Crossing the River with Dogs

Problem Solving for College Students

Second Edition

Ken Johnson
SIERRA COLLEGE

Ted Herr
ROSEVILLE HIGH SCHOOL

Judy Kysh
SAN FRANCISCO STATE UNIVERSITY

WILEY

John Wiley & Sons, Inc.

Vice President/Publisher:	Laurie Rosatone
Project Editor:	Jennifer Brady
Marketing Manager:	Jonathan Cottrell
Marketing Assistant:	Patrick Flatley
Photo Researcher:	Sheena Goldstein
Senior Production Manager:	Janis Soo
Associate Production Manager:	Joyce Poh
Assistant Production Editor:	Pauline Tan
Cover Designer:	Seng Ping Ngieng
Illustrator:	Dan Piraro
Cover Photo Credit:	©Kris Holland/iStockPhoto

This book was set in 10.5/14 StempelSchneidler by MPS Limited, a Macmillan Company and printed and bound by RR Donnelly (Von Hoffman).

This book is printed on acid free paper.

Founded in 1807, John Wiley & Sons, Inc. has been a valued source of knowledge and understanding for more than 200 years, helping people around the world meet their needs and fulfill their aspirations. Our company is built on a foundation of principles that include responsibility to the communities we serve and where we live and work. In 2008, we launched a Corporate Citizenship Initiative, a global effort to address the environmental, social, economic, and ethical challenges we face in our business. Among the issues we are addressing are carbon impact, paper specifications and procurement, ethical conduct within our business and among our vendors, and community and charitable support. For more information, please visit our website: www.wiley.com/go/citizenship.

Evaluation copies are provided to qualified academics and professionals for review purposes only, for use in their courses during the next academic year. These copies are licensed and may not be sold or transferred to a third party. Upon completion of the review period, please return the evaluation copy to Wiley. Return instructions and a free of charge return mailing label are available at www.wiley.com/go/returnlabel. If you have chosen to adopt this textbook for use in your course, please accept this book as your complimentary desk copy. Outside of the United States, please contact your local sales representative

Library of Congress Cataloging-in-Publication Data

Johnson, Ken, 1956–
    Crossing the river with dogs : problem solving for college students / Ken Johnson, Ted Herr, Judy Kysh. — 2nd ed.
        p. cm.
    Includes index.
    ISBN 978-0-470-46473-1 (pbk.)
        1. Problem solving. 2. Mathematics. I. Herr, Ted, 1955– II. Kysh, Judy, 1940–
III. Title.
QA63.H37 2012
510—dc23   2011039269

Printed in the United States of America
10 9 8 7 6 5 4 3 2 1

To Janie, Allyson, and Armand

Contents

Preface

What Is Problem Solving?

Problem solving has been defined as knowing what to do when you don't know what to do. Being one of the core human activities, it covers many daily functions such as going to the store, buying ingredients, and cooking a nutritionally sound meal. Beyond just grocery shopping, problem solving is used in every profession from teaching to engineering to politics, and being able to solve problems is essential to survival.

Typically, problem solving involves communicating, gathering information, organizing information, and implementing a plan. Imbedded within this context are specific strategies that aid in big-picture problem solving. Many of these strategies have a mathematical basis. This book focuses on strategies and provides opportunities to practice, integrate, and connect them to become a more skilled problem solver.

Now that most students have access to calculators and computers and can use math software, the time that once was used to practice computation can be used to learn problem solving. The hard part of "word problems" in mathematics is deciding what to do. Once a plan has been made, the rest—the computation part—is often relatively easy. Similarly, as students face jobs, deal with family, and work through various life situations, they will find that they need to use their problem-solving skills to determine what to do.

The Genesis of *Crossing the River with Dogs*

This book has had its genesis in many places: George Pólya's *How to Solve It;* Carol Meyer and Tom Sallee's *Make It Simpler; Topics of Problem Solving* by Randall Charles and colleagues; and the Lane County Problem Solving series; as well as several other diverse works.

These earlier works inspired us to develop a coherent course in mathematical problem solving. The topic was a natural for collaborative learning, and it also provided for extensive communication as a unifying theme.

Incorporating Problem Solving into College-Level Courses

All students can benefit from this book, whether majoring in math or not. This book is appropriate for a problem-solving course with an intermediate algebra prerequisite that could be taken as a general education math class. It is also appropriate for a liberal arts mathematics course. In addition, it works well in teacher credential programs for future elementary or secondary math teachers. Because much of the course is taught using groups, the course can provide teacher candidates a particularly valuable experience with this mode of learning and encourage them to incorporate cooperative learning in their own classes.

What's New in the Second Edition?

For this edition, we did the following:

- Rewrote parts of several chapters to sharpen explanations.

- Added margin notes in several chapters to clarify main points.

- Added new problems to Problem Set A in several chapters.

- Added a More Practice section in each chapter with answers in the back of the book to assist students in solidifying their grasp on each strategy.

- Included many problems written by students.

Expected Student Outcomes

From our experiences in our own high school and college classrooms, we find that this book fits best in courses that emphasize reasoning, communication, collaboration, and problem-solving strategies. The curriculum, the course content, and the delivery of instruction are uniquely consistent: Students learn best by working in groups, and the skills required for real workplace problem solving are those skills of collaboration.

By the end of the course, students will see a marked improvement in their writing, oral communication, and collaboration skills. They will acquire and internalize mathematical problem-solving strategies and increase their repertoire of learning strategies.

The students' own reflections on this course speak to its power. Students believe they are better equipped to continue their educational programs and courses of study at the end of a problem-solving course and report that it is the most useful course they have ever taken. Furthermore, students have increased their ability to think about and evaluate their thinking process (meta-cognition) while solving problems.

We believe this kind of learning is rare and special. We proudly present to you the new edition of *Crossing the River with Dogs: Problem Solving for College Students.*

Ken Johnson, Ted Herr, and Judy Kysh

Instructor Resources

Instructor resources are available to qualified adopters. Each printed *Instructor Resources* is packaged with a CD-ROM of Microsoft® PowerPoint slides. These two supplements work together to provide instructors with the following:

- Teaching suggestions, such as organizing a class structure, incorporating collaborative learning, and dealing with unique assessment issues.

- Chapter summaries and frequently asked questions.

- Notes on the chapter problems.

- Alternate Problem Sets A and B, which allow instructors to assign different problems each semester.

- Answers to the problem sets that occur in the book as well as the alternate versions that appear in the *Instructor Resources.*

- PowerPoint presentations that guide students step-by-step through the chapter problems.

For more information on these teaching resources, please visit www.wiley.com.

Acknowledgments

The support for this book has come in many forms. We wish to publicly thank the people who supported and inspired us and those who contributed problems and solutions:

Jen Adorjan, Reina Alvarez, Kevin Alviso, Allyson Angus Stewart, Sally Appert, Alan Barson, Michelle Bautista, Dave Beauchamp, Susan Behl, Dan Bennett, Sarah Bianchet, John Bingham, Paul Bohanna, Monica Borchard, Brianne Brickner, Josh Bridges, Judi Caler, Raul Carbajal, Barbara Carmichael, Gary Cerar, Randall Charles, Jeremy Chew, Evelyn Click, Cory Craig, Bob Daniel, Lisa DaValle, James Davis, Ginny Day, Carolyn Donohoe, Tab Dorando, Brandon Duckett, Don Elias, Halee Epling, Jessica Erickson, Carina Euyen, Katie Evenson, Katie Fitzpatrick, Russell Fredy, Ryan Gaffney, Greg Garretson, Nancy Gatzman, Don Gernes, Ed Gieszelmann, Michelle Goal, Joanne Gold, Aletta Gonzales, Suzanne Goodell, Gary Haas, Leah Hall, Rachel Hamilton, Kerry Harrigan, Nancy Harrigan, Tim Harrison, Alyssa Haugen, Haley Haus, Kyle Heckey, Holly Herr, Joe Herr, Ryan Hietpas, Alicia Hupp, Kelly Iames, Steve Jackson, Erica Johnson, Jennifer Johnson, Lucille Johnson, Rick Johnson, Heather 'Fonzie' Jordan, Erick Kale, Nick Landauer, David Lee, Steve Legé, Ashley Lien, Eric Lofholm, Chris Lowman, Irving Lubliner, Brad Marietti, Alyse Mason-Herr, Melissa McConnell, Jonathan Merzel, Amanda Mizell, Jacque Moncrief, Peter Murdock, Matthew Nelson, Ed Patriquin, Patrick Pergamit, Grover Phillips, Alex Ramm, Steve Rasmussen, Alyssa Reindl, Jason Reynolds, Travis Roberts, Kristen Rogers, Tom Sallee, Shari Seigworth, Sherrie Serros, Dave Seymour, Alan Shanahan, Robin Shortt, Tanja Slomski, Cheyenne Smedley-Hanson, Alyssa Spagnola, Gary Steiger, Danny Stevenson, Max Stockton, Brian Strand, Kristen Sullivan, Matt Swearingen, Tony Tani, Eric Theer, Andrea Trihub, Joanne Weatherly, Steve Weatherly, Stephanie Wells, Brandy Whittle, Amy Wilcox, Russell Wilcox, Joe Yates, Barbara Ybarra, Byron Young, the Northern California Mathematics Project, and the students enrolled in the problem-solving courses at Luther Burbank High School, Sacramento, California, and Sierra College, Rocklin, California.

We also owe thanks to our colleagues who were the sources for some of these problem ideas. You'll recognize in this book some classic problems (many with new twists) and familiar themes from the lore of the problem-solving community at large. It would be impossible to even guess at, much less acknowledge, a source for each of those problems—but we're grateful to be in a profession that values sharing ideas.

In particular, we would like to thank the authors of the Classic Problems that appear in this book. These people were some of the forefathers of problem solving who were inspirational to us: Henry Dudeney, James Fixx, Martin Gardner, Boris Kordemsky, Sam Loyd, Julio Mira, George Pólya, Raymond Smullyan, Ian Stewart, George Summers, and David Rock and Doug Brumbaugh and the other contributors to the Problem of The Week Website at Ole Miss.

We are very appreciative of the comments and criticisms of the reviewers of this edition of the text:

- Richard Borie, University of Alabama

- Maria Brunett, Montgomery College

- Michael Coco, Lynchburg College

- Stuart Moskowitz, Humboldt State University

- Christopher Reisch, Jamestown Community College

- Carla Tayeh, Eastern Michigan University, reviewer of the first and second editions

We'd also like to acknowledge the other reviewers of the first edition:

- Phyllis Chinn, Humboldt State University

- Joyce Cutler, Framingham State University

- Sandra Bowen Franz, University of Cincinnati

- Kathy Nickell, College of DuPage

- Laurie Pines, San Jose State University

- Joan Weinstein, Pine Manor College

- John Wilkins, California State University, Dominguez Hills

And thank you to Jim Stenson, the math checker of the first edition.

We would like to thank Jennifer Brady, Jonathan Cottrell, Sheena Goldstein, Laurie Rosatone, Pauline Tan, and the rest of the editorial staff at John Wiley and Sons.

A special acknowledgment is reserved for Tom Sallee. For each of us he played similar, overlapping roles. As a professor of mathematics at the University of California, Davis, he was an instrumental part of our development in mathematics, education, and mathematical problem solving and as mathematics teachers. For all of this, we are grateful.

And finally, thanks to Ken and Ted's children: Daniel, Gary, and Will Johnson and Alyse, Jeremy, and Kevin Mason-Herr.

Introduction

Letter to the Student

A mathematics course based on this book will be different from other math classes you've encountered. You may solve some equations, but mostly you'll be asked to think about problems, solve them, then write about your solutions. You may know students who have survived math classes without understanding the material, but that won't happen in this class. Our goal for you is that when you finish this course, you will be able to understand the mathematics you are doing and explain your reasoning in writing and to other students.

This course is based on the idea that you'll learn the strategies people in the real world use to solve problems. You will develop specific problem-solving strategies, communication skills, and attitudes. Learning problem-solving strategies will help you to do well on standardized tests used by colleges, graduate schools, and some employers. These tests often assess problem-solving skills more than the ability to solve equations. You will also learn to have fun doing mathematics. For centuries people have considered challenging problems, often called puzzles or brain teasers, to be a source of entertainment.

In this course you will be asked to solve some tough problems. You will be able to solve most of them by being persistent and by talking with other students. When you come across an especially difficult problem, don't give up. Use the techniques you've already learned in the course, such as drawing diagrams, asking other people for help, looking at your notes, and trying other approaches.

You will be expected to talk to your classmates! Your instructor will ask you to get help from one another. Thus, not all of the learning you'll do will be "book learning." You'll also learn how to work with others. Research has shown that even the best students learn from working with their peers. The communication skills you learn in this class will help you throughout your lifetime.

What you get out of this course will depend on how much you are willing to invest. You have a chance to take an active role in your education.

Enjoy the journey!

Answers to Questions That Students Usually Ask

Some people have said that America is not ready for a math book with the word *dogs* in the title, but we think the country can handle it. This book is different from many other mathematics books, from the title page to the last page. For one thing, this book is meant to be enjoyable to read. It is also meant to teach problem-solving **strategies,** and it incorporates research on how students best learn mathematics.

This book was written to take advantage of the strength of cooperative learning and the benefits of communicating your math work to others. You've probably attended classes in which your instructor encouraged you to work with others. Your instructor used this approach because research shows that students learn more when they work together. In addition to working with your fellow students, reading the book, and learning from your instructor, you will be expected to communicate about your work and your mathematical thinking. You will do this by presenting your **solutions** to the entire class and by writing up complete solutions to problems. You will do presentations and write-ups because talking and writing allow you to develop your thinking.

What is problem solving?

Problem solving has been defined as figuring out what to do when you don't know what to do. In some of your math courses, you probably learned about mathematical ideas by first working on an example and then practicing with exercises. An **exercise** asks you to repeat a method you learned from a similar example. A **problem** is usually more complex than an exercise. It is harder to solve because you don't have a preconceived notion about how to solve it. In this course you will learn general, wide-ranging strategies for solving problems. These strategies, many of them popularized by George (György) Pólya's classic book *How to Solve It* (Princeton University Press, 1988), apply to many different types of problems.[1]

Many of the problems you do in this course will be new to you. That is, you won't have seen a similar example in class or have a "recipe" to follow. To solve the problems, you will use the broad **heuristic,** or discovery-based, strategies you'll learn in this book. You may sometimes find that your first approach to a problem doesn't work. When this happens, don't be afraid to abandon your first approach and try something else. Be persistent. If you get frustrated with a problem, put it aside and come back to it later. Let your subconscious work on it—you may find yourself solving problems in your sleep! But don't give up on any problem.

Why should we work together? Can't we learn just as well on our own?

You will have many opportunities to work with other students in class. You should also try to get together with other students outside of class. When you work with other students, you are free to make conjectures, ask questions, make mistakes, and express your ideas and

---

[1]First published in 1945, *How to Solve It* was evidence that Pólya was far ahead of his time in his approach to mathematical problem solving.

opinions. You don't have to worry about being criticized for your thoughts or your wrong answers. You won't always proceed down the correct path. Support one another, question one another, ask another person to explain what you don't understand, and make sure the other members of your group understand, too. If others in your group make mistakes, don't berate them. Instead, help them to see why their approach is wrong.

Are there any limits to working together? What about searching for solutions on the internet?

The point of this class is for you to learn new problem-solving strategies, and use your skills to solve a variety of problems. While we encourage you to work together, you have to be reasonable about the amount of help you accept. It is fine to collaborate to reach a solution, with each member of the group having input to the final solution. It is not OK to copy someone else's work, with none of your own thinking. It is also not OK to search for solutions on the internet. This is a class about becoming a powerful problem solver and discovering the power of collaboration to solve difficult problems. This is not a class about getting an answer by any means necessary.

How should we study? How should we read the book?

The book is organized into chapters. Each chapter introduces a new problem-solving strategy and presents several problems within the text, which are identified by an icon of an attentive dog. You are asked to first solve each problem, then read its sample solution. To get the maximum benefit from this book, work the problems before reading the solutions. Even if you successfully solved a problem, read the solution anyway because it may differ from your own or bring up some points you hadn't thought about. Remember, your purpose is not just to get answers, but to learn more solution strategies. Although you'll learn the most by trying the problems yourself, if you don't have time to solve all of them, you will still get a lot out of at least reading the problems and their solutions. You'll learn the least if you don't read the book at all. Be willing to read the text slowly and carefully.

Sometimes you will see a problem that you saw earlier in another chapter. You'll solve this same problem again, but you'll use a different strategy. Solving the same problem in many different ways will help you become a better problem solver. You'll also often see sample solutions that actual students came up with as they worked through the problems in this book. Their work shows that there are many valid ways to tackle a problem, even if one particular approach doesn't result in a correct answer.

Some issues that cause confusion occur occasionally in the problem discussions. When these points of confusion come up, they are considered further and clarified in the discussion following the problem analysis. These points of possible confusion are marked by an icon of a confused dog.

Margin notes will clarify important steps in solutions to problems. An oar icon marks these spots.

Every chapter concludes with a summary of the main points of the strategy. In addition, important problem-solving advice is given in the body of the text. A pointing dog icon marks these spots.

There are many useful references in the back of the book. In the text there are many words in **bold** print. You will find the definition for these words in the glossary. There is a page of geometry formulas on area, volume, and triangle properties. You will find information on metric conversions, as well as common abbreviations for different units of measure. There is also a reminder on how to work with fractions. There is a section on divisibility rules and how to determine if a number is prime. There is an answer section for the More Practice problems. Finally, there is an index of problem titles and a general index.

What other problems will we do besides those discussed in the text?

Each chapter ends with a Problem Set A. The problems in these sets can always be solved with the strategy presented in that particular chapter. Some of the problems could also be solved by using a strategy you learned in an earlier chapter. However, because you're learning a new strategy with each new chapter, solve the Set A problems with their chapter-specific strategy. (To get more benefit from this class, you may also want to try solving many of the problems using other strategies.)

In most chapters, at the end of Problem Set A you will find one or two classic problems. There have been many famous puzzle writers through the years, beginning with Henry Dudeney in England and Sam Loyd in the United States at the end of the nineteenth century. In the twentieth century, Martin Gardner, Raymond Smullyan, George

Summers, Boris Kordemsky, and George Pólya joined them. These individuals, along with several others, created problems that are now considered classics by the mathematics community. In this book, we give you the opportunity to solve some of these famous problems. In most cases you can solve them with the strategy you learned in the chapter where you find them.

After the Classics section, there is a section called More Practice. The problems that appear here are relatively straightforward and you should use them to study from. Do the problems in this section after finishing the chapter to solidify your grasp of the strategy. Answers for these problems appear in the back of the book as one way to check your solution.

Beginning in Chapter 3, a Problem Set B ends the chapter. The five Set B problems in each chapter can be solved with any strategy you've already learned, and it's up to you to pick an appropriate strategy. The Set B problems are more difficult than the Set A problems. In fact, toward the end of the book many of the Set B problems are extremely difficult. Many instructors use Problem Set B as a week-long (or longer) assignment. They use the sets as take-home tests and allow students to work together, with each student turning in his or her own work. Each student is expected to provide answers and to explain their reasoning for each solution.

What is the role of the instructor in this course?

In many courses, the instructor is the final authority who determines whether the student is right or wrong. In this course your instructor will play that role at times, such as when she or he grades your work. But there will probably be times when the instructor will not play that role. For example, during student presentations several people may have different answers to the same problem. When this happens, it's natural to ask your instructor who is right. In this course your instructor may let you make up your own minds. Your in-class groups can discuss which answer they think is correct and why. Explaining why is a very important part of this course. Not relying on the instructor to verify your work will help you become a better problem solver. You will learn to carefully evaluate your own work as well as the work of others.

Why aren't there any other answers in the back of the book?

Answers appear in the back of the book for the More Practice problems. There are no other answers in the back of the book. When you are learning a strategy for the first time, an answer given too early will cut off the thinking processes that you need to develop. When working in a group of students in class, the group needs to come to a conclusion about approaches and answers. If there were answers in the back of the book, much of this process would be inhibited.

Will the skills I learn in this class help me to get a job?

What do employers want in their employees? The internet is full of top-ten lists that explore this question. There are many common themes to such lists. Employers want their employees to exhibit:

a. Problem-solving skills

b. Creativity in finding solutions to complex problems

c. The ability to analyze solutions to determine their workability

d. The ability to work well with other people, especially people different from yourself

e. The ability to apply previous knowledge to new situations

f. Good communication skills—both oral and written

g. Determination and persistence

h. A strong work ethic

i. Honesty and integrity

Every skill on this list will be greatly developed by studying and actively working on the problems in this book. Look at this course as a challenge—as an opportunity to learn new skills and show them off. You will use your creativity and problem-solving skills to solve a variety of different problems. We are sure that those who really buy into this will emerge from the course with an enriched toolbox of problem-solving skills.

Some Comments on Answers

When you turn in written work, you should write your answer to each problem in the form of a sentence, including any appropriate units. Be sure that you answer the question that is being asked, and make your answer entirely clear. Don't expect the reader to dig through your work to find your answer.

Think carefully about what your answer means, and make sure the form of the answer makes sense and is reasonable given the circumstances of the problem. For example, if the answer to a question is a certain number of people and your answer is a fraction or a decimal, think about what the question's answer should be. Does it make sense to round your answer up or down, or to leave it the way it is? Consider the following situation.

The Vans

There are 25 people going on a trip. They are traveling by van, and each van has a capacity of 7 people.

Some people might think the answer being sought is $3^4/7$ vans, but the answer depends on what question is posed. Here are some possible questions:

a. How many vans will be needed to transport all 25 people?

b. How many vans can be filled to capacity?

c. How many vans will have to be filled to capacity?

d. What is the average number of people in a van?

e. Must any van have 7 people in it?

f. How many more people could fit into the vans that will be required?

The answer to each of these questions is different, even though the situation is the same. The difference is in the question asked. For example, the van problem looks like it could be solved by dividing 25 by 7, but only one of the questions above looks like a division problem. In the van problem, we're working with units that are generally considered to be indivisible (vans or people) as opposed to units that are clearly divisible (pizzas). The answer to our division problem is reasonable only if our answer's units are also reasonable. That is, no matter what arithmetic is done with the numbers in this problem, the answer must still apply to human beings going somewhere in vans. Keep these issues in mind when you work problems.

Some Introductory Problems

During this course you will learn many problem-solving strategies and use them to solve many different problems. Solve the problems in this introduction with whatever strategy you wish. You will have an opportunity to share your solutions to some or all of these problems with a small group or the whole class. You can solve these problems with a variety of different strategies. In fact, you may want to solve each problem several times, using a different strategy each time. (The solutions to the example problems in this book are shown following those problems, but the solutions to these introductory problems are not shown.)

I.   SOCCER GAME

At the conclusion of a soccer game whose two teams each included 11 players, each player on the winning team "gave five" to (slapped hands with) each player on the losing team. Each player on the winning team also gave five to each *other* player on the winning team. How many fives were given?

2.   ELEVATOR

The capacity of an elevator is either 20 children or 15 adults. If 12 children are currently in the elevator, how many adults can still get in?

3.   THEATER GROUP

There are eight more women than men in a theater group. The group has a total of 44 members. How many men and how many women are in the group?

4. DUCKS AND COWS

Farmer Brown had ducks and cows. One day she noticed that the animals had a total of 12 heads and 32 feet. How many of the animals were ducks and how many were cows?

5. STRANGE NUMBER

If you take a particular two-digit number, reverse its **digits** to make a second two-digit number, and add these two numbers together, their sum will be 121. What is the original number?

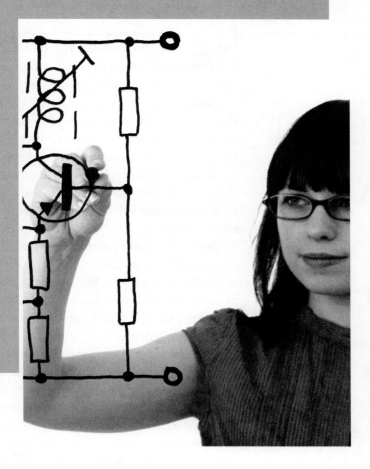

Draw a Diagram

Diagrams are often the key to getting started on a problem. They can clarify relationships that appear complicated when written. Electrical engineers draw diagrams of circuit boards to help them visualize the relationships among a computer's electrical components.

Y ou've probably heard the old saying "One picture is worth a thousand words." Most people nod in agreement when this statement is made, without realizing just how powerful a picture, or a **diagram,** can be. (Note that words in **bold** type are terms that are defined in this book's glossary.) A diagram has many advantages over verbal communication. For example, a diagram can show positional relationships far more easily and clearly than a verbal description can. To attempt to clarify ideas in their own minds, some people talk to themselves or to others about those ideas. Similarly, a diagram can help clarify ideas and solve problems that lend themselves to visual representations.

One of the best examples of a diagram in the professional world is a blueprint. An architect's blueprint expresses ideas concisely in a visual form that leaves little to interpretation. Words are added only to indicate details that are not visually evident. A blueprint illustrates one of the strengths of diagrams: the ability to present the "whole picture" immediately.

Problem solving often revolves around how information is organized. When you draw a diagram, you organize information spatially, which then allows the visual part of your brain to become more involved in the problem-solving process. In this chapter you will learn how you can use diagrams to clarify ideas and solve a variety of problems. You'll improve your diagramming abilities, and you'll discover that a diagram can help you understand and correctly interpret the information contained in a problem. You'll also see the value of using diagrams as a problem-solving strategy.

Solve this problem by drawing a diagram.

VIRTUAL BASKETBALL LEAGUE

Andrew and his friends have formed a fantasy basketball league in which each team will play three games against each of the other teams. There are seven teams: the (Texas A&M) Aggies, the (Purdue) Boilermakers, the (Alabama) Crimson Tide, the (Oregon) Ducks, the (Boston College) Eagles, the (Air Force) Falcons, and the (Florida) Gators. How many games will be played in all? Do this problem before reading on.

As you read in the Introduction, you'll see many different problems as you work through this book. The problems are indicated by an icon of an attentive dog. To get the maximum benefit from the book, solve each of the problems before reading on. You gain a lot by solving problems, even if your answers are incorrect. The *process* you use to solve each problem is what you should concentrate on.

You could use many different diagrams to solve the Virtual Basketball League problem, but you could also solve this problem in ways that do not involve diagrams. As you also read in the Introduction, throughout this book you will see some of the same problems in different chapters and solve them with different strategies. You will become a better problem solver in two ways: by solving many different problems and by solving the same problem in many different ways. In this chapter, the solutions involve diagrams. If you solved the Virtual Basketball League problem without using a diagram, try solving it again with a diagram before reading on.

What comes next is a solution process that is attributed to a student. The people mentioned in this book are real students who took a problem-solving class at either Sierra College in Rocklin, California, or at Luther Burbank High School in Sacramento, California. In those classes, the students presented their solutions on the board to their classmates. Ted Herr and Ken Johnson, two of the authors of this book, taught these classes. Our students presented their solutions because we felt that the other students in class would benefit greatly from seeing many different approaches to the same problem. We didn't judge each student's solution in any way. Rather, we asked each member of the class to examine each solution that was presented and decide which approach or approaches were valid or, perhaps, better. The purpose behind shifting this responsibility from the instructor to the students is to give the students practice in evaluating problem solving.

We have tried to re-create the same learning atmosphere in this book. Sometimes you'll see several different approaches to a problem in this book, but for the most part those approaches and the resulting solutions won't be judged. You are encouraged to evaluate the quality of the approaches. You may have been led to believe that there is always one right way—and many wrong ways—to solve problems. This notion couldn't be further from the truth. There are many right ways to solve problems, and you are encouraged to solve the problems in this book more than once, using different methods.

Here's how Rita solved the Virtual Basketball League problem: She drew a diagram that showed the letters representing each team arranged in a circle.

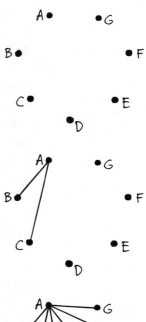

She then drew a line from A to B to represent the games played between the Aggies and the Boilermakers. Then she drew a line from A to C to represent the games played between the Aggies and the Crimson Tide.

She finished representing the Aggies' games by drawing lines from A to D, E, F, and G.

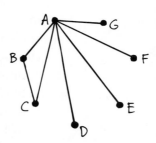

Next she drew the lines for the Boilermakers. She'd already drawn a line from A to B to represent the games the Boilermakers played against the Aggies, so the first line she drew for the Boilermakers was from B to C.

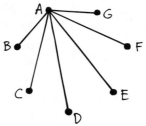

She continued drawing lines to represent the games that the Boilermakers played against each other team.

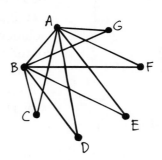

From C she drew lines only to D, E, F, and G because the lines from C to A and from C to B had already been drawn. She continued in this way, completing her diagram by drawing the lines needed to represent the games played by the rest of the teams in the league. Note that when she finally got to the Gators, she did not need to draw any more lines because the games the Gators played against each other team had already been represented with a line.

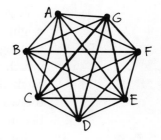

She then counted the lines she'd drawn. There were 21. She multiplied 21 by 3 (remember that each line represented three games) and came up with an answer of 63 games. Finally, Rita made sure that she'd answered the question asked. The question was "How many games will be played in all?" Her answer, "Sixty-three games will be played," accurately answers the question.

Mirka solved this problem with the diagram below. She also used the letters A, B, C, D, E, F, and G to represent the teams. She arranged the letters in a row and, as Rita did, she drew lines from team to team to represent games played. She started by drawing lines from A to the other letters, then from B to the other letters, and so on. She drew 21 lines, multiplied 21 by 3, and got an answer of 63 games.

MODEL TRAIN

Esther's model train is set up on a circular track. Six telephone poles are spaced evenly around the track. The engine of Esther's train takes 10 seconds to go from the first pole to the third pole. How long would it take the engine to go all the way around the track? Solve the problem before reading on.

If you read the problem quickly and solved it in your head, you might think the answer is 20 seconds. After all, the problem states that the engine can go from the first pole to the third pole in 10 seconds, which is three poles out of six and apparently halfway around the track. So it would take the engine 2 times 10, or 20 seconds, to go all the way around the track. But this answer is wrong. The correct answer becomes apparent when you look at a diagram.

Rena's diagram is shown at right. Rena explained that the train goes one-third of the way around the track in 10 seconds, not halfway around the track. So the train goes around the entire track in 3 times 10 seconds, or 30 seconds.

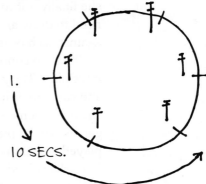

Phong drew the same diagram, but he interpreted it differently. He explained that if it takes 10 seconds to go from the first pole to the third pole, then it takes 5 seconds to go from the first pole to the second pole. So it takes 5 seconds to go from pole to pole. There are six poles, so it takes the train 30 seconds to go all the way around the track.

Pete interpreted the problem as Phong did, but he didn't draw a diagram. Thus, he neglected the fact that the train must return from the sixth pole to the first pole in order to travel all the way around the track. Therefore, he got the incorrect answer 25 seconds.

～～～～～

The diagram helped Rena and Phong solve the Model Train problem. If you used a diagram to solve the problem, you probably got the correct solution. If you were able to get the correct solution without drawing a diagram, think back on your process. You probably visualized the train track in your mind, so even though you didn't actually draw a diagram, you could "see" a picture.

Do you get the picture? Do you see why diagrams are important? Research shows that most good problem solvers draw diagrams for almost every problem they solve. Don't resist drawing a diagram because you think that you can't draw, or that smart people use only equations to solve problems, or whatever. Just draw it!

THE POOL DECK

Curly used a shovel to dig his own swimming pool. He figured he needed a pool because digging it was hard work and he could use it to cool off after working on it all day. He also planned to build a rectangular concrete deck around the pool that would be 6 feet wide at all points. The pool is rectangular and measures 14 feet by 40 feet. What is the **area** of the deck? As usual, solve this problem before continuing.

Jeff drew the diagram below to show the correct dimensions of the deck and pool, which together are 12 feet longer and 12 feet wider than the pool alone.

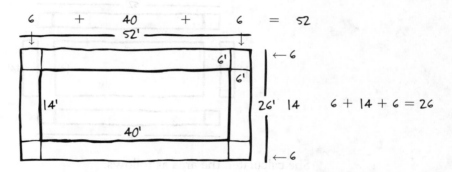

The diagram helps show the difficult parts of the problem. However, Jeff solved the problem incorrectly by finding the outside **perimeter** of the pool and the deck together, then multiplying the perimeter by the width of the deck.

52 feet + 26 feet + 52 feet + 26 feet = 156 feet
156 feet × 6 feet = 936 square feet

His approach is incorrect because it counts each corner twice.

Rajesh used the same diagram, but he solved the problem by first computing the area of the deck along the sides of the pool, then adding in the corners of the deck.

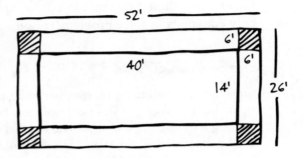

Two lengths: 40 ft x 6 ft x 2 = 480 sq ft
Two widths: 14 ft x 6 ft x 2 = 168 sq ft
Four corners: 6 ft x 6 ft x 4 = <u>144</u> sq ft
Total 792 sq ft

May's diagram shows the corners attached to the length of the deck.

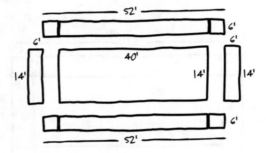

She calculated the area as follows:

52 ft x 6 ft = 312 sq ft
312 sq ft x 2 = 624 sq ft for extended lengths
14 ft x 6 ft = 84 sq ft
84 sq ft x 2 = 168 sq ft for widths
Total = 624 sq ft + 168 sq ft = 792 sq ft

Herb solved this problem by first computing the area of the pool and the deck together, then subtracting the area of the pool, leaving the area of the deck.

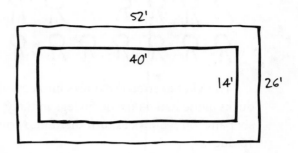

Area of entire figure = 52 ft x 26 ft = 1,352 sq ft
Area of pool alone = 40 ft x 14 ft = 560 sq ft
Area of deck = 1352 ft – 560 ft = 792 sq ft

FARMER BEN

Farmer Ben has only ducks and cows. He can't remember how many of each he has, but he doesn't need to remember because he knows he has 22 animals and that 22 is also his age. He also knows that the animals have a total of 56 legs, because 56 is also his father's age. Assuming that each animal has all legs intact and no extra limbs, how many of each animal does Farmer Ben have? Do this problem, and then read on.

Trent drew the following diagram:

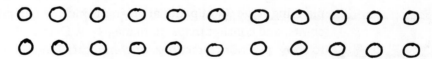

He then explained his thinking: "These 22 circles represent the 22 animals. First, I made all of the animals into ducks." (Trent is not much of an artist, so you just have to believe that these are ducks.) "I gave each animal two legs because ducks have two legs."

"Then I converted the ducks into cows by drawing extra legs. The ducks alone had 44 of the 56 legs initially, so I drew 12 more legs, or six pairs, on 6 ducks to turn them into cows. So there are 6 cows and 16 ducks."

Of course, Farmer Ben might have a problem when his father turns 57 next year.

Draw a Diagram

If you can visualize it, draw a diagram.

Any idea that can be represented with a picture can be communicated more effectively with that picture. By making visible what a person is thinking, a diagram becomes a problem-solving strategy. A diagram clarifies ideas and communicates those ideas to anyone who looks at it. Diagrams are used in many jobs, especially those that require a planning stage. Occupational diagrams include blueprints, project flow charts, and concept maps, to name a few. Diagrams are often necessary to show position, directions, or complicated multidimensional relationships, because pictures communicate these ideas more easily and more clearly than words.

Problem Set A

You must draw a diagram to solve each problem.

1. WORM JOURNEY

A worm is at the bottom of a 12-foot wall. Every day the worm crawls up 3 feet, but at night it slips down 2 feet. How many days does it take the worm to get to the top of the wall?

2. UPS AND DOWNS OF SHOPPING

Roberto is shopping in a large department store with many floors. He enters the store on the middle floor from a skyway and immediately goes to the credit department. After making sure his credit is good, he goes up three floors to the housewares department. Then he goes down five floors to the children's department. Then he goes up six floors to the TV department. Finally, he goes down ten floors to the main entrance of the store, which is on the first floor, and leaves to go to another store down the street. How many floors does the department store have?

3. FOLLOW THE BOUNCING BALL

A ball rebounds one-half the height from which it is dropped. The ball is dropped from a height of 160 feet and keeps on bouncing. What is the total vertical distance the ball will travel from the moment it is dropped to the moment it hits the floor for the fifth time?

4. FLOOR TILES

How many 9-inch-square floor tiles are needed to cover a rectangular floor that measures 12 feet by 15 feet?

5. **STONE NECKLACE**

Arvilla laid out the stones for a necklace in a big circle, with each stone spaced an equal distance from its neighbors. She then counted the stones in order around the circle. Unfortunately, before she finished counting she lost track of where she had started, but she realized that she could figure out how many stones were in the circle after she noticed that the sixth stone was directly opposite the seventeenth stone. How many stones are in the necklace?

6. **DANGEROUS MANEUVERS**

Somewhere in the Mojave Desert, the army set up training camps named Arachnid, Feline, Canine, Lupine, Bovine, and Thirty-Nine. Several camps are connected by roads:

Arachnid is 15 miles from Canine, Bovine is 12 miles from Lupine, Feline is 6 miles from Thirty-Nine, Lupine is 3 miles from Canine, Bovine is 9 miles from Thirty-Nine, Bovine is 7 miles from Canine, Thirty-Nine is 1 mile from Arachnid, and Feline is 11 miles from Lupine. No other pairs of training camps are connected by roads.

Answer each of the following questions (in each answer, indicate both the mileage and the route): What is the shortest route from

Feline to Bovine?   Canine to Thirty-Nine?

Lupine to Thirty-Nine?   Lupine to Bovine?

Canine to Feline?   Arachnid to Feline?

Arachnid to Lupine?

7. **RACE**

Becky, Ruby, Isabel, Lani, Alma, and Sabrina ran an 800-meter race. Alma beat Isabel by 7 meters. Sabrina beat Becky by 12 meters. Alma finished 5 meters ahead of Lani but 3 meters behind Sabrina. Ruby finished halfway between the first and last women. In what order did the women finish? What were the distances between them?

8. **A WHOLE LOTTA SHAKIN' GOIN' ON!**

If six people met at a party and all shook hands with one another, how many handshakes would be exchanged?

9. HAYWIRE

A telephone system in a major manufacturing company has gone haywire. The system will complete certain calls only over certain sets of wires. So, to get a message to someone, an employee of the company first has to call another employee to start a message on a route to the person the call is for. As far as the company can determine, these are the connections:

Cherlondia can call Al and Shirley (this means that Cherlondia can call them, but neither Al nor Shirley can call Cherlondia). Al can call Max. Wolfgang can call Darlene, and Darlene can call Wolfgang back. Sylvia can call Dalamatia and Henry. Max can get calls only from Al. Carla can call Sylvia and Cherlondia. Shirley can call Darlene. Max can call Henry. Darlene can call Sylvia. Henry can call Carla. Cherlondia can call Dalamatia.

How would you route a message from

Cherlondia to Darlene?	Shirley to Henry?
Carla to Max?	Max to Dalamatia?
Sylvia to Wolfgang?	Cherlondia to Sylvia?
Henry to Wolfgang?	Dalamatia to Henry?

10. ROCK CLIMBING

Amy is just learning how to rock climb. Her instructor takes her to a 26-foot climbing wall for her first time. She climbs 5 feet in 2 minutes but then slips back 2 feet in 10 seconds. This pattern (up 5 feet, down 2 feet) continues until she reaches the top. How long will it take her to reach the very top of the wall?

_____

This problem was written by Jen Adorjan, a student at Sierra College in Rocklin, California.

11. CIRCULAR TABLE

In Amanda and Emily's apartment, a round table is shoved into the corner of the room. The table touches the two walls at points that are 17 inches apart. How far is the center of the table from the corner?

12. THE HUNGRY BOOKWORM

Following is an expansion of a well-known problem:

The four volumes of *The World of Mathematics* by James R. Newman are sitting side by side on a bookshelf, in order, with volume 1 on the left. A bookworm tunnels through the front cover of volume 1 all the way through the back cover of volume 4. Each book has a front cover and a back cover that each measure $1/16$ inch. The pages of each book measure $1\frac{1}{8}$ inches. How far does the bookworm tunnel?

13. BUSING TABLES

Brian buses tables at a local café. To bus a table, he must clear the dirty dishes and reset the table for the next set of customers. One night he noticed that for every three-fifths of a table that he bused, another table of customers would get up and leave. He also noticed that right after he finished busing a table, a new table of customers would come into the restaurant. However, once every table was empty (no diners were left in the restaurant), nobody else came into the restaurant. Suppose there were six tables with customers and one unbused table. How many new tables of customers would come in before the restaurant was empty? After the last table of customers had left, how many tables were unbused?

This problem was written by Brian Strand, a student at Sierra College in Rocklin, California.

14. WRITE YOUR OWN PROBLEM

In each chapter you'll be given the opportunity to write your own problem that can be solved by using the strategy you studied in that chapter. The book will give you suggestions for how to go about writing these problems. Each time you write your own problem, solve it yourself to be sure that it's solvable. Then give it to another student to solve and, as needed, to help you with the problem's wording.

Create your own Draw a Diagram problem. Model it after either this chapter's Worm Journey problem or Ups and Downs of Shopping problem.

CLASSIC PROBLEMS

15. THE WEIGHT OF A BRICK

If a brick balances with three-quarters of a brick and three-quarters of a pound, then how much does the brick weigh?

Adapted from *Mathematical Puzzles of Sam Loyd,* Vol. 2, edited by Martin Gardner.

16. THE MOTORCYCLIST AND THE HORSEMAN

A motorcyclist was sent by the post office to meet a plane at the airport. The plane landed ahead of schedule, and its mail was taken toward the post office by horse. After half an hour, the horseman met the motorcyclist on the road and gave him the mail. The motorcyclist returned to the post office 20 minutes before he was expected. How many minutes early did the plane land?

Adapted from *The Moscow Puzzles* by Boris Kordemsky.

MORE PRACTICE

1. APARTMENT BUILDING

Joden just moved into a 12-story apartment building, and he is still having trouble finding which floor he lives on. He knows that he lives in the first apartment on the floor, but doesn't know which floor. He starts by going to the first floor and knocks on the door. Mrs. Smith answers and tells him to go up 8 floors and ask Mr. Jones. Joden does that and asks Mr. Jones where he lives. Mr. Jones doesn't know, but he says, "Go down 2 floors to Bryn's apartment and ask him." Bryn didn't know either. He told Joden to go up 5 floors to see his friend Trudie, because she knows where everyone lives. Trudie had no clue who Joden was. She said that the only one who might know where he lived was the new guy 7 floors below her. Joden goes down 7 floors and knocks on the new guy's apartment. No one answers. He stands there thinking for a while and finally realizes that he is the new guy. He opens the door and walks into his apartment. Which floor does Joden live on?

This problem was written by Jeremy Chew, a student at Sierra College in Rocklin, California.

2. MOVIE LINE

A bunch of people were standing in line for a movie. Averi got there late, and realized that she knew every person in line. She decided not to get in line until she figured out who to cut in with. She first stopped and talked to Jake, who was at the back of the line. Then she moved forward by passing 3 people and talked to Alexandra. She then moved forward again by passing 9 people and talked to Walter. Then she moved backward by passing 4 people and talked to Annie. Then she moved forward by passing 12 people and talked to Katie. Finally she moved backward by passing 2 people and joined Carli in line. Carli was originally the person in the exact middle of the line. Including Averi (who was now in line) how many people are in the line?

Note: Moving forward refers to moving toward the front of the line and moving backward refers to moving toward the back of the line.

3. BACKBOARD

Ei liked to play tennis. One day she didn't have anyone to play with, so she took her racket and tennis ball and began to hit the ball against the backboard. She hit it at the backboard, the ball bounced off the backboard and came back to her, and she hit it again, and so on. She started out 40 feet from the backboard. But she didn't hit the ball hard enough—the bounce came back only 90% as far, so she had to run up to hit it again. Again she didn't hit it hard enough, and it again only came back 90% as far as she had hit it. This continued for two more hits, each time the ball coming back 90% as far. Finally on the fifth hit she hit it really hard and it came back five times as far as she had hit it, going way over her head and hitting the fence. She got frustrated, and picked up her ball and went home. How far was the distance from the backboard to the fence? What was the total horizontal distance that the tennis ball traveled?

4. WORKING OUT

At the gym, there are 12 weight machines that Gina liked to use: 6 upper body machines and 6 lower body machines. All 12 machines were in one row, with 8 feet between each upper body machine and 8 feet between each lower body machine. The upper body machines were separated from the lower body machines by 16 feet. The lower body machines were numbered 1 through 6, and the upper body machines were numbered 7 through 12. Gina started at machine 6 (lower body) and then walked to machine 7 (upper body), then to machine 5 (lower body), then to machine 8 (upper body), and so on, alternating between lower body and upper body. After each 3 lower body machines, she would walk to the mat that was 20 feet past machine 1 and do some stretches. After each 3 upper body machines, she would walk to the wall that was 30 feet past machine 12 and do some stretching. When she finally finished lifting and stretching (the last thing she did was upper body stretching on the wall) how many feet had she walked in all?

2

Make a Systematic List

W hen you make a systematic list, you reveal the structure of a problem. Sometimes the list is all you need to solve it. Train schedules are systematic lists that help travelers find information easily and quickly.

Leslie has 25¢ in her pocket but does not have a quarter. If you can tell her all possible combinations of coins she could have that add up to 25¢, she will give you the 25¢. Solve this problem before continuing.

Many people start solving this problem as follows: "Let's see, we could have 5 nickels, or 2 dimes and 1 nickel. We might have 25 pennies. Oh yeah, we could have 10 pennies, 1 dime, and 1 nickel. Perhaps we could have . . . ." Solving the problem this way is extremely inefficient. It could take a long time to figure out all the ways to make 25¢, and you still might not be sure that you'd thought of all the ways.

A better way to solve the problem is to make a **systematic list.** A systematic list is just what its name says it is: a list generated through some kind of system. A **system** is any procedure that allows you to do something (like organize information) in a methodical way. The system used in generating a systematic list should be understandable and clear so that the person making the list can verify its completeness quickly. Additionally, another person should be able to understand the system and verify the solution without too much effort.

Many systematic lists are in the form of a table whose columns are labeled with the information given in a problem. The rows of the table are used to indicate possible combinations. As you read the following solutions for the Loose Change problem, make your own systematic list. Label the columns of the list Dimes, Nickels, and Pennies, and then fill in the rows with combinations of coins that add up to 25¢.

Brooke started her list in the first row of the Dimes column by showing the maximum number of dimes Leslie could have: two. In the Nickels column, she showed the maximum number of nickels possible with two dimes: one. In the second row, she decreased the number of nickels by one because it's possible to make 25¢ without using nickels. She then filled in the Pennies column by showing how

Dimes	Nickels	Pennies
2	1	0
2	0	5
1	3	0
1	2	5
1	1	10
1	0	15

and so on

many pennies she had to add to her dimes and nickels to make 25¢. After finding all the ways to make 25¢ with two dimes, Brooke continued filling in her list with combinations that include only one dime. In the third row, she showed the maximum number of nickels possible in one-dime combinations: three. As she did for the two-dime combinations, she decreased the number of nickels by one in each row until she ran out of nickels.

Brooke's completed list is shown at right. It includes all the possible zero-dime combinations. Finish your own list before reading on.

Brooke's systematic list is not the only one that will solve this problem. Heather used a different system. Before you look at her entire solution, which follows, cover all but the first three rows of the table at bottom right with a piece of paper. Look at the uncovered rows to figure out her system, and then complete the list yourself.

Heather explained her system like this: "I started with the largest number of pennies, which was 25. Then I let the pennies go down by fives and filled in the nickels and dimes to make up the difference."

Many people find Heather's system to be more difficult than Brooke's system. What do you think?

Dimes	Nickels	Pennies
2	1	0
2	0	5
1	3	0
1	2	5
1	1	10
1	0	15
0	5	0
0	4	5
0	3	10
0	2	15
0	1	20
0	0	25

Pennies	Nickels	Dimes
25	0	0
20	1	0
15	2	0
15	0	1
10	3	0
10	1	1
5	4	0
5	2	1
5	0	2
0	5	0
0	3	1
0	1	2

There are even more ways to do this problem as well. Kaitlyn made her list as shown on the right.

Kaitlyn explained, "I wrote 20 in the 10 cent column because two dimes is 20 cents. Then I wrote 5 in the five cent column because one nickel is 5 cents. In this way each row adds up to 25 cents. I also made sure that I only used 20, 10, or 0 in the 10 cent column, and 25, 20, 15, 10, 5, or 0 in the 5 cent column, because obviously you can't have half a nickel or something like that. I also froze the number in the first column and played with the other two

10¢	5¢	1¢
20	5	0
20	0	5
10	15	0
10	10	5
10	5	10
10	0	15
0	25	0
0	20	5
0	15	10
0	10	15
0	5	20
0	0	25

columns figuring out all possible ways before I changed the number in the first column. So there were 2 ways to put 20 in the first column, 4 ways to put 10 in the first column, and 6 ways to put 0 in the first column. That's a nice pattern. There are a total of 12 ways."

Mo made her list in another way (shown on the next page). "I saw what the others had done, but I didn't like the chart set up with the three columns. So I looked at the problem differently. I simply listed each coin, using 10 for a dime, 5 for a nickel, and 1 for a penny. I started off with two dimes and one nickel so I wrote 10-10-5. Then I traded the nickel in for five pennies so I wrote 10-10-1-1-1-1-1. Then I changed it to one dime and used three nickels. Continuing, I froze the 10, and changed one of the 5's to five 1's. Then I changed another 5 to five 1's, and finally the third 5 to five 1's. Finally I got rid of the 10, and started with five 5's. Then four 5's and five 1's, then three 5's, etc."

Mo went on, "The triangular pattern of the 1's was really cool. It helped me see that I hadn't missed any. I also noticed that my list was in exactly the same order as Brooke's list and Kaitlyn's list, but we thought about the problem in totally different ways. I liked Kaitlyn's explanation of freezing a column, and then unfreezing it and freezing it

10	10	5		
10	10	11111		
10	5	5	5	
10	5	5	11111	
10	5	11111	11111	
10	11111	11111	11111	
5	5	5	5	5
5	5	5	5	11111
5	5	5	11111	11111
5	5	11111	11111	11111
5	11111	11111	11111	11111
11111	11111	11111	11111	11111

again. I did the same thing, but without the columns. First I froze two 10's, then I froze one 10, and finally I froze zero 10's. I got the same answer of 12 ways."

Making systematic lists is a way to solve problems by organizing information. In this chapter you'll make systematic lists to organize information in tables and charts. You will also learn a little about using a special type of diagram called a **tree diagram.** Many of the strategies you'll explore later in this book involve organizing information in some sort of table or chart, and you'll learn other strategies that involve organizing information spatially.

There is often more than one correct approach to solving a problem.

Remember that there is often more than one useful approach to solving a problem. This is often true with devising systematic lists. When you solved the Loose Change problem, you may have used a different list from what was shown. The four students here—Brooke, Heather, Kaitlyn, and Mo—all used systematic lists to solve the problem, but each list was different. They all incorporated the idea of **freezing** an entry in the list, and working with the other entries until that possibility was exhausted. Then they unfroze that entry, changed it, and froze it again. This idea of freezing and unfreezing is often a key element of systematic lists. Any list is fine so long as you have a system that you understand and can use effectively. If you find that your original system is too confusing, scrap it and start over with a different system.

Just as you can use the *same* strategy, such as making a list, to solve a problem in different ways, you will also often find that you can use *more than one strategy* to solve a given problem. In Chapter 1 you solved the Virtual Basketball League problem with a diagram. Solve the problem again, but this time use a systematic list. Don't refer back to the diagram solution!

VIRTUAL BASKETBALL LEAGUE

Andrew and his friends have formed a fantasy basketball league in which each team will play three games against each of the other teams. There are seven teams: the (Texas A&M) Aggies, the (Purdue) Boilermakers, the (Alabama) Crimson Tide, the (Oregon) Ducks, the (Boston College) Eagles, the (Air Force) Falcons, and the (Florida) Gators. How many games will be played in all? Do this problem before reading on.

Michael is a basketball player, and he's always interested in the matchups. In this problem there are seven teams, which Michael quickly assembled into pairs of teams for games:

Aggies vs Crimson Tide   Crimson Tide vs Ducks

Boilermakers vs Gators   Gators vs Aggies

Falcons vs Aggies   Crimson Tide vs Gators

Ducks vs Eagles   Boilermakers vs Aggies

Crimson Tide vs Gators   Falcons vs Eagles

Eagles vs Boilermakers   Ducks vs Gators

Crimson Tide vs Gators   Crimson Tide vs Aggies

Eagles vs Ducks   Ducks vs Boilermakers

Boilermakers vs Eagles   Gators vs Eagles

Don't list the same combination twice.

Is Michael's list systematic? Are all possible matchups represented? Does the list contain omissions or duplications?

Instead of trying to verify the accuracy of Michael's nonsystematic list, look at the first two columns of Monica's systematic list, at right.

Monica represented each of the teams by the first letter of its name. For example, AB represents a matchup between the Aggies and the Boilermakers. She started her list by showing the matchups between the Aggies and the other six teams. In the second column of her list, she showed the matchups between the Boilermakers and the other teams. Note that she didn't include the matchup between the Aggies and the Boilermakers because she'd already shown it in the first column.

AB	BC
AC	BD
AD	BE
AE	BF
AF	BG
AG	

Look for patterns within your list.

She continued by listing, in order, the opposing teams for each remaining matchup. The complete list is shown below.

AB	BC	CD	DE	EF	FG
AC	BD	CE	DF	EG	
AD	BE	CF	DG		
AE	BF	CG			
AF	BG				
AG					

There are 21 different pairs of teams, and each pair played 3 games against each other. So to answer the question "How many games will be played in all?" multiply 21 by 3. The answer is 63 games.

Now compare Monica's solution to this problem's diagram solutions in Chapter 1. You can see that the diagram lines, which represent games, were drawn systematically so that they'd be easy to understand and follow. Diagrams are often systematic. Notice also that the diagram lines correspond exactly to the pairs in Monica's list.

PENNY'S DIMES, PART I

Nick's daughter Penny has 25 dimes. She likes to arrange them into three piles, putting an odd number of dimes into each pile. In how many ways could she do this? Solve this problem before continuing.

Randy solved this problem by making a systematic list of the possible combinations. He made three columns for his list and called them

Pile 1, Pile 2, and Pile 3. In the first row of the list, he indicated the first combination of dimes. He put 1 dime in the first pile and 1 dime in the second pile. This left 23 dimes for the third pile. In the second row he started again with 1 dime in the first pile, then increased the second pile by 2 and decreased the third pile by 2. (Remember that each pile contains an *odd* number of dimes.) He continued in this way for a while, as shown above.

Pile 1	Pile 2	Pile 3
1	1	23
1	3	21
1	5	19
1	7	17
1	9	15
1	11	13
1	13	11

At this point in his list, Randy needed to decide whether or not 1, 13, 11 is a repeat of 1, 11, 13. In other words, is 13 in one pile and 11 in the other the same as 11 in one pile and 13 in the other? Randy decided that the piles were indistinguishable and therefore that these two combinations were indeed the same. He realized that crossing out repeats would save him a lot of work and make his list a lot shorter. So he crossed out the row with 1, 13, and 11. The next combination would be 1, 15, 9, which is a repeat of 1, 9, 15. So he concluded that he'd exhausted the combinations for 1 dime in the first pile.

Pile 1	Pile 2	Pile 3
1	1	23
1	3	21
1	5	19
1	7	17
1	9	15
1	11	13
~~1~~	~~13~~	~~11~~
~~3~~	~~1~~	~~21~~
3	3	19

Next he began finding combinations that started with 3 dimes in the first pile. The first combination he wrote down was 3, 1, 21. He quickly crossed out this combination because he realized that 3, 1, 21 was a repeat of the second combination in the list, 1, 3, 21. So he started with 3, 3, 19. He continued listing combinations with 3 dimes in the first pile until he reached 3, 11, 11. He stopped at this combination because he knew the next combination would be 3, 13, 9, which again would be a repeat.

Randy then moved on to listing combinations with 5 dimes in the first pile. To avoid repeating 5, 1, 19 and 5, 3, 17, he started his combinations with 5, 5, 15. He realized that when he changed the number in the first pile, he had to use that *same number* in the second pile to avoid repeating an earlier arrangement. He also noticed that he began to get repetitious combinations after the number in the second pile became as high as the number in the third pile. For example, when he reached 1, 13, 11, he had a repeat of 1, 11, 13. So here is the primary pattern present in this list: When moving from the first pile to the second pile to the third pile, the numbers cannot decrease. The second pile must be equal to or greater than the first pile, and the third pile must be equal to or greater than the second pile. This type of pattern can appear in many systematic lists.

Pile 1	Pile 2	Pile 3
1	1	23
1	3	21
1	5	19
1	7	17
1	9	15
1	11	13
3	3	19
3	5	17
3	7	15
3	9	13
3	11	11
5	5	15
5	7	13
5	9	11
7	7	19
7	9	9

Randy continued with his list, using the pattern he'd discovered. When he began listing combinations with 9 dimes in the first pile, his first combination was 9, 9, 7. At that point his list was complete, because 9, 9, 7 is a repeat of 7, 9, 9.

Randy's complete list is shown at right. There are 16 ways to form three piles of dimes.

You can solve this problem differently by experimenting with other systems—we encourage you to do so. One possible system would begin with 23 dimes in the first pile. You might also decide to solve the problem again, but this time assume that the three piles are distinguishable, which leads to a much longer list that has 78 possibilities. You should make this list, too. You'll have to modify the system that Randy used, because it will no longer be true that 1, 3, 21 is the same as 3, 1, 21.

FRISBIN

On a famous episode of *Star Trek,* Captain Kirk and the gang played a card game called Phisbin. This problem is about another game, called Frisbin. The object of Frisbin is to throw three Frisbees at three different-sized bins that are set up on the ground about 20 feet away from the player. If a Frisbee lands in the largest bin, the player scores 1 point. If a Frisbee lands in the medium-sized bin, the player scores 5 points. If a Frisbee lands in the smallest bin, the player scores 10 points. Kirk McCoy is playing the game. If all three of his Frisbees land in bins, how many different total scores can he make? Make a systematic list for this problem before reading on.

You can make two different types of systematic lists for this problem. An example of each follows.

Derrick set up a list with columns titled 10 Points, 5 Points, 1 Point, and Total. He began by indicating the maximum number of 10-point throws: 3. He continued by indicating the other possible 10-point throws: 2, 1, and 0. In each row he adjusted the 5-point and 1-point throws so that three throws were always accounted for. After calculating all the point totals, Derrick concluded that Kirk McCoy can make ten different total scores.

10 POINTS	5 POINTS	1 POINT	TOTAL
3	0	0	30
2	1	0	25
2	0	1	21
1	2	0	20
1	1	1	16
1	0	2	12
0	3	0	15
0	2	1	11
0	1	2	7
0	0	3	3

Notice the system in the list. The 10 Points column starts with the highest possible number of throws, then decreases by 1. The column entry stays at each particular possible number of throws (3, then 2, then 1, and finally 0) as long as it can. The 5 Points column follows a similar process: It starts with the highest possible number of 5-point throws for each particular score and decreases by 1 each time. The 1 Point column makes up the difference in the scores.

Derrick made this list very quickly, and anyone seeing the list for the first time should immediately be able to follow the system. To help ensure that the system is evident, we have provided an explanation of it. In this course, when you write solutions to problems that you'll turn in to your instructor, you'll be asked to also provide a written explanation of your work. By explaining your work, you'll not only become a better problem solver, but you'll also become proficient at explaining your reasoning, which is a very valuable skill.

Julian used a different method, shown next. He labeled each column with the number of the three possible throws. Then he wrote down the points for each throw. Describe Julian's system.

THROW 1	THROW 2	THROW 3	TOTAL
10	10	10	30
10	10	5	25
10	10	1	21
10	5	5	20
10	5	1	16
10	1	1	12
5	5	5	15
5	5	1	11
5	1	1	7
1	1	1	3

Julian started by freezing the first two throws at 10 points. He then adjusted the third throw to include each possibility. Then he unfroze the second throw, changed it to 5 points, and froze it again while he adjusted the third throw to include each remaining possibility. (Note that he did not list 10, 5, 10 as a possibility, because that would be a repeat of the second entry in the list, 10, 10, 5.) Then he changed the second throw to 1, and this finished the possibilities where the first throw was 10. He then changed the first throw to 5, froze it again, and adjusted the other two columns. (Again, note that he did not list 5, 10, 10 as a possibility here, because that was included earlier as 10, 10, 5. A rearrangement of the same three numbers is not considered a new possibility, unless the order of the throws makes a difference.) Finally he finished the list with 1, 1, 1. The point totals came out exactly the same as those in Derrick's list, but Julian's approach made it easier to add up the total scores.

Take another look at each entry in Julian's list. Do you notice anything? Study the list before reading on. You should notice a special property about each entry in the list. Reading left to right: the numbers in each row either remain the same or decrease. The numbers never get bigger going from left to right. So, for example, you will never see an entry like 10, 1, 5 or 5, 5, 10 in your list. Why? Think about it before reading on. Julian designed the system to prevent repeated entries. He froze the larger numbers at the beginning of each row, and adjusted the other columns downward. Starting with 10, 10 in the first two columns, he adjusted column three going from 10 to 5 to 1. Then he continued to freeze column one at 10, while he changed column two to 5, but he could not begin column three with 10 because he would repeat an earlier entry. So he used only 5 and then 1 for column three.

This idea of a row never increasing is a property of many systematic lists of this type. You may remember that Randy found a similar pattern in his systematic list for Penny's Dimes Part 1 on page 35, but his list never decreased. For Frisbin you could also have created a list where you never decreased going from left to right. For example, try making a systematic list in a similar style to Julian's list, but begin with the entry 1, 1, 1. Then freeze the first two 1's, and change the third column to 5 and then to 10. Finish the list in this way, and you will notice that reading each row from left to right, you will never decrease.

Finally, let's look at this problem one more time, with a different system. Cali's list does not employ the idea described above of never decreasing or never increasing in each row. What system did Cali use in the list below? Before reading further, study her list to figure out her system.

Throw 1	Throw 2	Throw 3	Total
10	10	10	30
5	5	5	15
1	1	1	3
10	10	5	25
10	10	1	21
5	5	10	20
5	5	1	11
1	1	10	12
1	1	5	7
1	5	10	16

Cali started by listing those situations where all three throws landed in the same bin. Then she listed the situations where two throws landed in the same bin. Finally she listed the one possibility where all three landed in different bins.

AREA AND PERIMETER

A rectangle has an area measuring 120 square centimeters. Its length and width are whole numbers of centimeters. What are the possible combinations of length and width? Which possibility gives the smallest perimeter? Work this problem before continuing.

Tuan explained his solution for this problem: "I read that the area of the rectangle was 120 square centimeters. The first thing I did was to draw a picture of a rectangle.

"I had no idea whether this rectangle was long and skinny, or shaped like a square. But I did know that the area was supposed to be 120 square centimeters. So I made a list of whole-number pairs that could be multiplied to get 120.

"I knew I was done at this point because the next pair of factors of 120 is 12 and 10, which I'd already used. A 12-by-10 rectangle is the same as a 10-by-12 rectangle turned on its side, and I saw no need to list it twice. I also realized that neither 7 nor 9 would work for the width, because they don't divide evenly into 120.

WIDTH	LENGTH	AREA
1 cm	120 cm	120 cm²
2 cm	60 cm	120 cm²
3 cm	40 cm	120 cm²
4 cm	30 cm	120 cm²
5 cm	24 cm	120 cm²
6 cm	20 cm	120 cm²
8 cm	15 cm	120 cm²
10 cm	12 cm	120 cm²

"Now I had to find which possibility gives the smallest perimeter. I knew that the perimeter of a rectangle is the distance around the rectangle, so I needed to add up the length and width. But this would only give me half of the perimeter, so I would have to double the sum of the length and width. I added the Perimeter column to my chart."

WIDTH	LENGTH	AREA	PERIMETER
1 cm	120 cm	120 cm^2	242 cm
2 cm	60 cm	120 cm^2	124 cm
3 cm	40 cm	120 cm^2	86 cm
4 cm	30 cm	120 cm^2	68 cm
5 cm	24 cm	120 cm^2	58 cm
6 cm	20 cm	120 cm^2	52 cm
8 cm	15 cm	120 cm^2	46 cm
10 cm	12 cm	120 cm^2	44 cm

"Now I can see from my chart that the rectangle measuring 10 centimeters by 12 centimeters (which does have an area of 120 cm^2) gives the smallest perimeter of 44 cm."

WHICH PAPERS SHOULD KRISTEN WRITE?

For her Shakespeare course, Kristen is to read all five of the following plays and choose three of them to write papers about: Richard III, The Tempest, Macbeth, A Midsummer Night's Dream, and Othello. How many different sets of three books can Kristen write papers about? Do the problem before continuing.

Li explained her systematic list, shown at right: "I decided to abbreviate the names of the books so I wouldn't have to write out the whole names each time. I used R3, TT, Mac, AMND, and Oth. Then I just made a list. I made my list by letting R3 stay in front as long as it could, and rearranged the other four books into the remaining two spots. Once I had all the combinations that include R3, I dropped it from the list. Then I used TT in the first spot and listed all the combinations that included it. Then I dropped TT, and finally I used Mac in the first spot. I listed the

R3	TT	Mac
R3	TT	Oth
R3	TT	AMND
R3	Mac	Oth
R3	Mac	AMND
R3	Oth	AMND
TT	Mac	Oth
TT	Mac	AMND
TT	Oth	AMND
Mac	Oth	AMND

combination that included Mac, but by that time there was only one more way to do it. There are ten ways altogether."

Travis used a different systematic list to solve the problem: "I made columns for the different books, and then I checked off three in each row. There are ten ways."

	R3	TT	Mac	Oth	MND
1	X	X	X		
2	X	X		X	
3	X	X			X
4	X		X	X	
5	X		X		X
6	X			X	X
7		X	X	X	
8		X	X		X
9		X		X	X
10			X	X	X

A **tree diagram** is another type of systematic list and is used to organize information spatially. A tree diagram's name reflects the fact that it looks like the branches of a tree. (Note that tree diagrams will be discussed further in Chapter 17: Visualize Spatial Relationships.)

After Hosa solved the Which Papers Should Kristen Write? problem, he wondered how many different orders Kristen could write the papers in once she had chosen the books. Suppose she chose *Richard III, The Tempest,* and *Macbeth.* Hosa solved the problem with a tree diagram.

Hosa's tree diagram of the different orders in which the papers could be written for R3, TT, and Mac is shown at right.

R3 — TT — Mac
R3 — Mac — TT
TT — R3 — Mac
TT — Mac — R3
Mac — R3 — TT
Mac — TT — R3

Hosa explained: "The first branch of the tree shows the paper written first. The second branch shows the paper written next. In the second branch I didn't repeat the paper that was written in the first branch. Finally, the third branch shows the paper written last."

You can solve some systematic list problems with tree diagrams. However, sometimes a tree diagram would be too confusing or cumbersome. You will need to decide when a tree diagram would be more useful than a standard systematic list.

Make a Systematic List

Making a systematic list is a great way to organize information. Your first attempt at a list will probably not be the one you end up using.

Plan your list, but don't hesitate to change your plan.

- Start with a messy list or several lists you are willing to give up. They will help you think more carefully about your planning.

- When you make your list, be sure you thoroughly understand your system.

- Continue to monitor your system. When you reach a logical break point, think carefully about the next entry so the next part of your list will continue the patterns you established earlier. Many mistakes are made at transition points in lists, so pay special attention to every transition point.

- Look for the chance to freeze a column or several columns, while you adjust the others accordingly, exhausting the possibilities. Then unfreeze, change, and freeze again.

- If the system doesn't seem to be working, don't be afraid to revise it or to start over.

Enjoy this strategy. Solving problems with systematic lists can be a lot of fun.

General Tips for Problem Solvers

As you work through this book, one major challenge you will face is choosing an appropriate strategy to solve a problem. Often you will find that the best strategy is the first one you chose, but sometimes you'll have to experiment with different strategies to see which one is most effective. As you work on the problems, keep the following thoughts in mind:

Don't give up—try a different strategy.

- Sometimes you'll need to use two or more strategies to solve a problem.

- Being persistent as you try different strategies will often pay off.

- On the other hand, you'll need to develop a sense of when to try something completely different. *Take a risk!*

The most important thing to know about problem solving is that most problems can be solved. As you solve increasingly difficult problems, your confidence and your abilities will increase.

Problem Set A

Solve each problem by making a systematic list.

1. CARDS AND COMICS

Charmaign's daughter has $6.00 she wants to spend on comic books and superhero cards. Comic books cost 60¢ each, and deluxe packages of superhero cards cost $1.20 each. List all the ways she can spend all of her money on comic books, superhero cards, or both.

2. TENNIS TOURNAMENT

Justin, Julie, Jamie, Matt, Ryan, and Roland are the six players in a round-robin tennis tournament. Each player will play a set against each of the other players. List all the sets that need to be played.

3. FREE CONCERT TICKETS

Alexis, Blake, Chuck, and Dariah all called in to a radio show to get free tickets to a concert. List all the possible orders in which their calls could have been received.

4. APARTMENT HUNTING

A management company offers two payment plans for leasing an apartment for one year. Plan A is designed so that a tenant's entry cost is low, and Plan B is designed so that there are more gradual price increases:

PLAN A	PLAN B
12-month lease	12-month lease
$400 first month	$500 first month
$30 per month increase each month	$15 per month increase each month

Which plan costs more for only the ninth month of tenancy? Which plan costs more for the entire year?

5. STORAGE SHEDS

Andre's company manufactures rectangular storage sheds. The sheds are made with aluminum side panels that measure 8 feet, 10 feet, 12 feet, and 15 feet along the bottom edge. For example, one possible shed measures 10 feet by 10 feet. Another possible shed measures 12 feet by 15 feet. List the measurements of all the possible sheds.

6. MAKING CHANGE

Ms. Rathman has lots of nickels, dimes, and quarters. In how many ways can she make change for 50¢?

7. FINISHED PRODUCT

The product of two **whole numbers** is 360, and their sum is less than 100. What are the possibilities for the two numbers?

8. BASKETBALL

Yolanda scored 10 points in a basketball game. She could have scored with one-point free throws, two-point field goals, or three-point field goals. In how many different ways could she have scored her 10 points?

9. THREE ERRANDS

Wes has six errands to run (picking up the dry cleaning, going to the grocery store, getting gas, going to the bank, going to the library, and going to the video store). He only has time to run three of the errands this afternoon. How many different combinations of three errands could he choose to run?

10. PRODUCT

The product of three whole numbers is 120. How many different possibilities are there for the three numbers?

11. AMERICAN IDIOT

At the end of the Broadway musical *American Idiot*—featuring a score by the rock group Green Day—the entire cast comes out to sing the song "Good Riddance (Time of Your Life)". (You can watch a live recording session of this song on Youtube.) For a new recording of the song, the 19 members of the cast arranged themselves into groups of size 2, 3, or 4. How many different ways can they arrange themselves in this way? (For example, on the video on Youtube, where there are only 18 cast members present, there are 2 groups of 3 and 6 groups of 2.)

12. TWENTY-FOUR

How many ways are there to add four positive even numbers to get a sum of 24?

13. TARGET PRACTICE

In a target shooting game, Spencer had four arrows. He hit the target with all four shots. With each shot he could have scored 25 points, 10 points, 5 points, or 1 point. How many total scores are possible?

14. TANYA'S TERRIFIC T-SHIRTS

Tanya is visiting New Orleans, and she wants to bring back T-shirts for all her friends. She's found T-shirts she likes for $5, $10, and $15. She has budgeted $40 for the gifts. List all the ways Tanya can spend $40 *or less* on T-shirts.

15. WRITE YOUR OWN

Create your own systematic-list problem.

CLASSIC PROBLEMS

16. ARCHERY PUZZLE

A target shows the numbers 16, 17, 23, 24, 39, and 40. How many arrows does it take to score exactly 100 on this target?

Adapted from *Mathematical Puzzles of Sam Loyd,* Vol. 2, edited by Martin Gardner.

17. WHICH BARREL WAS LEFT?

There are six barrels, containing 15 gallons, 8 gallons, 17 gallons, 13 gallons, 19 gallons, and 31 gallons. Each barrel contains either oil or vinegar. The oil sells for twice as much per gallon as the vinegar. A customer buys $14 worth of each, leaving one barrel. Which barrel was left?

———

Adapted from *Mathematical Puzzles of Sam Loyd,* Vol. 2, edited by Martin Gardner.

MORE PRACTICE

1. PIZZA

Zaida is throwing a pizza party for her extended family. She has narrowed down her pizza choices to pepperoni for $6, Canadian bacon and pineapple for $10, and deluxe veggie for $15. She has $90 she can spend. In how many different ways can she spend her $90 on some combination of those three pizza types?

2. TOMATO SAUCE

Yesenia plans to make some spaghetti sauce, so she goes to the store to buy some tomato sauce. Tomato sauce comes in three different size containers: 29 oz, 15 oz, and 8 oz. She needs about 60 oz of tomato sauce, but it's OK if she gets 3 or 4 ounces more or less than that. In how many ways can she buy between 56 and 64 ounces of tomato sauce?

3. ODDS TO 22

How many ways are there to add four positive odd numbers to get a sum of 22?

4. BLACKJACK

In blackjack, cards are worth between 1 and 11 points. The object of the game is to get 21 points. In how many numerically different ways can a blackjack player get 21 points with three cards?

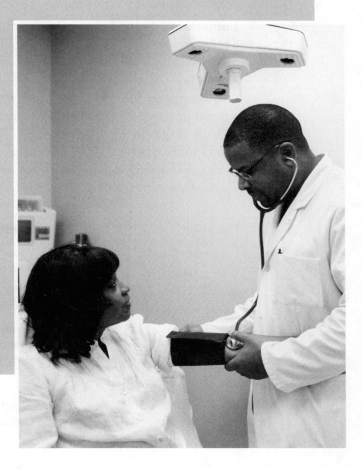

3

Eliminate Possibilities

This strategy forms the basis of logical deduction. Consider all the possibilities, and then eliminate those that lead to contradictions. Health care professionals must conduct tests to eliminate possibilities when diagnosing an illness.

Eliminating possibilities is a powerful problem-solving strategy. Sherlock Holmes used this strategy quite extensively. In his words, "Once you have eliminated the impossible, then whatever is left, no matter how improbable, must be the solution." In this chapter you will solve some problems by eliminating possibilities. Let's start with a made-up game of Twenty Questions.

Two people are playing this game. One player says, "I'm thinking of a number between 1 and 100, inclusive. You may ask whatever twenty questions you like, and I will respond with yes or no answers. To keep the game fair, I will always respond truthfully."

Question	Response	Comments
Is it greater than 50?	Yes	Good question. All numbers less than 51 are eliminated.
Is it 62?	No	Wasted guess. It eliminates only one number (62).
Is it 13?	No	Really wasted guess. The number 13 was eliminated earlier because the number we want is greater than 50.
Is it 3842?	No	
Is it greater than 75?	Yes	Good question. Eliminates about half the remaining numbers.
Is it odd or even?	Yes	
Is it odd?	No	Eliminates about half the remaining numbers.
Is it 83?	No	Bad guess. We know the number is not odd.
Is it less than 85?	Yes	Good; a further narrowing down.
Is it less than 80?	Yes	Again good.
Is it 76?	No	
Is it 78?	Yes	

You probably use this strategy in your daily life. For example, you consider and eliminate possibilities when you decide what to watch on TV or what to have for dinner. Restaurant customers often use the strategy of eliminating possibilities when ordering dinner. Many people decide what to eat based not on what they want but on what they don't want. After eliminating foods they obviously don't like, they often settle on choosing among three or four items. The decision may come about something like this: Imagine that Artie is out to dinner with his family. "Let's see, the chicken sounds good, but so do the pasta, the steak, and the salmon. Well, the salmon comes poached with a white wine cream sauce, and I really would rather have it broiled with lemon, so I'll skip that. The chicken sounds great, but it comes with artichokes, and the thought of artichokes scares me. The steak sounds really good, but my wife says I've been eating too much red meat lately. (I think what she really means is, it's too expensive.) So I'll go with the pasta."

When eliminating possibilities, it helps to remember the possibilities that have already been eliminated. Solve the next problem by eliminating possibilities.

PENNY'S DIMES, PART 2

Penny's favorite coin is the dime, as we saw in Chapter 2. Since we last saw Penny, she has spent some of her dimes and has acquired some more. She doesn't know how many she has now, but she knows she has fewer than 100. One day she was arranging them on her desk in different ways. She found that when she put them into piles of 2, there was 1 left over. When she put them into piles of 3, again there was 1 left over. The same thing happened when she put them into piles of 4. She then tried putting them into piles of 5 and found that there were none left over. How many dimes does Penny have? Solve this problem before continuing. (There is more than one correct answer.)

Make a list of clues from the problem.

Five clues are given in this problem:

1. When divided by 2, the remainder is 1.

2. When divided by 3, the remainder is 1.

3. When divided by 4, the remainder is 1.

4. When divided by 5, the remainder is 0.

5. There are fewer than 100 coins.

This problem will clearly demonstrate that it's important to consider all possibilities before eliminating some. Here's how a group of students—Marli, James, Dennis, and Troy—solved this problem. (Note: You might want to review the divisibility rules in the appendix.)

JAMES: Let's start with the first clue. I'll list all the numbers that give a remainder of 1 when divided by 2. (He wrote down the list shown below.)

All odds

$$3, 5, 7, 9, 11, 13, 15, 17, 19, 21, 23, 25, 27, 29, 31, 33, 35, 37, 39, \ldots$$

MARLI: (interrupting him) Wait a minute, this can be done more efficiently. I like your list because it's systematic, but I think we can improve it. Instead of just considering one clue at a time, I think we can compress this by using two of the clues from the beginning.

DENNIS: Good idea, Marli. Let's try using clues 1 and 2.

TROY: I think we should use clues 4 and 5 instead.

DENNIS: Why not 1 and 2?

Use the clue that gives the shortest list.

TROY: I chose 4 because it involves the fewest numbers, and 5 because it is so easy. So here is a list of all the numbers less than 100 that are divisible by 5. (He then wrote the list below.)

Multiples of 5

$$5, 10, 15, 20, 25, 30, 35, 40, 45, 50, 55, 60, 65, 70, 75, 80, 85, 90, 95$$

Compress the list by applying more than one clue.

JAMES: Now we can go back and reconsider clue 1. My initial list showed us that the number we're looking for is odd. Because we're trying to eliminate possibilities, let's cross off all the even numbers in Troy's list. (He crossed off the numbers as shown.)

$$5, \cancel{10}, 15, \cancel{20}, 25, \cancel{30}, 35, \cancel{40}, 45, \cancel{50}, 55, \cancel{60}, 65, \cancel{70}, 75, \cancel{80}, 85, \cancel{90}, 95$$

Cross out the evens

MARLI: Now let's continue on through the clues. Clue 2 allows us to cross off any number that is not 1 greater than a multiple of 3. Let me explain that. Clue 2 says, "When she put them into piles of 3, again there was 1 left over." That means if you divide the number of dimes by 3, the remainder would be 1. So we have to keep each number that is 1 more than a multiple of 3.

DENNIS: Oh, I see. That means we have to eliminate each number that is *not* 1 more than a multiple of 3.

JAMES: I don't get it. Can you give me an example?

MARLI: Okay. For example, 5 is 2 more than a multiple of 3, so we can cross it out. We can also cross out 15 because it's a multiple of 3. On the other hand, we can't cross out 25 because it's 1 more than 24, which is a multiple of 3. Thus, we can eliminate the multiples of 3—15, 45, and 75—and numbers that are 2 more than a multiple of 3—5, 35, 65, and 95. (She crossed off those numbers in the revised list.)

5, ~~15~~, 25, ~~35~~, ~~45~~, 55, ~~65~~, ~~75~~, 85, ~~95~~

TROY: That leaves us with 25, 55, and 85. Finally, we apply clue 3 and cross out any number that is not 1 greater than a multiple of 4. That means we can cross off 55 because it's 3 more than a multiple of 4 ($52 = 4 \times 13$).

JAMES: That leaves us with the numbers 25 and 85. Each of these satisfies all the clues. Which is the correct answer?

MARLI: I guess we can't tell for sure.

Some problems have more than one correct answer.

This problem has more than one possible answer. You may be accustomed to math problems having only one answer. This is true for many types of equations and problems, but it won't always be true for the problems in this book. In this respect, this book mirrors life, that is, there isn't always one correct answer for a problem, just as there isn't only one correct approach to finding an answer. In this course, when you solve a problem ask yourself whether you have the only answer or there are others to consider. We warned you that this problem had more than one answer, but we may not *always* warn you!

Eliminating possibilities is a way of organizing information. That is, you can eliminate certain possibilities after you organize the information given in a problem. It often helps to consider the possibilities systematically, as in the last problem. The strategy of eliminating possibilities also contains aspects of the guess-and-check strategy, which you'll explore in Chapter 6: Guess and Check.

One particular form of eliminating possibilities is a strategy that is often used in problem solving and in formal mathematics. This strategy is called **proof by contradiction** or **indirect proof.** In this book we will refer to this strategy as **seeking contradictions.** The diagram below shows the process of seeking contradictions:

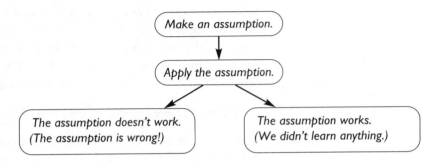

To seek contradictions, (1) make an assumption about the problem and (2) see what happens to the problem when you apply the assumption. This strategy has two possible outcomes: Either it works or it doesn't work. "It doesn't work" is the contradiction we're seeking because it allows us to eliminate the assumption as a possibility. Although "it works" seems to be a good sign, all it really does is confirm that an existing possibility is still an existing possibility.

Seeking contradictions is a common strategy for eliminating possibilities, especially when other possible eliminations are not evident. You'll use this strategy in this chapter, especially when you solve the cryptarithmetic puzzles. You'll also see this strategy in Chapter 4: Use Matrix Logic.

The strategy of seeking contradictions is very useful in problems that involve truth tellers and liars. Variations of this type of problem have been around for years. Solve the next problem by seeking contradictions.

WHO IS LYING?

Jim tells lies on Fridays, Saturdays, and Sundays. He tells the truth on all other days. Freda tells lies on Tuesdays, Wednesdays, and Thursdays. She tells the truth on all other days. If they both say "Yesterday I lied," then what day is it today? Solve this problem before continuing.

Danyell solved this problem by seeking contradictions. First she organized the information by setting up a chart of the two liars and the days of the week.

	MON	TUE	WED	THUR	FRI	SAT	SUN
JIM	T	T	T	T	L	L	L
FREDA	T	L	L	L	T	T	T

Danyell said, "I decided to solve this problem one day at a time. I'll assume that it is a certain day, and see if I can get to a contradiction. To start, assume today is Tuesday. Jim tells the truth on Tuesday. So when he makes the statement, 'Yesterday I lied,' it is supposed to be true. But yesterday was Monday, and I can see by looking at the chart that Jim tells the truth on Monday. This is a contradiction, which is what I was looking for. So today can't be Tuesday. I can make the exact same argument for Wednesday and Thursday using Jim. I can also make the exact same argument using Freda for Saturday and Sunday. So today can't be Tuesday, Wednesday, Thursday, Saturday, or Sunday.

"What about Monday? Assume today is Monday. Both people tell the truth on Monday, which is the only day of the week that happens. When Jim truthfully says on Monday, 'Yesterday I lied,' then he must have lied on Sunday. When I look at the chart, that works—Jim did lie on Sunday. What about Freda? When Freda on Monday says truthfully, 'Yesterday I lied,' then she must have been lying on Sunday. But when I looked at the chart, I see that Freda told the truth on Sunday. That's a contradiction. I can conclude that today cannot be Monday.

"Since I have reached contradictions for every other day, then today must be Friday. I need to check to make sure that it works. I look at the chart and I see that Jim lies on Friday. So if today is Friday his statement, 'Yesterday I lied,' is false. That works, because the chart shows that on Thursday Jim tells the truth. Now consider Freda. She tells the truth on Friday, so her statement, 'Yesterday I lied,' must be the truth. That works because the chart shows that on Thursday Freda lies. So having reached contradictions for every other day, and having shown that Friday worked, I conclude that today is Friday."

Carefully analyze false compound statements.

When dealing with problems involving truth tellers and liars, you must take special care in analyzing compound statements. Consider the statement, "Today Montego ate a hamburger in Memphis." If that statement is false, exactly what does that mean? If that statement is true, exactly what does that mean? If the statement is true, then Montego had to do both things today, she had to eat a hamburger, and she had to be in Memphis when she did so. If she ate a hamburger somewhere other than Memphis, than that statement is false. If she was in Memphis, but did not eat a hamburger, then that statement is false. And if she neither ate a hamburger, nor was in Memphis, then that statement is false. Or if she ate a hamburger in Memphis three days ago but not today, then that statement is false. False compound statements are really tricky.

Let's look at one more case. Consider the following statement, "Yesterday I went to the movies and I played miniature golf." The only way for that statement to be true is if both things happened. If I went to the movies but didn't play miniature golf then the statement is false. If I played miniature golf but did not go to the movies, then the statement is false. If I did neither, then the statement is false. So if a statement like that is a lie, you have to think carefully about what that means. It is natural to assume that if the above statement is a lie, then the person making the statement did not go to the movies and did not play miniature golf. But only one part of a statement involving the word "and" needs to be false in order for the whole statement to be false.

In the next problem you'll eliminate possibilities, seek contradictions, and maybe guess a little bit. Problems like this are sometimes called **cryptarithms.** Doing math problems for pleasure may seem strange to you, but over the years many people have enjoyed cryptarithmetic problems just for fun. We hope you enjoy them, too.

DOWN ON THE FARM

Gordon and Pearl lived on a farm in Concordia, Kansas, with their father, Emil, and their mother, Olive. One day Gordon asked his father, "Dad, what happened to that cat I used to have?" Pearl, overhearing this, said, "Yeah, and I used to have a horse. Where is she?" Emil replied, "Gordon's tomcat and Pearl's old nag were not much use. I traded them for my new goat."

Olive then said, "Hey, that sounds like a good cryptarithmetic problem. Let's see if we can solve it." She wrote down GTOM + PNAG = EGOAT. Each letter stands for a different digit, 0 through 9. No two letters stand for the same digit. Determine which digit each letter represents.

There are two clues. G = 5, and A represents an odd digit. Do this problem before reading on.

Ms. Pieracci's class discussed this problem together.

SILAS: The first thing we can do is fill in that every G = 5.

$$
\begin{array}{ccccc}
 & G & T & O & M \\
 & 5 & & & \\
+ & P & N & A & G \\
 & & & & 5 \\
\hline
E & G & O & A & T \\
 & 5 & & & \\
\end{array}
$$

Use a list to keep track of what you know.

MS. PIERACCI: Let's make two lists. One list for all the digits, and another list for the letters. Notice in this problem there are only eight letters, so we won't need to use all ten digits.

0 I 2 3 4 5 6 7 8 9   G T O M P N A E
            G   5

SIOBHAN: E has to be 1. When you add two numbers together, even if they are 9,999 plus 9,999, the answer is at most 19,998. That means E has to be 1.

MS. PIERACCI: That's good. Does anyone see anything else?

BRENT: I see something kind of weird. In the first column G + P = G. That makes me think that P must be zero, because G + 0 = G. So I think P = 0.

LEXIE: Wait a minute. That also happens in the third column, where O + A = A. So O must be zero too.

MS. PIERACCI: But no two letters can represent the same digit, so O and P can't both be zero. Besides, P can't be zero because no number can start with a zero. If it started with a zero, it would be a three-digit number and not a four-digit number.

LEXIE: That's right. Something is messed up here.

SIOBHAN: P can't be zero anyway. If P were zero, then the answer could not possibly be a five-digit number. You can't add 5 thousand something to a three-digit number and end up with 15 thousand something.

KATSU: What if P is 9?

LEXIE: How could P be 9? If you add 5 and 9, you get 14, not 15. The G in EGOAT wouldn't be a G, it would be something else.

TYLER: I think I see what Katsu means. If there were a carry into the thousands column, then you could add 1 (carry) + 5 + 9 = 15. The two G's would both be 5.

KATSU: Right. You have to have a carry.

MS. PIERACCI: What's the largest digit you can carry when you add two numbers?

SILAS: 1. Even if you had 9 + 9 + a carry, it's only equal to 19. So the biggest number you can carry is 1.

KATSU: So P is 9, and there is definitely a carry into the thousands column.

LEXIE: So does that mean that O has to be zero?

CLASS: Yes.

This is what the solution looks like so far.

```
carries    /

           G  T  O  M          0  I  2  3  4  5  6  7  8  9
           5     0             O  E           G           P

        +  P  N  A  G          G  T  O  M  P  N  A  E
           9        5          5  0     9        1
           ─────────────
           E  G  O  A  T
           /  5  0
```

LUCY: Now what do we do?

MS. PIERACCI: How about figuring out which other columns will have carries in them?

LUCY: The tens column won't. We have that zero in the tens column and the two A's are the same, so we can't have a carry into the tens column because that would make the two A's different.

BRENT: The hundreds column can't have a carry either. The tens column can't add up to be high enough.

MS. PIERACCI: That's great. Finding out that a column does not have a carry in it is just as important as finding out that it does have a carry in it. Now what?

NATALIE:	I think we need to make a list. The hundreds column adds up to 10, because O is zero, and there is a carry into the next column. Let's make a list of all the digits that add up to 10, like 7 and 3, 8 and 2, and so on.
CLINT:	Shouldn't we make the list systematic?
MS. PIERACCI:	Good idea. Let's set up the list. (She wrote it on the board.)

```
T:  9  8  7  6  5  4  3  2  1
N:  1  2  3  4  5  6  7  8  9

carries   1     0     0
          G     T     O     M
          5           0
        + P     N     A     G
          9                 5
        _____
          E     G     O     A     T
          1     5     0
```

MS. PIERACCI:	Is there anything we can eliminate from this list?
SIOBHAN:	Yes. Let's eliminate all the numbers we used already: 0, 1, 5, and 9. And we couldn't use 5 and 5 anyway, because T and N have to be different.
MS. PIERACCI:	Great. (She crossed out some numbers. These were left.)

```
T:  8  7  6  4  3  2
N:  2  3  4  6  7  8
```

MS. PIERACCI:	What else can we eliminate?
EMMA:	What if we look at M? In the ones column, we have $M + G = T$. We already know that G is 5. Since our list has T in it, couldn't we add a row for M too?
SHARON:	I get it. If T is 8, then M has to be 3, because $3 + 5 = 8$.
BJ:	Are we looking for 8 or 18?
KEINAN:	We're looking for 8. We know that we don't have a carry into the tens column.
MS. PIERACCI:	Let's add the possible M's to the list and then look for contradictions. Which possibilities will not work?

```
T:  8  7  6  4  3  2
N:  2  3  4  6  7  8
M:  3  2  1  X  X  X
```

TAYA: Only the first three are going to work. When T is 4, for example, M would have to be −1 in order for −1 plus 5 to equal 4, because we can't have a carry. If we could have a carry, then M could be 9, since 9 + 5 = 14. But, since we can't use negative digits M can't be −1, so this possibility doesn't work.

MS. PIERACCI: So we have contradictions for all but three possibilities.

BRANDY: M can't be 1, because E is already 1.

RODRIGO: So there are two answers that work. And any digit will work for A since 0 + A = A. Zero plus something equals itself. What do we do now?

BJ: Rodrigo is right. Like 0 + 4 = 4 or 0 + 7 = 7. A can be any digit. I think we are stuck.

SARA: Aren't we forgetting something? A has to be an odd digit, because that's what the problem says.

CORIE: Yes, so let's look at the two possible answers.

If T is 8, then N is 2 and M is 3: If T is 7, then N is 3 and M is 2:

```
carries   1  0  0              carries   1  0  0

          G  T  O  M                     G  T  O  M
          5  8  0  3                      5  7  0  2

       +  P  N  A  G                  +  P  N  A  G
          9  2     5                      9  3     5
         _____                    _____
          E  G  O  A  T               E  G  O  A  T
          1  5  0     8               1  5  0     7
```

```
0  1  2  3  4  5  6  7  8  9       0  1  2  3  4  5  6  7  8  9
O  E  N  M     G        T  P       O  E  M  N     G     T     P
```

SARA: Now let's look at the letters that are left for A. In the second case, there are no odd digits left for A.

CORIE: Right, but in the first case the odd digit 7 is left for A. So the first case has to be the answer, with A equal to 7. The digits 6 and 4 don't get used.

carries	1	0	0	
	G	T	O	M
	5	8	0	3
+ P	N	A	G	
9	2	7	5	
E	G	O	A	T
1	5	0	7	8

0	1	2	3	4	5	6	7	8	9
O	E	N	M		G		A	T	P

G	T	O	M	P	N	A	E
5	8	0	3	9	2	7	1

MS. PIERACCI: Great job, everyone.

Solving cryptarithmetic problems is a terrific example of eliminating possibilities. The class solving this problem made many useful points that will help you solve problems of this type. The main points are summarized here.

- Write down the problem with large letters, keeping a space underneath for the digits as you figure them out.

- Carries are important; pay attention to whether a column carries or doesn't carry. Record both carries and noncarries.

- Make a list of the digits from 0 to 9 and write in the letters as you find them.

- Make a list of the letters in the problem and write in the digits as you find them. It also helps to count the letters to discover whether you will use all ten digits.

- The maximum carry when adding two numbers is 1.

This situation occurs often.

- Look for situations like those in this problem, where a digit plus a different digit equals the first digit. In this problem, we had G + P = G. This can happen only if P is equal to 0, or if P is equal to 9 with a carry. This situation occurs often in cryptarithmetic problems.

- Make lists of possibilities that add to a known digit.

- Eliminate things from your lists and then add other digits to your list for each remaining possibility.

- Try alternative possibilities. There may be more than one solution.

- Seek contradictions as a way to eliminate possibilities.
- Don't give up.

The Benefits of Working in Groups

The solution to the Down on the Farm problem also illustrates some important points about working in groups and about class discussions. Each person who worked on this problem contributed something to its solution. Students who contributed useful ideas didn't necessarily know that their ideas would lead to a solution. Share your thinking with one another, even if you haven't solved the problem or your work turns out to be wrong. Mistakes often trigger good ideas. You and other people will benefit from the exchange of information and the thought process. This solution had the benefit of an instructor to facilitate the process. When you work in groups on your own, having someone summarize the group's findings every once in a while may help.

Mistakes often trigger good ideas.

Some people resent the idea that they might require a second or even a third brain to solve a problem. Their perception is that accepting the help of another means they are not smart enough to solve the problem on their own. This perception usually passes quickly as the group works toward a solution. Although it is probable that you can eventually solve most problems by yourself, access to the ideas of other people is very beneficial: Solutions are usually achieved more quickly, and seeing how other people think will make you a better thinker.

Work together.

Studies have shown that the number-one reason people are fired from jobs is their inability to work well with others. Most jobs require significant group interaction. You will be more successful if you are comfortable in the group environment. Cooperation and communication are essential to all social organizations, including families, and solid friendships are based on good communication. Working with others is a skill that cannot be overemphasized. Problem solving offers a good opportunity to further develop this skill.

Remember that many problems cannot be solved by using only one strategy. As you work through this book, you'll come across more and more problems that may require several strategies to solve. The next problem requires that you use two strategies together: making a systematic list and eliminating possibilities. The problem is tough and a little tricky. Take your time, and be sure you consider all possibilities before eliminating some.

Seymour owns his own business. He makes deli sandwiches, which he wraps to retain their freshness and then distributes to several convenience stores for resale. One of his favorite sandwiches is the Sausage and Meatball Combo, but it has a very low distribution. In fact, only three stores take deliveries of the Sausage and Meatball Combo: two Fast Stop stores and one Circle B store.

One morning, Seymour suffered an unfortunate accident. He slipped on the floor and banged his head. He seemed to be fine, except that when he was out on his delivery route, he couldn't remember which streets the three Sausage and Meatball Combo stores were on. The streets were numbered from 1st Street up to 154th Street, and he remembered that the two Fast Stop stores were on streets whose numbers added up to 50. He also remembered that the Circle B store was two streets away from one of the Fast Stop stores, and he remembered that he called the Sausage and Meatball Combo his "prime" favorite because all three stores were on prime-numbered streets. Unfortunately, the information he remembered wasn't enough to get him to the stores.

He called his friend Gus, and Gus remembered that Seymour had told him the **product** of the numbers of the streets the stores were on, but Gus could remember only the last digit of the product. This proved to be enough for Seymour, who promptly double-parked, whipped out a pencil, made a systematic list, and eliminated possibilities to find the answer.

Now it's your turn to re-create Seymour's heroics.

Make a systematic list.

Pairs add to 50.

Richard solved this problem as follows. First he considered the streets that the Fast Stop stores could be on. He made a systematic list of all the pairs of **odd numbers** that add to 50. He used odd numbers because **even numbers** cannot be **prime.** (The exception to this is 2, but 2 would be paired with 48, which obviously is not prime.) He knew some of the odd numbers weren't prime, but he listed them anyway because he wanted to be careful not to accidentally eliminate any possibilities before he considered them.

FS #1	FS #2
1	49
3	47
5	45
7	43
9	41
11	39
13	37
15	35
17	33
19	31
21	29
23	27
25	25

Next he eliminated all the numbers that weren't prime: 1, 49, 45, 9, 39, 15, 35, 33, 21, 27, and 25. He also eliminated the numbers they were paired with, even if they were prime. He was left with four pairs: 3 and 47, 7 and 43, 13 and 37, and 19 and 31. Now he had a better idea of the possible combinations of streets the Fast Stop stores could be on. However, he didn't know specifically which store was on which street. For instance, was the first store on 3rd Street and the second on 47th Street, or was it the other way around? To make sure he accounted for all possible combinations, he added his four pairs to the list in reverse order: 47 and 3, 43 and 7, 37 and 13, and 31 and 19. His revised list is shown above.

FS #1	FS #2
3	47
47	3
7	43
43	7
13	37
37	13
19	31
31	19

Next he added a third column to represent the street the Circle B store was on. He knew that the Circle B store was two streets away from one of the Fast Stop stores. He didn't worry about *which* store the Circle B store was near because he'd already listed all the possible combinations of Fast Stop stores, but he did want to consider whether or not the Circle B store was two streets *up* or two streets *down* from a Fast Stop store. This meant that each third number he added to his list would have to be 2 greater than or 2 less than the second number. He revised his list as shown at right, adding all possible combinations.

FS #1	FS #2	CB
3	47	45
3	47	49
47	3	1
47	3	5
7	43	41
7	43	45
43	7	5
43	7	9
13	37	35
13	37	39
37	13	11
37	13	15
19	31	29
19	31	33
31	19	17
31	19	21

Only primes remain.

However, Richard needed to consider only third numbers that were prime. For example, the first set of Fast Stop numbers was 3 and 47. The two possible third numbers for this pair were 45 and 49, but neither of these numbers is prime so he didn't have to consider them. When he eliminated all the third numbers that were not prime, he was left with the list at right.

FS #1	FS #2	CB
47	3	5
7	43	41
43	7	5
37	13	11
19	31	29
31	19	17

Now consider the last clue of the problem. When Gus told Seymour the last digit of the product of the three numbers, Seymour was able to figure out the answer. Richard didn't know the product, but it is enough for him to know that if *Seymour* knew the product, Seymour could figure it out. See if you can figure out the answer before reading on.

There's a shortcut for finding the last digit of a product.

For each set of numbers in the list, Richard added the last digit of their product to his list. Note that he needed only the last digit of the product. For instance, the product $47 \times 3 \times 5 = 705$ ends in 5. To save himself from having to multiply all the numbers, he used a multiplication shortcut.

FS #1	FS #2	CB	Last Digit
47	3	5	5
7	43	41	1
43	7	5	5
37	13	11	1
19	31	29	1
31	19	17	(3)

He simply multiplied the digits in the ones places of the numbers to figure out what the last digit of the total product would be. For example, for $7 \times 43 \times 41$ he multiplied $7 \times 3 \times 1 = 21$, which ends in 1, the last digit of the product of $7 \times 43 \times 41$.

Richard noticed that 3 appears in the Last Digit column only once, whereas 5 appears twice and 1 appears three times. Thus, because Seymour was able to deduce the correct combination once he knew only the last digit of the product, the last digit must be 3. Otherwise, without more information he wouldn't have been able to figure out the combination. Richard concluded that the streets were numbered 31, 19, and 17. This meant that the Fast Stop stores were located on 31st Street and 19th Street and that the Circle B store was on 17th Street.

This problem contains a key element that appears in a lot of puzzle problems: You didn't have all the information that Seymour had. However, it was enough to know that if Seymour had the information, then he could solve the problem. This bit of knowledge allows you to solve the problem, too.

Look for hidden clues.

Eliminate Possibilities

Use a systematic list to reduce the number of possibilities.

Sometimes seeking contradictions is a valuable problem-solving strategy. By listing all possibilities and eliminating the obviously incorrect ones, you can narrow down the list of possible right answers. Sometimes not all of the possibilities can be eliminated right away.

Detectives use the process of elimination as they try to solve crimes. When they are able to eliminate some of the possible perpetrators quickly, they can concentrate on investigating the remaining suspects. In highly circumstantial cases, they may follow up on their assumptions by sifting through the evidence looking for contradictions. The person for whom there is no contradictory evidence becomes the leading suspect. One potential pitfall of this strategy is that there may be others, who were not investigated, for whom there is a motive but no contradictory evidence.

Beware of assumptions that lead directly to a solution. More than one solution may be possible.

Similarly, when you use the strategy of eliminating possibilities, you may have to make an assumption and follow that line of thinking until you reach either a contradiction or a solution. If you reach a contradiction, you know you can go back and eliminate that assumption. On the other hand, if you reach a solution, you need to go back and check all remaining possible assumptions, because they may also lead to a solution.

Problem Set A

1. SQUARE ROOTS

The square root of 4,356 is an **integer.** Without a calculator, determine what that integer is by eliminating possibilities. Do the same for 8,464.

2. SQUARE WITH MISSING DIGITS

The five-digit number 5abc9 is the square of an integer, where a, b, and c each represent a missing digit. If all five digits are different, what is the five-digit number?

3. PARKING LOT

There are between 75 and 125 cars in a parking lot. Exactly 25% of them are red. Exactly one-ninth of them have out-of-state license plates. How many cars are in the lot?

4. HOW MANY LINES?

Stu counted the lines of a page in his book. Counting by threes gave a remainder of 2, counting by fives also gave a remainder of 2, and counting by sevens gave a remainder of 5. How many lines were on the page?

5. EGGS IN A BASKET

If the eggs in a basket are removed two at a time, one egg will remain. If the eggs are removed three at a time, two eggs will remain. If the eggs are removed four, five, or six at a time, then three, four, and five eggs will remain, respectively. But if they are taken out seven at a time, no eggs will be left over. Find the least number of eggs that could be in the basket.

6. DARTBOARD

Juana threw five darts at a dartboard. The possible scores on the target were 2, 4, 6, 8, and 10. Each dart hit the target. Which of these total scores can be *quickly* identified as "not possible": 38, 23, 58, 30, 42, 31, 26, 6, 14, or 15? (Don't spend more than 1 minute on this problem.)

7. FIND THE NUMBER

If you multiply the four-digit number *abcd* by 4, the order of digits will be reversed. That is, *abcd* × 4 = *dcba*. The digits *a, b, c,* and *d* are all different. Find *abcd*.

8. WOW, WOW, SO COOK!

Denée was having an argument with her roommate, Frankie, about whether or not Frankie could cook. After arguing for a while, Denée said, "Wow, wow, so cook!" Frankie, who was a math teacher, noticed that what Denée said might be a cryptarithm. She sat down to work on it, and Denée ended up cooking dinner. Each letter in the cryptarithm stands for a different digit. (Hint: K = 9.)

```
    W  O  W
    W  O  W
 +     S  O
 ----------
 C  O  O  K
```

9. NELSON + CARSON = REWARD

A story from the Old West tells the tale of two famous outlaws named Nelson and Carson. The wanted poster calling for their arrest indicated that a substantial reward would be offered to the person who caught up to both of them and brought them in for trial. Amazingly, it turned out that the poster contained a great cryptarithm. All who saw the poster realized this and spent their time solving the puzzle rather than looking for Nelson and Carson! When Nelson and Carson heard all the ruckus about the poster, they also tried to solve the problem. However, they weren't too bright and ended up visiting their local sheriff for a clue. He told them that N = 5, and then he arrested them. They solved the problem during the time they spent in jail. Each letter in the cryptarithm stands for a different digit. Find the digits that the other letters represent.

```
   N  E  L  S  O  N
 + C  A  R  S  O  N
 ------------------
   R  E  W  A  R  D
```

10. THE THREE SQUARES

Three cousins, Bob, Chris, and Phyllis, were sitting around watching football on TV. The game was so boring that they started talking about how old they were. Bob (the oldest) noticed that they were all between the ages of 11 and 30. Phyllis noticed that the sum of their ages was 70. Chris (the youngest) pointed out, "If you write the square of each of our ages, all the digits from 1 to 9 will appear exactly once in the digits of the three squares." How old was each person?

11. TO TELL THE TRUTH

Many puzzle books contain puzzles that involve people or creatures who are either liars or truth tellers. You no doubt know from experience, for instance, that many talking dogs are notorious liars. Imagine that you have just encountered three talking dogs and you ask them whether they are liars or truth tellers. Dog 1 says something that you do not understand. Dog 2 says, "He said he was a truth teller." Dog 3 says, "Joe is lying." You then ask, "Which one of you is Joe?" Dog 2 says, "I am the only Joe." Dog 3 points to Dog 1 and says, "He is the only Joe." Determine whether each dog is a liar or a truth teller, or whether it can't be determined.

12. FOUR COLLEGE ROOMMATES

Thuy (the tallest) is older than Miguel (the lightest). Jerel (the oldest) is shorter than Nate (the heaviest). No one has the same rank in any category. For example, if someone is the second tallest, he can't also be the second heaviest or the second oldest. Rank the four roommates in each category: age, height, and weight.

13. THE LETTER FROM COLLEGE

The story goes that a young man away at college needed some extra cash. He sent his mother this plea. He wanted her to send the amount indicated by the following sum: SEND + MORE = MONEY. Each letter stands for a different digit, 0 through 9. No two letters stand for the same digit. How much money did the young man want? (Assume that there is a decimal point between N and E because his mother is probably not willing to send ten thousand or so dollars to her son on request.)

14. WRITE YOUR OWN

Create your own problem that has to be solved by eliminating possibilities. In your problem you may want to somehow use the systematic-list problem you created in Chapter 2: Make a Systematic List. You also may want to try making up a cryptarithmetic problem.

CLASSIC PROBLEMS

15. TURKEYS

Among Grandfather's papers a bill was found:

72 turkeys $_67.9_

The first and last digits of the number that obviously represented the total price of those fowls are replaced here by blanks, for they have faded and are now illegible. What are the two faded digits and what was the price of one turkey?

Adapted from *How to Solve It* by George Pólya.

16. SCRAMBLED BOX TOPS

Imagine that you have three boxes, one containing two black marbles, one containing two white marbles, and one containing one black marble and one white marble. The boxes were labeled for their contents— BB, WW, and BW—but someone has switched the labels so that every box is now incorrectly labeled. You are allowed to take one marble at a time out of any box, without looking inside, and by this process of sampling you are to determine the contents of all three boxes. What is the smallest number of drawings needed to do this?

Adapted from *The Scientific American Book of Mathematical Puzzles and Diversions* by Martin Gardner.

17. THE TRIAL

A king tried his political prisoners based on Frank Stockton's story "The Lady or the Tiger?" The prisoner must choose one of two doors. Behind each door is a room containing either a lady or a tiger. It is possible that both rooms contain ladies or both rooms contain tigers, or that one room contains a lady and the other room contains a tiger. If the prisoner chooses the room with the lady, he goes free. If he chooses the room with the tiger, he is eaten. There is a sign on each door. The king tells the prisoner that the signs on the doors are either both true or both false.

The sign on door I says "At least one of these rooms contains a lady." The sign on door II says "A tiger is in the other room." Which room should the prisoner pick?

———

Adapted from *The Lady or the Tiger? and Other Logic Puzzles* by Raymond Smullyan.

18. THE CONSPIRATORS' CODE

A correspondent sends this interesting puzzle: Two conspirators had a secret code. Their letters sometimes contained little arithmetical sums relating to some quite plausible discussion and having an entirely innocent appearance. In their code each of the ten digits represented a different letter of the alphabet. Thus, on one occasion, a letter contained a little sum in simple addition that, when the letters were substituted for the figures, read as follows:

$$\begin{array}{r} F\ L\ Y \\ F\ O\ R \\ +\ Y\ O\ U\ R \\ \hline L\ I\ F\ E \end{array}$$

It will be found an interesting puzzle to reconstruct the addition sum with the help of the clue that the letters I and O stand for the figures 1 and 0, respectively.

———

Adapted from *536 Puzzles and Curious Problems* by Henry Dudeney, edited by Martin Gardner.

MORE PRACTICE

1. MANDARINS

Adrian grabbed a bag of mandarin oranges off the shelf and spilled them all out on the counter. He divided them up into four piles, but there were 3 left over. He rearranged them into five piles, but this time there were 2 left over. Then he tried putting them into six piles, but there were 3 left over. When he tried seven piles there were 3 left over again. Finally he put them into three piles and he didn't have any left over. If the number of mandarins in the bag was less than 100, how many mandarins were in the bag?

2. BLOCKS

Caden was playing with his blocks. He first put all of his blocks into piles of 3 (there were 3 blocks in each pile) but he had 2 blocks left over. He thought that was annoying, so he then tried putting all of his blocks into piles of 6 and again there were 2 left over. He started to get frustrated. He then put his blocks in piles of 5, and this time there was 1 left over. He was about to throw his blocks at his sister, Brooklyn, but he decided to try one more time. So finally he put them in piles of 7, and there weren't any blocks left over. He was ecstatic and Brooklyn was saved! He has fewer than 100 blocks. How many blocks does Caden have?

3. MISSING DIGITS

The sum below is missing some digits. The missing digits have been replaced by *. Determine what the missing digits are.

$$
\begin{array}{r}
*\ 6\ 9\ 4 \\
8\ *\ *\ 8 \\
+\ 9\ 4\ 9\ * \\
\hline
1\ *\ 9\ 1\ 5
\end{array}
$$

4. INTERESTING NUMBER

If you multiply the five-digit number *abcde* by 4, the order of the digits will be reversed. That is, *abcde* × 4 = *edcba*. The digits *a, b, c, d,* and *e* are all different. Find *abcde*.

Problem Set B

1. THE SIDEWALK AROUND THE GARDEN

We have a garden that measures 17 feet by 20 feet. We want to pour concrete for a 3-foot-wide sidewalk around the garden. To make the forms for the concrete, we will need to buy some lumber. How many feet of lumber will we need just for the perimeter of the walk? (Consider both the inside and the outside perimeter.)

2. A NUMBER OF OPTIONS

Dusty Rhodes is planning to buy a new mountain bike. The different available options are as follows.

a. She can choose either regular tires or extra-heavy-duty tires.

b. She can choose to get plastic, vinyl, cloth, or leather for the seat.

c. She can choose a paint color: brown, silver, or black.

In how many different ways can she order her new bike?

3. GOOD DIRECTIONS?

I stopped at a street corner and asked for directions to Burger Jack. Unfortunately, the person I asked was Larry Longway, whose directions are guaranteed to be too complicated. He said, "You are now facing north. Go straight for two blocks. Turn left. Go straight for one block. Turn right. Go straight for three blocks. Turn right. Go straight for five blocks. Turn right. Go straight for three blocks. Turn left. Go straight for one block. Turn right. Go straight for four blocks. Turn left. Go straight for two blocks. Turn left. Go straight for one block. Turn left. Go straight for five blocks, and you are there." By the time I arrived, I was out of breath and Burger Jack was closed. Please give me the directions for the shortest path from my original spot to Burger Jack. (Assume no streets have dead ends.)

HIGH SCORERS

The five starters for the Seaside Shooters scored all the team's points in the final basketball game of the season. Regina Reporter covered the game, but later her notes were accidentally destroyed. Fortunately she had taped some interviews with the players, but when she played them back only a few quotes seemed relevant to the scoring. She knew the final score was 95 to 94. Using the players' observations, determine how many points each of them scored.

KELLENE: Everybody's totals were odd.

SHAUNTAE: Donna was fourth highest with 17 points. I scored 12 more points than Kellene.

MARTINA: Kellene and I scored a total of 30 points. I outscored her.

HALEY: The last digit in everybody's score was different.

DONNA: Our highest scorer had 25 points.

5. **WAYS TO SCORE**

The Chicago Bears score 18 points in a football game. In how many different ways can the Bears score these points? Points are scored as follows: A safety is 2 points, a field goal is 3 points, and a touchdown is 6 points. After a touchdown is scored, the team may get 1 extra point for a kick, 2 extra points for a 2-point conversion, or 0 points if they fail on either of the above.

4

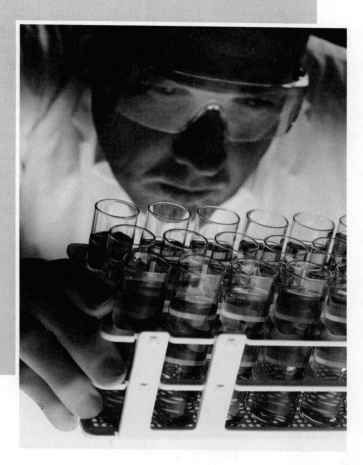

Use Matrix Logic

If you use a matrix, or two-dimensional chart, to organize the information in a problem, you can more easily use the process of elimination in a systematic way. In their study of the chemical processes of living organisms, such as the regulation of metabolism, biochemists use matrix logic to organize various chemical compounds and reactions.

O ne way to improve your logical reasoning ability is to solve logic problems. These problems are fun and challenging. Most grocery stores and drugstores carry booklets or magazines that are full of logic problems. Three such magazines are *Games* and *World of Puzzles,* published by Kappa Publishing Group (kappapublishing.com), and *Dell Math and Logic Problems,* published by Penny Publications (pennydellpuzzles.com). The basic idea behind logic problems is to solve them by matching items in various categories. For example, you might match the first and last names of a list of people, their favorite foods, their favorite colors, and the pets they own, or you might match each person's name with a kind of car and an occupation. Generally speaking, these problems get harder as more people or more categories are added.

Most logic problems can be solved by using a chart or a table that we will call a **matrix.** A matrix helps organize the information in a problem in a useful way and facilitates eliminating possibilities, which is an important strategy for solving logic problems. Many people use matrices to solve problems in their work. Using a **logic matrix** sets up a **one-to-one correspondence** between the elements in a problem. This is a standard feature of most logic problems. In the next problem, for example, each person has one favorite sport.

FAVORITE SPORTS

Ted, Ken, Allyson, and Janie (two married couples) each have a favorite sport: running, swimming, biking, and golf. Given the following clues, determine who likes which sport.

1. Ted hates golf. He agrees with Mark Twain that golf is nothing but a good walk spoiled.

2. Ken wouldn't run around the block if he didn't have to, and neither would his wife.

3. Each woman's favorite sport is featured in a triathlon.

4. Allyson bought her husband a new bike for his birthday to use in his favorite sport.

Use the following matrix to work this problem before continuing. Use an X to represent no and an O to represent yes.

SPORTS

	Running	Swimming	Biking	Golf
NAMES Ted				
Ken				
Allyson				
Janie				

Notice that the top and left side of the matrix are labeled with the main categories given in the problem: names and sports. This matrix labels sports on top and names on the left side, but it could have been set up the other way around. If the problem featured more categories, you'd need more matrices. (You will see examples of this later in the chapter.)

Here's how Damon solved the Favorite Sports problem.

"First I set up my chart [matrix], with the names on the side and the sports up on top. Then I read through the clues. Clue 1 said that Ted didn't like golf. So that eliminated Ted from being the golfer. I put an $\times$ in the Ted-golf space. I also put a 1 next to this $\times$ to show that I had eliminated this possibility by using clue 1.

"Clue 2 said that Ken wouldn't run around the block. So I figured that Ken wasn't the runner, and I put an $\times_2$ in the Ken-running space. The clue also said Ken's wife wasn't the runner, but I didn't know who his wife was, so I left this part of the clue for later.

"The third clue said that the women like sports featured in the triathlon. The only sport in the list that isn't featured in the triathlon is golf, so I put $\times_3$'s in the Allyson-golf and Janie-golf spaces."

SPORTS

	running	swimming	biking	golf
NAMES Ted				$\times_1$
Ken	$\times_2$			
Allyson				$\times_3$
Janie				$\times_3$

"Now Ken is the only person who can like golf, because all other people have been eliminated for golf. So I put an O in the Ken-golf space. This meant that Ken couldn't like any of the other sports, so I put X's in the Ken-swimming and Ken-biking spaces."

SPORTS

	running	swimming	biking	golf
Ted				X_1
Ken	X_2	X	X	O
Allyson				X_3
Janie				X_3

(NAMES)

"Next I read clue 4. It said Allyson's husband likes biking. This meant that Ted had to like biking, since he was the only man left once I knew Ken liked golf. So Ted likes biking and is Allyson's husband. So I put an O_4 in the Ted-biking space. This meant that Ted could not be the runner or the swimmer, so I put X's in the Ted-running and Ted-swimming spaces. The Ted-golf space already had an X in it. I also put X's in the Allyson-biking and Janie-biking spaces, since if Ted was the bicyclist, then nobody else could be."

SPORTS

	running	swimming	biking	golf
Ted	X	X	O_4	X_1
Ken	X_2	X	X	O
Allyson			X	X_3
Janie			X	X_3

(NAMES)

Reread the clues.

"Now I had to figure out who was the runner and who was the swimmer. But I didn't have any more clues. I read through all the clues again. Clue 2 said that Ken's wife wouldn't run around the block. I had skipped that part of the clue when I first read it, but now I thought I could use it. I knew that Allyson bought her husband a bike and that her husband turned out to be Ted. This meant that Ken is married to Janie, so Janie was the one who won't run around the block. I substituted the name Janie for Ken's wife in clue 2 and put an X in the Janie-running space. This left Allyson as the runner and Janie as the swimmer."

SPORTS

NAMES	running	swimming	biking	golf
Ted	X	X	O_4	X_1
Ken	X_2	X	X	O
Allyson	$O_{2,4}$	X	X	X_3
Janie	X	O	X	X_3

"The answers are as follows: Allyson is the runner, Janie is the swimmer, Ted is the bicyclist, and Ken is the golfer. The chart really helped to solve this problem."

Think negatively to eliminate possibilities.

Here are a few things to remember about matrix logic problems:

First, although most people schooled in psychology will tell you to think positively, that's the wrong attitude for this kind of problem. Negatives help eliminate possibilities. **Matrix logic** is a strategy based on things that cannot be.

Second, as soon as you know a connection, correct or incorrect, list it immediately and make the needed mark in your matrix. For example, in the Favorite Sports problem, as soon as you knew that Ted and Allyson were married to each other, you could have started a list with "Ted-Allyson (married)," or you could have put a little heart symbol in the matrix by their names so that you could easily remember their connection. Listing connections for a problem this simple isn't a big deal, but for a more difficult problem keeping a careful list can be an asset. A list like this is called an **adjunct list.** You'll use adjunct lists for some problems later in this chapter.

Finally, Damon had to reread a clue while solving the Favorite Sports problem. He initially skipped the part of clue 2 about Ken's wife and came back to it later when he knew who Ken's wife was. Rereading clues is helpful in solving matrix logic problems.

The Favorite Sports problem was concerned with only two categories: the first name of each person and his or her favorite sport. We needed only one matrix to solve the problem.

The next problem, Outdoor Barbecue, is about four men and uses three categories: the men's first names, their occupations, and what type of meat each one brought to a barbecue. To solve this problem, you need to match up each first name with the correct occupation and the correct meat.

As with the Favorite Sports problem, you'll have to use matrices to compare each category with each other category. A good way to see how many matrices you'll need is to make a diagram. Here is a diagram for the Outdoor Barbecue problem:

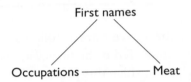

Each trait is matched with another trait in each of the categories. The diagram shows that you'll need three matrices, one for matching first names with occupations, one for matching first names with meats, and one for matching occupations with meat.

Making another diagram helps you organize your matrices into a chart for solving the problem. The three squares in the following diagram represent the three matrices you'll need. Each is labeled with the proper categories.

Next match up like categories, as in the diagram below. You may have to rearrange some of the labels for the trait boxes so categories match up. In this diagram, the labels on the Names-Occupations box were switched so that the two occupations categories and the two names categories could match up.

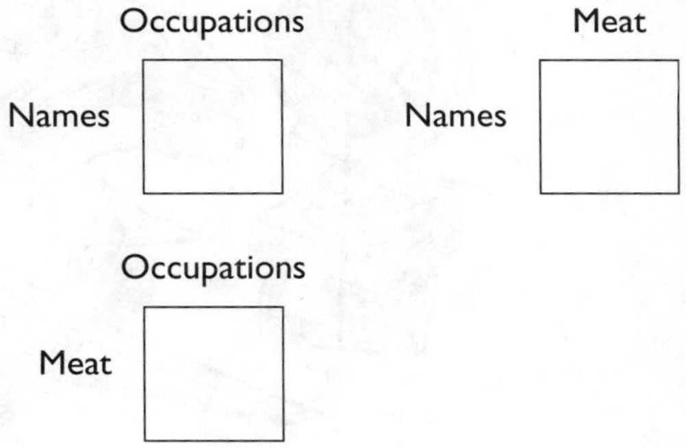

Occupations   Meat

Names   ☐   Names   ☐

Occupations

Meat   ☐

Now put your boxes together, making them into one figure, as shown. Delete the category names that are "squeezed" by putting the boxes together.

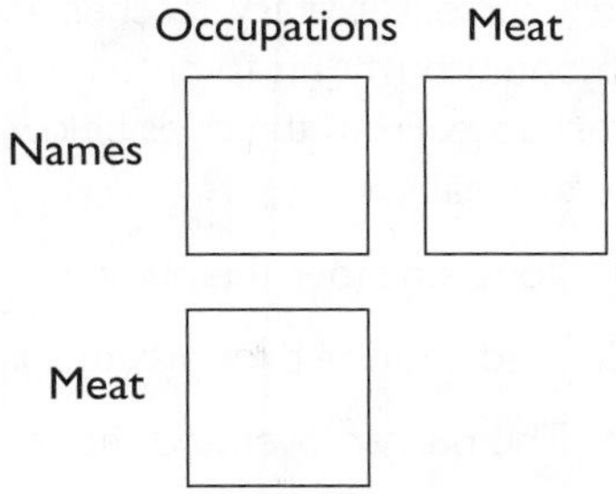

Occupations   Meat

Names   ☐   ☐

Meat   ☐

Now you have an idea of how your Names/Occupations/Meat matrix will be organized. This structure is a good way to organize your chart. As you'll see, it helps you keep track of the information in the problem. Look at the final Names/Occupations/Meat chart in the Outdoor Barbecue problem. Notice that it contains all the categories given in the problem and that its labels correspond to the labels in your final diagram.

OUTDOOR BARBECUE

Tom, John, Fred, and Bill are friends whose occupations are (in no particular order) nurse, secretary, teacher, and pilot. They attended a picnic recently, and each one brought his favorite meat (hamburger, chicken, steak, and hot dogs) to barbecue. From the clues below, determine each man's occupation and favorite meat.

1. Tom is neither the nurse nor the teacher.
2. Fred and the pilot play in a jazz band together.
3. The burger lover and the teacher are not musically inclined.
4. Tom brought hot dogs.
5. Bill sat next to the burger fan and across from the steak lover.
6. The secretary does not play an instrument or sing.

Use the following chart to work this problem before continuing.

		Nurse	Scty	Tchr	Pilot	Burg	Chkn	Steak	Hdog
		OCCUPATIONS				MEAT			
NAMES	Tom								
	John								
	Fred								
	Bill								
MEAT	Burg								
	Chkn								
	Steak								
	Hdog								

Millissent and Tami worked on this problem.

MILL: Let's read the clues and start eliminating. The first clue says that Tom is neither the nurse nor the teacher. So put X's in the Tom-nurse and Tom-teacher boxes.

TAMI: Put little 1's next to the X's so we know we eliminated a box because of clue 1.

MILL: Okay. The next clue says that Fred and the pilot play in a jazz band together. That means that Fred is not the pilot, so put X_2 in the Fred-pilot space.

TAMI: Then from clue 3 we know that the burger lover and the teacher are not musically inclined. So put X_3 in the burger-teacher box.

MILL: Maybe we should start one of those adjunct lists.

TAMI: What are those?

Adjunct lists can organize information.

MILL: When you have other things referred to in the clues, it helps to list them separately. Here we have two people who play in the jazz band and two people who don't. Listing information like this will help us keep everything straight. (She wrote down the adjunct list shown on the next page.)

	In jazz band		Not in jazz band
Fred			burgers
	pilot		teacher

TAMI: How come you wrote Fred and the pilot down on different lines?

MILL: Because we know Fred is not the pilot, so "Fred" and "pilot" describe two different people. And I might get some more information about jazz musicians that I could add next to "Fred" or "pilot."

TAMI: Oh, I see. And the burger lover and the teacher are written on different lines because they're different people as well. I'll bet we can use that list later, too. Wait a second. Fred can't like burgers or be the teacher.

MILL: Why not?

TAMI: Fred plays in the jazz band, and the burger lover and the teacher don't. So Fred can't be either one of them. In the Fred-burger and Fred-teacher spaces, put $X_{2,3}$ because we used clues 2 and 3.

MILL: That's right. And by the same reasoning we can cross off pilot-burger also.

		OCCUPATIONS				MEAT		
		Nurse	Scty	Tchr	Pilot	Burg	Chkn	Steak	Hdog
NAMES	Tom	X_1		X_1					
	John								
	Fred			$X_{2,3}$	X_2	$X_{2,3}$			
	Bill								
MEAT	Burg			X_3	$X_{2,3}$				
	Chkn								
	Steak								
	Hdog								

TAMI: Okay, let's go on. Clue 4 says "Tom brought hot dogs." That's easy. Put an O in the Tom–hot dog space.

MILL: And put X's in the rest of that row and column because Tom couldn't have brought anything else and nobody else brought hot dogs. But watch out, don't put X's past the bold lines by accident.

		OCCUPATIONS				MEAT			
		Nurse	Scty	Tchr	Pilot	Burg	Chkn	Steak	Hdog
N A M E S	Tom	X_1		X_1		X_4	X_4	X_4	0
	John								X_4
	Fred			$X_{2,3}$	X_2	$X_{2,3}$			X_4
	Bill								X_4
M E A T	Burg			X_3	$X_{2,3}$				
	Chkn								
	Steak								
	Hdog								

Cross-correlate with negative information.

MILL: Hey, look at this! We know that Tom brought hot dogs. We also know that Tom is not the nurse or the teacher. So the nurse and the teacher didn't bring hot dogs. So put an X in the nurse–hot dog and teacher–hot dog spaces. We just used cross-correlating with negative information.

① Tom ≠ nurse
Tom ≠ tchr

② Tom = hdog

		OCCUPATIONS				MEAT			
		Nurse	Scty	Tchr	Pilot	Burg	Chkn	Steak	Hdog
N A M E S	Tom	X_1		X_1		X_4	X_4	X_4	0
	John								X_4
	Fred			$X_{2,3}$	X_2	$X_{2,3}$			X_4
	Bill								X_4
M E A T	Burg			X_3	$X_{2,3}$				
	Chkn								
	Steak								
	Hdog	$\times$		$\times$					

Therefore
③ Nurse ≠ hdog
Tchr ≠ hdog

TAMI: That was good, Millissent. The next clue says Bill is not the burger fan or the steak lover, because he was sitting in different spots from them. So put X_5 in Bill-burger and Bill-steak.

MILL: Look, now we can fill in that Bill must be chicken. All the other possibilities are eliminated. Bill brought chicken, so nobody else could have brought chicken. Cross off the rest of the Chicken column.

TAMI: That means that John brought burgers and Fred brought steak.

MILL: Great, one of our matrices is done.

		OCCUPATIONS				MEAT			
		Nurse	Scty	Tchr	Pilot	Burg	Chkn	Steak	Hdog
N A M E S	Tom	X_1		X_1		X_4	X_4	X_4	0
	John					0	X	X	X_4
	Fred			$X_{2,3}$	X_2	$X_{2,3}$	X	0	X_4
	Bill					X_5	0	X_5	X_4
M E A T	Burg			X_3	$X_{2,3}$				
	Chkn								
	Steak								
	Hdog	X		X					

Chicken is the only thing left for Bill. All other meat was eliminated.

MILL: Now we can do more cross-correlating with negative information. Look across Fred's row. Fred is the steak lover, and he isn't the teacher or the pilot. So the steak lover is not the teacher or the pilot. Put X's in the steak-teacher and steak-pilot spaces.

TAMI: And now we see that the teacher likes chicken.

MILL: It also shows that the pilot likes hot dogs, after we cross off the rest of the Chicken row and the Teacher column.

	Nurse	Scty	Tchr	Pilot	Burg	Chkn	Steak	Hdog
Tom	X_1		X_1		X_4	X_4	X_4	0
John					0	X	X	X_4
Fred			$X_{2,3}$	X_2	$X_{2,3}$	X	0 ←	X_4
Bill					X_5	0	X_5	X_4
Burg			X_3	$X_{2,3}$				
Chkn	X	X	0	X				
Steak			X	X				
Hdog	X	X	X	0				

(2) Fred ≠ tchr
Fred ≠ pilot
Therefore
(3) Steak ≠ tchr
Steak ≠ pilot

(1) Fred = steak

Cross-correlate with positive information.

TAMI: Now let's do some cross-correlating with positive information. We know that Tom likes hot dogs. We also know that the pilot likes hot dogs. So Tom must be the pilot. Put an O in the pilot-Tom space.

MILL: Great, Tami. We can do the same thing with Bill. We know that Bill likes chicken and the teacher likes chicken, so Bill must be the teacher.

	Nurse	Scty	Tchr	Pilot	Burg	Chkn	Steak	Hdog
Tom	X_1	X	X_1	0	X_4	X_4	X_4	0
John			X	X	0	X	X	X_4
Fred			$X_{2,3}$	X_2	$X_{2,3}$	X	0	X_4
Bill	X	X	0	X	X_5	0 ←	X_5	X_4
Burg			X_3	$X_{2,3}$				
Chkn	X	X	0	X				
Steak			X	X				
Hdog	X	X	X	0				

(1) Bill = chkn

(2) Chkn = tchr

Therefore (3) Bill = tchr

Write in what we know now.

MILL: We're almost there. The last clue says the secretary does not play an instrument or sing. How are we supposed to use that?

TAMI: We can use that. Remember that adjunct list we made before? Let's look at that again with the new information we've gained so far.

In jazz band			Not in jazz band		
Fred	steak		John	burgers	
Tom	hot dogs	pilot	Bill	chicken	teacher

MILL: Okay, so the secretary is not in the jazz band. That means the secretary can't be Fred. So the secretary has to be John.

TAMI: Right, so Fred is the nurse. Great job.

		OCCUPATIONS				MEAT			
		Nurse	Scty	Tchr	Pilot	Burg	Chkn	Steak	Hdog
N A M E S	Tom	X_1	X	X_1	0	X_4	X_4	X_4	0
	John	X	0	X	X	0	X	X	X_4
	Fred	0	X_6	$X_{2,3}$	X_2	$X_{2,3}$	X	0	X_4
	Bill	X	X	0	X	X_5	0	X_5	X_4
M E A T	Burg	X	0	X_3	$X_{2,3}$				
	Chkn	X	X	0	X				
	Steak	0	X	X	X				
	Hdog	X	X	X	0				

TAMI: Let's be sure to answer the question.

MILL: Okay.

Tom	pilot	hot dogs
John	secretary	burgers
Fred	nurse	steak
Bill	teacher	chicken

Subscripts

The X's and O's in the completed matrix include many subscripts. These subscripts indicate the numbers of the clues that allowed Millissent and Tami to mark them off in the matrix. The subscripts are useful for two reasons.

First, if you make a mistake while solving a problem, you will have to do part of the problem over again. The subscripts will help you remember the information you already had before you made the mistake. Without them, you'd have to start the whole problem over again. With them, you can probably erase just a few marks and start from where you made the mistake.

Second, the subscripts are extremely useful when you re-create your reasoning after you've solved the problem. If you saw only X's and O's in the matrix, you'd have to do the problem again to explain your reasoning. The subscripts remind you why you marked each X and O in the matrix and thus make your explanation easier to write.

Matrix Logic Substrategies

Millissent and Tami used a number of substrategies to solve the Outdoor Barbecue problem. Here is a brief explanation of these techniques.

Adjunct Lists

One approach to using extra information is to create an adjunct list based on all the clues. Millissent and Tami made a list of those who play in the jazz band and those who don't play in the jazz band.

In jazz band	Not in jazz band
Fred	burgers
pilot	teacher
	secretary

From this list it is clear that Fred, who is not the pilot, cannot be the teacher or the secretary because they are not in the band. Therefore Fred must be the nurse.

Marking Traits

The substrategy of marking traits could also be used in this problem. It allows us to mark off a number of squares as Millissent and Tami did when they made an adjunct list using the clues that mention music:

2. Fred and the pilot play in a jazz band together.

3. The burger lover and the teacher are not musically inclined.

6. The secretary does not play an instrument or sing.

On the chart below, each person listed in the clues about music has been marked with either a musical note symbol ♩ or a negated musical note symbol as appropriate.

		OCCUPATIONS					MEAT		
		Nurse	Scty	Tchr	Pilot	Burg	Chkn	Steak	Hdog
NAMES	Tom								
	John								
	Fred ♩								
	Bill								
MEAT	Burg								
	Chkn								
	Steak								
	Hdog								

The burger lover's lack of musical talent needs to be indicated twice, once on each of the two charts where "Burg" appears. We also know that two people play in a jazz band (and two people apparently don't). Because we have already identified the occupations of the two who are not in a band, we can determine that the nurse must be a member of the jazz band, even though the clues about music don't refer to the nurse directly.

Applying these music symbols, we can eliminate a number of boxes, as shown in the next chart.

		OCCUPATIONS				MEAT		
	Nurse ♩	Scty ♪	Tchr ♪	Pilot ♩	Burg ♪	Chkn	Steak	Hdog
Tom								
John								
Fred ♩		X	X		X			
Bill								
Burg ♪	X			X				
Chkn								
Steak								
Hdog								

NAMES (rows: Tom, John, Fred, Bill) · MEAT (rows: Burg, Chkn, Steak, Hdog)

The Outdoor Barbecue problem contained extra information about a jazz band. In other problems, information about a person's gender will often appear. A couple of clues might read "The dog owner sold his old truck yesterday," and "The skydiver gave birth to triplets last week." These clues make it evident that the dog owner and the skydiver are two different people, because they are different genders.

Substitution

After going through the first five clues of the Outdoor Barbecue problem, many people determine that John is the burger lover. This information can be substituted into a couple of the clues to eliminate more possibilities.

Specifically, clues 3 and 5 can be rewritten with the new knowledge:

3. The burger lover (John) and the teacher are not musically inclined.

5. Bill sat next to the burger fan (John) and across from the steak lover.

A **substitution** may lead to the one elimination you need in order to get the rest of the problem to fall into place.

Combining Clues

Clues that talk about the same people can be combined to eliminate more possibilities. Consider clues 1 and 4 from the Outdoor Barbecue problem:

1. Tom is neither the nurse nor the teacher.

4. Tom brought hot dogs.

These clues can be combined into "Tom, the person who brought hot dogs, is neither the nurse nor the teacher." The combined clue allows us to eliminate both the teacher and the nurse as the person who brought the hot dogs.

Be sure to account for all the people in the problem.

A word of caution about combining clues: Make sure your conclusion accounts for all the people in the problem. For example, clue 2 in the Outdoor Barbecue problem informs you that two people play in the same jazz band. Clue 3 describes the two people who are *not* musical, yet clue 6 also describes someone who is not musical. The problem itself has only four people in it, so among these three descriptions at least one person is described twice. Out of four people, we can see that two people are musical (from clue 2) and two people are not musical (from clue 3). Clue 6 characterizes the secretary as not being musical, so the overlap must occur with clues 3 and 6.

3. The burger lover and the teacher are not musically inclined.

6. The secretary does not play an instrument or sing.

The secretary cannot be the teacher, so the secretary must be the burger lover.

Combining clues can be done incorrectly. Consider these two clues:

3. The burger lover and the teacher are not musically inclined.

5. Bill sat next to the burger fan and across from the steak lover.

Does this mean that Bill is the teacher or that the teacher is the steak lover? We can't tell. The clues show an overlap in that the burger lover receives two references. However, we don't know if there is any additional overlap between the two people in clue 3 and the three people in clue 5. These two clues could refer to all four people in the problem, or they could refer to only three. We can't draw any conclusions from combining these two clues.

Cross-Correlations (or Bouncing)

As you work through the clues and gain knowledge about the people and their traits, you will notice more positive identifications of the people and traits. Suppose in a problem about people, favorite color, and home state, you had already found out that Jane is "red" and Jane is "Texas." It then makes sense that "Texas is red." This is called **cross-correlating** (or **bouncing**) with positive information. Cross-correlating can also be done with negative information. Examine the following charts. The chart on the left comes from a problem involving names, colors, and states. The chart on the right comes from a problem involving names, pets, and jobs.

Cross-correlating with positive information	*Cross-correlating with negative information*
Jane is "red."	*Mark is "C.E.O."*
+ Jane is "Texas."	*+ Mark is not "cat owner."*
therefore	*therefore*
"Texas is red."	*"C.E.O. is not cat owner."*

Another way to look at these correlations is with what we call "chart tools," as shown next. Understanding how they work will allow you to transfer information from one chart to another chart for recording different associations of characteristics.

Cross-Correlating with Positive Information

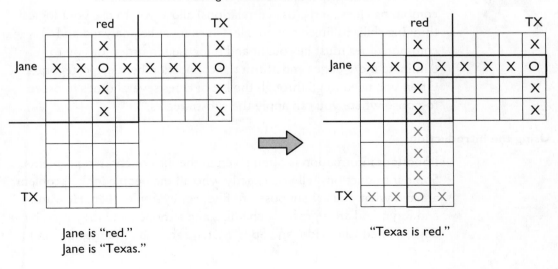

Jane is "red."
Jane is "Texas."

"Texas is red."

Using the positive connection between Jane and red and between Jane and Texas, we can make the connection between Texas and red.

Cross-Correlating with Negative Information

	C.E.O.							cat
Mark	X	O	X	X				X
		X						
		X						
		X						
cat								

Mark is "C.E.O."
Mark is not "cat owner."

	C.E.O.							cat
Mark	X	O	X	X				X
		X						
		X						
		X						
cat		X						

"C.E.O. is not cat owner."

Cross-correlations involve using the information marked on two matrices to transfer it in some form to a third matrix. From positive information, you are able to derive more positive information. When some of the information is negative (negated), you can create only more negated information.

These substrategies (adjunct lists, marking traits, substitution, combining clues, and cross-correlations) allow you to use your logical thinking skills to finish the puzzle. Clues must be analyzed and commonalities must be sought and refocused in order for you to eliminate more boxes and affirm the connection between categories. Often you need to go through the list of clues several times in order to notice where you can apply the substrategies.

Using the Introduction

Important information is often given in the introduction to problems. Some introductions tell you exactly who all the people in the problem are and what you are supposed to find. For example, "Len, Howard, and Raymond are a grocer, a chemist, and a model, and they each have dogs, named Clio, Fido, and Spot. Match each man's name, job, and dog." In other problems, introductions are a little vague. For example, "Three men (one is named Len) have different jobs (one is a model). Each also has a dog (one dog is named Fido). From the clues below, determine each man's name, job, and dog." In this case, you have to read the clues to find the names of the other men, the names of the other dogs, and the other occupations. You would then mark that information in your matrix. Of course, you would hope to be able to tell the difference between a man's name and a dog's name, but what would you do with the name Rex?

Introductions can have hidden information.

Sometimes introductions hide pieces of information you need to solve the problem. For example, "Ivan and three friends (one is a mechanic) like different sports." This sentence tells you that the problem is about four people, one of whom is named Ivan. As with other matrix logic problems, you can probably assume that the people have different jobs unless the problem states otherwise. What you may not notice right away is that the mechanic is not Ivan, because the mechanic is referred to as one of Ivan's friends. If you read the sentence carefully, you'll notice this kind of clue, and you can mark off the Ivan-mechanic space in your matrix.

When you are solving matrix logic problems, you need to be able to make certain assumptions. For example, if someone is described as being "unmusical," it is reasonable to assume that he or she doesn't play in a jazz band. As a general rule, we use the most common understanding of statements in order to interpret the information given. This in turn allows us to eliminate possibilities.

One area in particular that can cause problems is making assumptions about someone's gender. If you are given clues that state "*his* favorite meat . . . ," it is quite clear that the person is a male. Be careful about assigning gender based on names. Bob and Ray clearly are men, and Mary and Beth clearly are women. However, some problems deliberately include names used by either gender. For example, in a problem about four people named Taylor, Blair, Sam, and Casey, whether they were female or male wouldn't necessarily be clear. Determining genders might turn out to be an important part of solving the problem. If a problem includes names from a culture unfamiliar to you, you might have trouble determining gender from the names. The bottom line is to avoid assuming too much.

Making an Assumption

With that warning about assumptions, let's consider one more substrategy: making an **assumption.** Making an assumption is tricky because, in keeping with the idea of thinking negatively, the most important part of making an assumption is proving your assumption *incorrect.* In fact, this substrategy works *only* if you can prove you made an incorrect assumption. The solution to the next problem uses this substrategy.

COAST TO COAST

Four women live in different cities. One of the cities is San Francisco. Determine which city each woman lives in.

1. The woman from Charleston (South Carolina), the woman from Gainesville (Florida), and Riana are not related.

2. Wendy and the woman from Provo are cousins.

3. Neither Phyllis nor Wendy is from the West Coast.

4. Ann is from a coastal city.

Solve this problem before reading on.

		NAMES			
		Ann	Phyllis	Wendy	Riana
CITIES	SF		X_3	X_3	
	Gain	X_4			X_1
	Chrl				X_1
	Provo	X_4		X_2	

Make an assumption.

Right away, the problem's clues allow us to mark the X's in the matrix above. Now let's use the substrategy of making an assumption. We'll assume that Riana is from Provo. The new chart that follows has a light orange O marked in the Riana-Provo space. If Riana is from Provo, then she's not from San Francisco, so mark an X in the Riana–San Francisco space. Thus, Ann must be from San Francisco, because Ann's space is the only space left for San Francisco. We already know that Wendy is from either Charleston or Gainesville (clue 3: Wendy is not from the West Coast). The matrix below shows all of our original marks in blue and all of our new marks, based on our assumptions, in light orange.

		NAMES			
		Ann	Phyllis	Wendy	Riana
CITIES	SF	O	X_3	X_3	X
	Gain	X_4			X_1
	Chrl	X			X_1
	Provo	X_4	X	X_2	O

Seek a contradiction.

Now reread the clues to look for a contradiction. Clue 2 says that Wendy and the woman from Provo are cousins. We assumed that Riana is from Provo, which means that Wendy and Riana are cousins (clue 2). But clue 1 says that Riana and the women from Charleston and Gainesville are not related. Because Wendy is from either Charleston or Gainesville (clue 3), clue 1 tells us that she should not be related to Riana, but according to clue 2 Wendy and the woman from Provo (who we assumed was Riana) are cousins. This contradiction arose from our assumption, so our assumption that Riana is from Provo must be false. Proving an assumption wrong is cause for celebration: Hooray! Therefore, Riana is not from Provo. Knowing this allows us to mark off the Riana-Provo space.

A contradiction eliminates a possibility.

		NAMES			
		Ann	Phyllis	Wendy	Riana
CITIES	SF		X_3	X_3	
	Gain	X_4			X_1
	Chrl				X_1
	Provo	X_4		X_2	X

Now the matrix clearly shows that Riana is from San Francisco. From this point on, we can simply mark spaces in the matrix to solve the entire problem, as in the next matrix.

		Ann	Phyllis	Wendy	Riana
	SF	X	X₃	X₃	O
CITIES	Gain	X₄	X	O	X₁
	Chrl	O	X	X	X₁
	Provo	X₄	O	X₂	X

At this point, be sure to recheck who is related to whom. The women from Gainesville and Charleston are Wendy and Ann, and they are unrelated to Riana. Wendy's cousin is Phyllis (the woman from Provo), so that connection works.

The substrategy of making an assumption can be very dangerous. If your assumption does *not* result in a contradiction, then it proves nothing. Use this substrategy only as a last resort when you are completely stuck on a problem and need to try something drastic. First try the other substrategies in this chapter. You'll be able to solve most logic problems without making any assumptions.

Here's a tip for when you do want to use this substrategy: Before making an assumption, mark in pen every known connection in the matrix. Then, from the point at which you make the assumption, mark in light pencil all further connections. If you determine that the assumption is wrong, that's great. You can erase all your pencil marks, and your earlier connections will still be marked in pen. We illustrated this idea in the second matrix of the Coast to Coast problem by using light and dark colors.

Many logic problems feature more people or categories than you've seen so far in this chapter. For example, a problem may ask you to match up first names, last names, cities, and jobs. As with problems that include fewer categories, your first step in solving such a problem would be to determine how many matrices to combine into your final matrix. A diagram like the one you used for the Outdoor Barbecue problem can help.

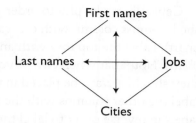

You must have a matrix that matches each category with each other category. From the diagram, you can see that you need the following six matrices:

First names/Last names

First names/Cities

First names/Jobs

Last names/Cities

Last names/Jobs

Cities/Jobs

Setting up the form of the chart is the next important step in solving the problem. Make a 1-by-3 chart comparing first names with each of the other categories: last names, cities, jobs.

	Last names	Cities	Jobs
First names			

The chart above contains three of the six matrices we need. We need to finish the chart to include the other three matrices. Reversing the order of cities and jobs allows us to build the chart without repetitions.

	Last names	Cities	Jobs ← Cities followed by jobs
First names			
Jobs			
Cities			

Jobs followed by cities → (pointing to Jobs/Cities rows)

Generally it's helpful to order your category labels like this: Label the left-most column with one category (First names in the preceding chart). Label the top row with another category (Last names in the chart). Note that when two of your categories are first and last names, they should always be placed in the first column and first row. Then label the other columns with the remaining categories, and repeat these labels in reverse order to label the other rows so that each category matches up with each other category. The repeated categories should always be in reverse order in the rows and columns.

Use Matrix Logic

Matrix logic problems provide great practice for producing logical arguments.

By using a two-dimensional chart, matrix logic combines the strategies of using systematic lists and eliminating possibilities. When you are confronted with a problem, be sure to experiment with your preliminary matrices before settling on a matrix to use. Graph paper makes it easier to draw the matrix. Keep the substrategies in mind:

- Adjunct lists

- Marking traits

- Substitution

- Combining clues

- Cross-correlations (bouncing)

- Using the introduction

- Making an assumption

Solving logic problems can improve your reasoning abilities. When you solve these problems step-by-step, you are using a process similar to that used to develop a written argument or proof. Logic problems can be fun and challenging, as people who buy recreational problem-solving magazines have discovered. Enjoy the challenge!

Problem Set A

1. THE FISHING TRIP

Several friends take a fishing trip every year. Each year they have a contest to see who catches the heaviest fish. The loser has to pay for all of the junk food they eat on the trip. (Second and third places are also expected to chip in token amounts.) Determine each friend's standings in this year's contest by using the following clues. By the way, in the tradition of fishing trips, every statement quoted here is a falsehood.

MARTA: Mickey was first. **WOODY:** I beat Sally.

SALLY: Marta beat Woody. **MICKEY:** Woody was second.

2. DIVISION I ATHLETES

Russ, Don, Pamela, and Stephanie are the first names of four friends who all received sports scholarships. Krieger actually has a full ride, because he is a star in two different sports. Use the clues to determine each person's full name.

1. Hicks and Russ play on the same men's volleyball team.

2. Drake and Braun have both set women's records in swimming.

3. Stephanie and Drake both went to the same high school.

3. A DAY ON THE LAKE

It was a beautiful summer day, and a group of senior citizens—John, Ellen, Armand, and Judy—decided to take the boat out for a day on the Lake. One person looked forward to driving the boat, and another wanted to water-ski. One person always enjoyed diving off of the rocks on Emerald Island, and another was just looking forward to a swim in the cool, clear water. Use the clues to match each person with his or her activity for the day.

1. Judy had been pulling weeds all week and complained of sore "gripping muscles."

2. Armand had forgotten to bring his bathing suit, so he would not get in the water.

3. Ellen had just had her hair done and avoided splashing so it wouldn't get wet.

4. BROTHERS AND SISTERS

Jennifer is on a rampage. Her four younger siblings are in big trouble. When she returned home from the concert, she found that her computer had crashed, her favorite nail polish was half gone, and someone had used her last drop of shampoo. To add insult to injury, her stuffed panda was hanging by one paw from the light fixture.

There was no doubt in her mind that she was the victim of a conspiracy and that all her siblings had been involved. She confronted each one individually, knowing full well that each and every one of them would lie. Although no one gave her enough information to give anyone else away, she was able to put it all together to match each misdeed with its perpetrator. Who was guilty of each crime?

Scott said, "Neither of the girls touched your computer."

Lynn said, "I saw Greg using the nail polish. I think he was gluing his model together."

Cheryl added, "One of the boys used your shampoo."

Greg said, "Lynn used the shampoo. She wanted to brighten up her pink hair."

5. STATE QUARTERS

Two waitresses, Robin and Jen, and their sons, Nicholas, Dustin, and Miles, just started collecting state quarters. To start their collection, they each acquired one quarter representing a different state: Rhode Island, Delaware, Massachusetts, New York, and Pennsylvania. Use the clues to match the names and the states.

1. Nobody got a state quarter from a state whose name started with the same letter as his or her name.

2. Robin and Jen work together. Their coworker Thelma gave one of them the Delaware state quarter.

3. Robin told her son that the Pennsylvania quarter was worth $2. Neither of them has it.

4. One of Jen's two sons has the Massachusetts quarter.

5. Miles got home from school and told his mom that he got the New York quarter. His brother Dustin wasn't home.

6. CABINET MEMBERS

The president was discussing some politics with the vice president and three of her cabinet members: the secretaries of state, education, and the treasury. Using the clues in the conversation below, determine which woman (one is named Norma) holds which position.

Paula said, "Ms. President, I don't think the secretary of state knows what she is talking about. I think our foreign policy has really deteriorated lately."

The secretary of state shook her head. So did the vice president.

The secretary of the treasury said, "I agree with Paula. We haven't even talked to Japan lately."

The vice president jumped in. "Will you two leave Inez alone? She is doing a fine job."

Georgianne, who had been silent so far, finally said, "Okay, let's get on to something else."

The secretary of education said, "I'm sorry, Inez. Nothing personal."

Colleen said, "I'm sorry too, Inez. I guess we just got carried away."

Inez replied, "That's okay. I know we've all been under a lot of stress lately."

7. ECOLOGY EXPERTS

Abbie, Bridget, Cynthia, and Dena are women whose professions are water quality engineer, soil contamination scientist, air pollution consultant, and biological diversity advocate. Match each woman to her expertise using the following four clues.

1. Bridget's expertise is not biological diversity.

2. Abbie loves to garden, but her expertise has nothing to do with soil or air pollution.

3. The air pollution expert, the water quality engineer, and Dena all met one another at a global warming conference.

4. Bridget has never met the person who works on air pollution.

8. **VOLLEYBALL TEAM**

Three friends—Elaine, Kelly, and Shannon—all start for their college volleyball team. Each plays a different position: setter, middle blocker, and outside hitter. Of the three, one is a freshman, one a sophomore, and the other a junior. From the clues below, determine each woman's position and year in school.

1. Elaine is not the setter.

2. Kelly has been in school longer than the middle blocker.

3. The middle blocker has been in school longer than the outside hitter.

4. Either Kelly is the setter or Elaine is the middle blocker.

9. **MUSIC PREFERENCES**

Two men (Jack and Mike) and two women (Adele and Edna) each like a different type of music (one likes jazz). Their last names are Mullin, Hardaway, Richmond, and Higgins. From the following clues, find each person's full name and favorite type of music.

1. Hardaway hates country-western music.

2. The classical-music lover said she'd teach Higgins to play the piano.

3. Adele and Richmond knew the country-western fan in high school.

4. Jack and the man who likes rock music work in the same office building.

5. Richmond and Higgins are on the same bowling team. There are no men on their team.

10. **GRADUATE SCHOOL APPLICATIONS**

Four college friends (one is named Cathy) were all accepted to graduate school. Their last names are Williams, Burbank, Collins, and Gunderson. Each will attend a different graduate school (one is a law school). From the following clues, determine each person's full name and the graduate school he or she plans to attend.

1. No student's first name begins with the same first letter as her or his last name. No student's first name ends with the same letter as the last letter of her or his last name.

2. Neither Hank nor Williams is going to the seminary.

3. Alan, Collins, and the student who is going to the veterinary school all live in the same apartment building. The other student shares a house two blocks away.

4. Gladys and Hank live next door to each other in the same apartment building.

5. The medical school accepted Hank's application, but he decided to study in a different field.

11. SUSPECTS

The police department arrested four suspects—two men and two women—on suspicion of petty theft. The sergeant on duty who processed the suspects was having a bit of a bad day. He produced this list of suspects and descriptions:

Dana Wilde: scar on left cheek

Cary Steele: purple hair

Morgan Fleece: tall and blonde

Connie Theeves: birthmark on left wrist

(problem continued on the next page)

When the list landed on the arresting detective's desk, he was furious. He went to the sergeant and said, "Jean-Paul, you might be having a bad day, but this list is full of mistakes. The first and last names are all mismatched. And none of the descriptions matches either the first or the last name it is listed with. Do you think you can fix this?"

The sergeant replied, "Sorry, Dick. I *am* having a bad day. But I think I need a little bit more information."

The detective answered, "Okay, Jean-Paul. Here's some more info. Connie has purple hair to match her purple high-tops. The men are Steele and Fleece. A woman has the scar. Do you think you can straighten out this mess now?"

The sergeant now determined the first and last names of each suspect, as well as their descriptions. You work it out too.

12. ANNIVERSARIES

Three couples are good friends. At a dinner party one night, they discovered that their anniversaries were in different months: May, June, and July. They also discovered that they had each been married a different number of years: 11, 12, and 13. From the clues below, match up each husband (one is Pierre) with each wife (one is Lorna), the month of their anniversary, and the number of years each couple has been married.

1. Jorge and his wife have three children. Their anniversary is not in July. They have not been married as long as Tara and her husband have.

2. Nylia and her husband have four children. Their anniversary is in June. They have been married longer than Ahmed and his wife.

13. PAYING THE BILLS

Oh, those bills! Kathleen, Ginny, Rosita, and Michelle are all friends in college. Each of them is stuck with bills, some because that's just the way life is and others because they have some financial help from their parents and can spend a little more on extras in life. Even though they're friends, each has a different favorite activity: scuba diving, swimming, rock climbing, or snorkeling. With the help of the following eight clues, determine which woman has which item (apartment, car, clothes, or dorm) as her biggest monthly bill and which activity is her favorite.

1. Neither the person who likes snorkeling, the person whose biggest bill is her apartment rent, nor Kathleen knew one another before college.

2. Ginny and the person who makes dorm payments have been friends since high school.

3. Michelle walks most places because she doesn't have a license. Ginny's grandfather bought her car for her.

4. Rosita left the dorms for good last year after she almost went crazy living there.

5. Rosita and the rock climber keep their clothing expenses to a minimum by making most of their own clothes.

6. The person whose biggest bill was the apartment rent practiced her activity in the Caribbean Sea last summer.

7. The person whose biggest bill is for clothing, the swimmer, and Kathleen went to the football play-offs together.

8. The swimmer, the snorkler, and Rosita all went to the opera together last week.

After filling in as much of your chart as you can using the clues and cross-correlations, you will have to make an assumption and seek contradictions. Try assuming that Kathleen's largest bill is for the dorm, and see where that assumption leads.

14. CLASS SCHEDULES

You are a counselor at River High School. The master schedule for River High School is shown next. The dots indicate the period(s) in which a class is offered. All students must take the classes marked with asterisks. The other classes are electives.

	1	2	3	4	5	6	7
*English	•		•			•	
*Math		•		•	•		
*Science			•	•			
*PE	•	•			•		
History		•					•
Drama			•				
Typing						•	•
Band	•						
*Lunch				•	•		

The following students come to you with schedule requests.

a. Make up a schedule for Jillian. She wants to take Band and History as her electives. She needs Lunch during fifth period because the science club meets then.

b. Make up a schedule for Todd. He wants to take Drama and Typing. He would like to have Lunch during fourth period, but he doesn't care that much.

c. Make up a schedule for Leanne. She wants to take Band and Drama.

d. Make up a schedule for Mea. She wants to take History and Typing.

e. Part 1: José wants to take Drama and History. Make up his schedule.

Part 2: José had his schedule all figured out, but when he went to sign up for first-period PE, he found out that the class was closed. However, the other two periods of PE were still open. What should he do?

Part 3: After making a new schedule that he wasn't too happy with, José found out that the school was opening a new sixth-period Science class. Now what should he do?

15. **WRITE YOUR OWN**

Create your own matrix logic problem. First come up with the people in the problem and what their characteristics are. In other words, start with the answer. We suggest that you use only one chart for your first made-up problem. Then write the clues and solve the problem as you are writing clues. In this way, you will know when you have enough clues.

CLASSIC PROBLEM

16. **THE ENGINEER'S NAME**

Three businessmen—Smith, Robinson, and Jones—all live in the Leeds-Sheffield district. Three railwaymen (guard, stoker, and engineer) of similar names live in the same district. The businessman Robinson and the guard live in Sheffield, the businessman Jones and the stoker live in Leeds, whereas the businessman Smith and the engineer live halfway between Leeds and Sheffield. The guard's namesake earns $10,000 per year, and the engineer earns exactly one-third of what the businessman living nearest to him earns. Finally, the railwayman Smith beats the stoker at billiards. What is the engineer's name?

Adapted from *536 Puzzles and Curious Problems* by Henry Dudeney, edited by Martin Gardner. This problem is one of Henry Dudeney's most popular puzzles and is said to be the prototype of all matrix logic problems.

MORE PRACTICE

1. BOOK CLUB

Four women—Teddi, Sherwood, Luann, and Kathi—have all been in a book club for several years. Each woman has a favorite book (*Water for Elephants* by Sara Gruen, *The Kite Runner* by Khaled Hosseini, *The Poisonwood Bible* by Barbara Kingsolver, and *The Little Book* by Selden Edwards). From the clues below, determine each woman's favorite book.

1. Teddi liked *The Poisonwood Bible*, but it was not her favorite of the four.

2. Sherwood loved the time travel aspects of *The Little Book*, but the last name of the author of her favorite book had a longer last name than Edwards.

3. Luann was too depressed by some of the scenes in Afghanistan in *The Kite Runner* and the African scenes in *The Poisonwood Bible*, so neither book was her favorite.

4. Kathi loved the way they staged the circus elements of her favorite book in the movie *Water for Elephants*.

2. MUSCLE CARS

Two men (Garrett and P-Dawg) and three women (Cigi, Tesha, and Karrin) each drive a muscle car (Chevy Camaro, Chevy Chevelle, Dodge Challenger, Dodge Charger, Ford Mustang). From the clues below, determine which person drives which car.

1. One woman drives a Dodge, one woman drives a Chevy, and the other woman drives the Ford.

2. Garrett and the Camaro owner both work at Denny's. Karrin has never been there.

3. P-Dawg and the two Chevy owners went to the movies together.

4. Cigi loves her Challenger.

3. LUNCH

Nattalie, Max, Andrea, Savanah, and Stephanie went out to lunch. In no specific order they ate spaghetti, steak, salad, chili, and cheese quesadilla. Find out what each person ate for lunch.

1. Nattalie sat across from the person who ordered steak and next to her new friend Savanah.

2. Max does not like vegetables or noodles.

3. Stephanie is lactose intolerant.

4. Savanah is related to the person who ate spaghetti.

5. Neither Stephanie, Nattalie, nor Max like beans.

6. Andrea ate the salad.

This problem was written by Andrea Trihub, a student at Sierra College in Rocklin, California.

4. SECOND GRADE RECESS

Mrs. Lily promised her second grade class (four boys and four girls) ten extra minutes at recess, as long as the students followed one simple rule: only one student is allowed to play one activity at one time. However, two students can play on the see-saw as well as the swings. After recess was over, Mrs. Lily wanted to know which students played what activity. The students thought it would be fun to play a game with their teacher to make her guess which activities they did. Mrs. Lily remembered Juanita got in trouble last week for climbing up the slide, and wasn't allowed to go back on the slide for two weeks. She knew that earlier today, one of the boys pushed his classmate, Alyssa, and received detention, and had to miss out on recess. Just before recess, while Mrs. Lily was helping Nicholas with his math homework, she overheard Jessica telling Juanita that she wasn't able to play with her at recess because she was going to play hopscotch. Pedro's sister told Mrs. Lily yesterday that she didn't want to play on the monkey bars anymore, because they were boring. Mrs. Lily took into account that Isabella and Sean are best friends and knew the two would do something together. Last week the boys had said the see-saw was only for girls, so she knew Pedro and Ryan were going to be playing outside, but not on the see-saw. During art class, Mrs. Lily noticed that Juanita's brother, Pedro, always has the opposite opinion of his sister. Which student went with what activity?

This problem was written by University of Wisconsin – Eau Claire students Jessica Erickson, Ryan Hietpas, Nick Landauer, and Alyssa Spagnola.

SUNSHINE CAFE

Sunshine runs a café in the Galleria. Her servers are named Mace, Shonda, Jaime, and Trixie. Servers' sections are numbered consecutively—1, 2, 3, 4—and each server works in a different section. Each server is wearing a different color shirt—purple, yellow, green, or blue. Match each server to their section and shirt color.

1. The person in section 4 is not wearing a purple shirt.

2. Mace is in section 2 and is not wearing a yellow shirt.

3. Shonda is wearing a blue shirt but is not working in section 3 or 4.

4. Jaime isn't wearing a purple shirt or a green shirt and is working in a section next to Mace's.

This problem was written by Michelle Bautista, a student at Sierra College in Rocklin, California.

Problem Set B

Introducing the Family Family

1. **THE PHONE NUMBER**

Ed, the eldest child of the **Family family,** met a new girl named Candy at the beginning of his senior year in high school. He really liked her, so he wanted her phone number. He knew the first three digits of the number were 492 because the town was so small that everyone had the same telephone prefix. She wouldn't give him the rest of the number at first, but he persisted. Finally, at the beginning of the lunch period, she handed him a piece of paper with several numbers on it.

"The last four digits of my phone number are on this page," she explained.

3257	4682	8824	0626	4608
8624	4632	6428	8604	8428
8064	3195	8420	4218	8240
7915	6420	4602	2628	4178
3281	2804	4002	4826	0846
4718	4680	6402	0428	2406

Ed protested, "But there must be thirty numbers here!"

Candy laughed. "That's right. But I'll give you some clues. If you really want my number, you'll figure it out."

"Okay," he said. "Shoot. I'm ready."

Candy listed her clues:

1. All the digits are even.

2. All the digits are different.

3. The digit in the tens place is less than the other digits.

4. The sum of the two larger digits is 10 more than the sum of the two smaller digits.

During lunch, Ed frantically worked away and tried to figure out which number was Candy's. He looked up with only a few minutes left in the lunch period. "I don't have enough information," he protested. The bell rang, so they walked out of the lunchroom together. She said, "Okay, I'll tell you the sum of all the digits." She whispered the information in his ear.

"Thanks!" Ed said, because this gave him enough information to figure out the number. "I'll call you tonight."

What was Candy's phone number?

2. THE BILLBOARD

Ed went to work for an advertising company after school and on Saturdays. For his first assignment, he hired himself to create a billboard. The billboard will be 20 feet high and will proclaim his love for Candy (the girl, not the sweets). He decided that each letter should be 2 feet high and that there will be a $1/2$-foot space between the bottom of one line of words and the top of the next line of words. In addition, he wants borders measuring $1^1/2$ feet at both the top and the bottom of the billboard. How many lines of words can Ed fit onto the billboard?

Lisa, the eldest daughter in the Family family, is a junior at the same high school that her brother attends. Ed told her about the incident with the phone number, and she decided that she might do something similar sometime. Her opportunity came the next day when Ernie asked her for a date. She liked Ernie, but she couldn't resist a good puzzle. She decided to test Ernie's skill as a puzzler.

"Okay, Ernie, I'd like to go out with you, but I need to see if we're compatible puzzlers. Can you find five ways to add up four even, **positive numbers** (not including zero) and get a **sum** of 16?"

Ernie replied, "Oh sure, I can do that." He whipped out a piece of paper and a pencil and wrote down

$$2 + 2 + 2 + 10 = 16$$
$$2 + 2 + 4 + 8 = 16$$
$$2 + 2 + 6 + 6 = 16$$
$$2 + 4 + 4 + 6 = 16$$
$$4 + 4 + 4 + 4 = 16$$

"Good job," said Lisa. "Now, if you can do this next problem, I'll go out with you. How many ways are there to add up six even, positive numbers (again not including zero) to get a sum of 26?"

Ernie did it. Now you do it too.

Papa Family was sitting at the kitchen table one day, looking at a piece of paper that had his children's names written on it. The paper read

LISA
JUDY
EDWARD

He said to Mama Family, "You know, the names of our kids almost make one of those cryptarithm word-arithmetic problems."

Mama said, "Let me see." She looked at the paper. "Papa, you forgot: Ed's name is Eduard with a *u*, not a *w*. And Lisa's name is really Elisa, but we don't call her that much anymore since she started high school and decided that Lisa is prettier. And don't you remember about Judy's name? It's really Ajudy."

"Oh, yeah," Papa said. "I forgot about that. The nurse came into the delivery room and said, 'Well, what do we have here: a Kathy?' And you said, 'No, a Judy.' What a surprise when the birth certificate actually had 'Ajudy' written on it."

Mama said, "Look, Papa, if you use their real names, it does work as a word-arithmetic problem." She wrote down

$$
\begin{array}{r}
E\ L\ I\ S\ A \\
+\ A\ J\ U\ D\ Y \\
\hline
E\ D\ U\ A\ R\ D
\end{array}
$$

Together, she and Papa solved the problem in a few minutes. You solve it, too. Each letter stands for one of the digits 0 to 9, and no two letters stand for the same digit.

5. THE SPORTING EVENTS

Papa Family just got a new job at the recreation department. His first task is to schedule the playing field for the afternoons of the upcoming week. His first priority in scheduling is to make his three children—Ed, Lisa, and Judy—happy. They all play various sports, and their teams are supposed to play after school during this week. Papa also wants to make his wife, Mama Family, happy. She is taking a golf class and also needs to use the field this week. Use the following information to help Papa figure out a schedule that will satisfy all the desires of all the members of his family.

1. The field is available only on Monday, Tuesday, Thursday, and Friday. One sport will be scheduled each day.

2. Lisa wants to watch the baseball game, but she's in the school play and has rehearsal after school on Tuesday and Thursday.

3. Lisa and Judy have no interest in seeing each other play.

4. One daughter wants her ultimate-Frisbee event to precede the soccer game and the golf class because soccer players and golfers tear up the field.

5. Ed is going to visit colleges on the weekend and wants to get an early start immediately after school on Friday.

5

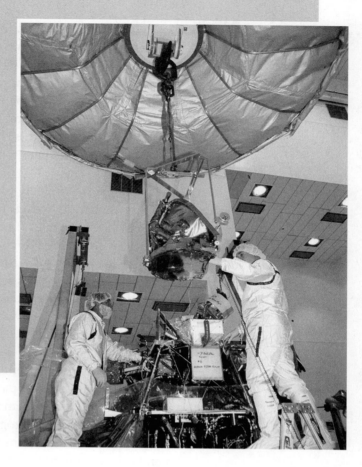

Look for a Pattern

Mathematics has been called the science of patterns. We make sense of our universe by inventing patterns to describe it. Practice will sharpen your pattern-seeking skills. An astronomer uses a spectrometer to record and observe patterns in the frequencies of light radiated from a distant star. The patterns enable her to determine what type of star it is by indicating its chemical composition.

T he study of mathematics is often called the study of **patterns.** Patterns show up everywhere around us. In everyday life, we see thousands of patterns: wallpaper, traffic, automobile designs, weekday afternoon TV schedules, cabinet door arrangements, swimming pool tiles, fence links, and much more. In mathematics, patterns can repeat and extend indefinitely. In this chapter you will develop your ability to recognize and extend patterns, which is a very valuable problem-solving skill.

Recognizing patterns is extremely useful for real-world problems. Detectives look for similarities between crimes to link them together in order to use evidence from one crime to bolster the evidence from another. The patterns established can help predict and prevent similar crimes and apprehend criminals. Researchers need to be able to detect patterns so they can isolate variables and reach valid conclusions. A child uses patterns to learn about the world. For example, children learn to differentiate between positive and negative behavior as they recognize patterns in their parents' reactions.

As a problem-solving strategy, recognizing patterns enables you to reduce a complex problem to a pattern and then use the pattern to find a solution. Often the key to finding a pattern is organizing information.

We begin this chapter with mathematical sequences. A **sequence** is an ordered string of numbers tied together by a consistent rule, or set of rules, that determines the next term in the sequence. A **term** is an individual member of a sequence.

Consider this sequence: 3, 7, 11, 15, ___, ___, ___, ___.

The rule that ties these numbers together appears to be "Add 4 to each term to generate the next term in the sequence," giving 3, 7, 11, 15, 19, 23, 27, 31.

Note that you could continue this sequence by copying the first four terms: 3, 7, 11, 15, 3, 7, 11, 15, ... However, we will not be using copied patterns in this text, nor should you in seeking solutions. In other words, don't take the easy way out while working on these sequence problems. Instead, find the rule and apply it to find subsequent terms in the sequence.

The following problem features examples of mathematical sequences.

Find the pattern and predict the next four terms. Then write a sentence that explains your pattern. Solve each problem before continuing.

A. 1, 2, 4, _____, _____, _____, _____

B. 1, 3, 5, 7, _____, _____, _____, _____

C. 1, 6, 11, 16, _____, _____, _____, _____

D. 1, 4, 9, 16, _____, _____, _____, _____

E. 1, 3, 6, 10, _____, _____, _____, _____

F. 3, 6, 5, 10, 9, 18, 17, 34, _____, _____, _____, _____

G. 1, 3, 4, 7, 11, 18, 29, _____, _____, _____, _____

H. 2, 3, 5, 9, 17, 33, _____, _____, _____, _____

I. 77, 49, 36, 18, _____ (This sequence ends here.)

A. Sequence A in this problem illustrates an important point. Here are two possible answers with the patterns explained:

- 1, 2, 4, 8, 16, 32, 64, . . . (Double each term to find the next term in the sequence.)

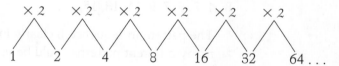

Use "hats" (∧) to show the relationship between successive terms.

- 1, 2, 4, 7, 11, 16, 22, . . . (Start by adding 1 to the first term, then add one greater number to each successive term. That is, add 1, then add 2, then add 3, and so on.)

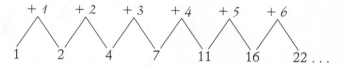

As you can see, there is more than one reasonable, correct answer to this sequence, and each answer follows a consistent rule. Therefore, from this point on check your answers carefully. If you come up with the same answers as the authors of this book did, you are obviously a genius. If you come up with different answers, double-check to make sure you're applying your rules consistently. It is quite possible that some of the sequences in this problem have more than one verifiable correct answer. It's also possible that you could use different patterns to arrive at the same answers, as you'll see in the solutions for sequences B and D.

Here are the most common answers to each problem, although you may be able to justify a different set of answers:

B. 1, 3, 5, 7, 9, 11, 13, 15, . . .

C. 1, 6, 11, 16, 21, 26, 31, 36, . . .

D. 1, 4, 9, 16, 25, 36, 49, 64, . . .

E. 1, 3, 6, 10, 15, 21, 28, 36, . . .

F. 3, 6, 5, 10, 9, 18, 17, 34, 33, 66, 65, 130, . . .

G. 1, 3, 4, 7, 11, 18, 29, 47, 76, 123, 199, . . .

H. 2, 3, 5, 9, 17, 33, 65, 129, 257, 513, . . .

I. 77, 49, 36, 18, 8 (no further terms)

Before you go on to read the explanations below, see if you can decipher the patterns that you didn't get.

B. 1, 3, 5, 7, 9, 11, 13, 15, . . .

The pattern in sequence B clearly involves odd numbers, but many different patterns could be used to arrive at this answer.

1. The pattern is a sequence of odd numbers.

2. Add 2 to each term to get the next term.

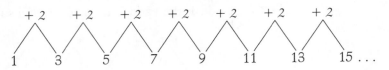

3. The table below shows each term in this sequence in its **position** in the sequence.

Position in the sequence	1	2	3	4	5	6	7	8
Term	1	3	5	7	9	11	13	15

Take the position of each term, double that number, and subtract 1. For example, term number 3 in this sequence is 5. We get 5 by doubling the term number, 3, and subtracting 1. That is, $2 \times 3 - 1 = 5$. Similarly, the eighth term is 15 because $2 \times 8 - 1 = 15$.

C. 1, 6, 11, 16, 21, 26, 31, 36, . . .

To get the next term in sequence C, add 5 to the previous term.

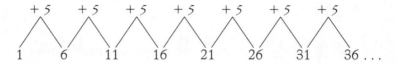

D. 1, 4, 9, 16, 25, 36, 49, 64, . . .

Most people see the pattern in sequence D in one of these ways.

1. To get from one term to the next, add an odd number. To get the term after that, add the next greatest odd number. Continue in this way, adding successive odd numbers. That is, first add 3, then 5, then 7, and so on, always adding the next odd number.

Express the relationship between consecutive pairs of numbers.

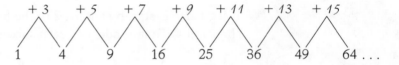

2. Begin by adding 3 to the first term. Then add 2 to each successive addition:

Find the difference of the differences.

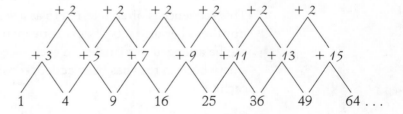

(Note that this pattern is essentially the same as adding successive odd numbers, but it is explained in a slightly different way.)

3. This sequence is a sequence of **square numbers.** The first term is 1×1, or 1^2, the second term is 2^2, the third term is 3^2, the fourth term is 4^2, and so on. **Note** that these numbers are called "square" because sets of dots corresponding to these numbers can be arranged in squares:

The pattern of adding odd numbers is shown in this diagram of squares. In each successive diagram, a new bottom row and right-most column have been added. The number of dots required for these additions always corresponds to the next odd number.

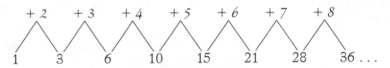

E. 1, 3, 6, 10, 15, 21, 28, 36, . . .

In sequence E, start by adding 2 to the first term. Then add 3. Then add 4. That is, add 1 to each successive addition.

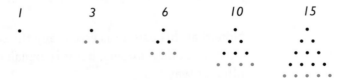

This sequence is often referred to as a sequence of **triangular numbers** because sets of dots corresponding to these numbers can be arranged in triangles. In each successive diagram below, a new bottom row has been added, consisting of one more dot each time.

1	*3*	*6*	*10*	*15*
•	•	•	•	•

F. 3, 6, 5, 10, 9, 18, 17, 34, 33, 66, 65, 130, . . .

Sequence F shows a different kind of pattern. First, multiply the first term by 2 to generate the second term. Then subtract 1 to generate the third term. Then again multiply by 2 to get the fourth term. Then again subtract 1 to get the fifth term. Continue in this way, alternately multiplying by 2 and subtracting 1. Thus, $2 \times 3 = 6$. Then $6 - 1 = 5$, then $5 \times 2 = 10$, and then $10 - 1 = 9$. After that, $9 \times 2 = 18$ and $18 - 1 = 17$. The pattern continues in this way.

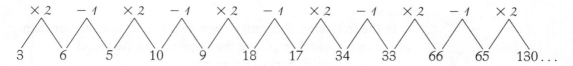

G. 1, 3, 4, 7, 11, 18, 29, 47, 76, 123, 199, . . .

To find the next term in sequence G, add the previous two terms: $1 + 3 = 4, 3 + 4 = 7, 4 + 7 = 11, 7 + 11 = 18, 11 + 18 = 29$. So $18 + 29 = 47, 29 + 47 = 76$, and so on.

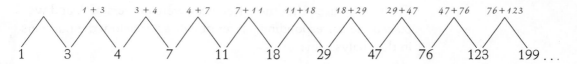

H. 2, 3, 5, 9, 17, 33, 65, 129, 257, 513, . . .

Write the difference between each pair of terms.

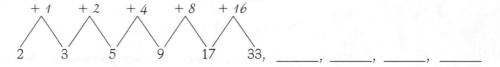

To get from one number to the next, double the previous difference. The first difference is 1, the second difference is $1 \times 2 = 2$, the third difference is $2 \times 2 = 4$, the fourth difference is $4 \times 2 = 8$, and so on. So to continue the sequence, we need to add 32, 64, 128, and 256. This gives us our answers: $33 + 32 = 65, 65 + 64 = 129$ and so on. So fill in the blanks with 65, 129, 257, and 513.

It may appear that we are done, but did you notice that the differences are all powers of 2? That means there may be a general rule for this sequence. So let's take a minute here to explore finding rules in sequences. Sequences are studied in many math courses, and being able to calculate the nth term of a sequence is an important mathematical skill. To calculate the nth term you need to figure out a rule that you can use to get the value of any term by starting with the term number and applying the rule. In the set of differences shown for sequence H, the first difference is 1, the second difference is 2, the third difference is 4, and so on.

In the following diagram, we wrote out 1, 2, 3, 4, etc. for each of the terms and then placed the sequence below it and the analysis of the differences below that.

Term Number	1	2	3	4	5	6
Sequence	2	3	5	9	17	33
Analysis		+1	+2	+4	+8	+16	+32

Then we noticed that the differences were powers of 2, and we placed the corresponding power of 2 directly below the numbers in the analysis line:

Term Number	1	2	3	4	5	6
Sequence	2	3	5	9	17	33
Analysis		+1	+2	+4	+8	+16	+32
Powers of 2		2^0	2^1	2^2	2^3	2^4	2^5

We noticed that the exponent for each power is one less than the term number, so the 1st term goes with 2^0, an exponent of zero, the 2nd term goes with 2^1, and so on. Following this pattern, the terms in the analysis row can be written as $2^{(n-1)}$, where n refers to the term number. Thus, the sixth term in the analysis row is $2^{(6-1)}$, which is $2^5 = 32$.

Looking back at the original sequence, each of the terms is one more than its corresponding term in the analysis row, so the original sequence can be generalized as $2^{(n-1)} + 1$ The sixth term for the original sequence can be calculated as $2^5 + 1 = 33$. The seventh term is $2^6 + 1 = 65$, and so on.

I. 77, 49, 36, 18, 8 (no further terms)

The pattern for sequence I may seem strange compared to the patterns you've looked at so far. This pattern illustrates the need to be mentally flexible when dealing with patterns. For this sequence, treat the digits in each term as separate numbers, then multiply them together. For the first term, $7 \times 7 = 49$. Then $4 \times 9 = 36$, $3 \times 6 = 18$, and $1 \times 8 = 8$. At this point the pattern is over. Try making up your own problem like this. Start with a two-digit number and see if you can come up with a longer sequence than sequence I.

Be mentally flexible.

Finding patterns in sequences is a skill that you can develop. Remember that patterns can be identified at different levels of complexity. Sometimes the pattern lies in the differences between the numbers, sometimes it lies in the differences between the differences, and so on. Sometimes you need to use other operations, such as multiplication or division. Following these procedures will help develop your pattern-detection skills.

Now that you've practiced finding patterns in number sequences, try using patterns to solve problems. Often problems are easier to solve after you find a pattern in them—patterns can reduce a complex problem to a simple one.

DODGER STADIUM

Radio broadcasters joke about the number of people who start leaving Dodger Stadium during the seventh inning of baseball games. One evening, during a particularly boring baseball game in which the Dodgers were trailing by six runs after six innings, the fans began to leave at a record pace. After the first out in the top of the seventh inning, 100 fans left. After the second out, another 150 fans left. After the third out, still another 200 fans left. The pattern continued in this way, with 50 more fans leaving after each out than had left after the previous out. The ridiculous thing was, the Dodgers tied the game in the bottom of the ninth inning, and people continued leaving early.

The game lasted ten innings (the Dodgers lost anyway), and the pattern continued through the bottom of the tenth inning. How many fans left early? Work this problem before continuing.

Chemene wrote the following solution.

"My first reaction was, 'Holy cow, there won't be any fans left at the end of the game.' But after the panic subsided, I was ready. And as I thought about my visit to Dodger Stadium a few summers ago when we visited L.A., I realized that the problem was probably accurate. Anyway, I made a chart [also called a table]."

Inning	Out	Fans leaving	Fans who've left so far
top 7th	1	100	100
top 7th	2	150	250
top 7th	3	200	450
bot 7th	1	250	700
bot 7th	2	300	1,000

"At this point, I realized that this was going to take forever and that there had to be a more efficient way. After I thought about it a little more, I realized that from the top of the seventh inning to the bottom of the tenth, there were going to be 24 outs. So in my next chart, I just counted outs instead of writing the inning too. This was my next chart."

Out	Fans leaving	Fans who've left so far
1	100	100
2	150	250
3	200	450
4	250	700
5	300	1,000

"Then I realized I could do this in a different way. I made a third chart. This time I set up the chart so that I could easily see the total number of fans leaving after each out. I broke down each out into the 100 fans that leave every time, plus the 50 more fans that leave for each out after the first out. For example, after the fourth out, 250 fans had left. This breaks down into 100 + 50 + 50 + 50, which is the way it is shown in the chart below. For each successive out, I just added another row of 50 fans leaving from that out until the end."

Out #	1	2	3	4	5	6	7	...	Total
Base fans leaving	100	100	100	100	100	100	100	...	2,400
Additional fans leaving		50	50	50	50	50	50	...	1,150
Additional fans leaving			50	50	50	50	50	...	1,100
Additional fans leaving				50	50	50	50	...	1,050

"To get the totals on the right, I just multiplied. There were 24 outs of 100 people leaving, then 23 outs of 50 people leaving, then 22 outs of 50 people leaving, and so on. I then totaled the amounts at the side, which, of course, showed a pattern after the first 2,400: 1,150 + 1,100 + 1,050 + 1,000 + ... I then totaled the subtotals to get my answer. So 16,200 people left early. But that includes the people who left after the last out, which technically isn't early. If we don't include the 1,250 people who left right after the last out, then 14,950 people left early."

Chemene realized that her first few attempts would have worked but would have taken forever. She didn't hesitate to abandon one chart in favor of another. Her third chart clearly showed a pattern, which she used to solve the problem.

PITTER PATTER RABBITS' FEET

Tessa wanted to buy a rabbit. She had liked the Easter bunny when she was a kid, so she decided to raise some bunnies of her own. She went to the store with the intention of buying one rabbit, but she ended up with two newborn rabbits, a male and a female. She named them Patrick and Susan. Well, rabbits being what they are (rabbits), it is fairly impossible to have just two rabbits for an extended period of time. She bought them on April 1, 2011. On June 1, she noticed that Patrick and Susan were the proud parents of two newborn rabbits, again one male and one female. She named these new arrivals Thomas and Ursula.

On July 1, Patrick and Susan again gave birth to a male and a female rabbit. She named these Vida and Wanda.

On August 1, Patrick and Susan again gave birth to a male and a female. But Tessa was really surprised to see that Thomas and Ursula also gave birth to a male and a female. Tessa was running out of names, so she didn't bother giving them any.

On September 1, Patrick and Susan gave birth to a male and a female, and so did Thomas and Ursula, and so did Vida and Wanda.[1]

Tessa noticed a pattern to the breeding. A pair of rabbits was born. Two months later they bred a pair of rabbits and continued to breed a pair of rabbits every month after that. Tessa wondered, "If this keeps up, how many pairs of rabbits am I going to have on April 1, 2012?"

Do this problem before continuing.

Four students—whose names just happened to be Pat, Sue, Tom, and Ula—worked on this problem.

PAT: Wow, this is weird. These rabbits have the same names that we do.

TOM: Let's try to do this problem, okay? We don't need to know whether the rabbits are named after us.

PAT: I just thought it was interesting.

ULA: How are we going to do this? I'm totally confused.

SUE: Let's try making a systematic list.

PAT: Okay, let's see. We have adult rabbits and baby rabbits.

SUE: Yeah, but we also have teenaged rabbits. After Thomas and Ursula were born on June 1, they didn't have babies until August 1. So in July they were just teenagers.

---

[1]Tessa also started trading young rabbits with other breeders, male for male and female for female, in order to maintain genetic diversity.

Make a table. Find a pattern.

ULA: Let's get all this down in a table. Then maybe we can find a pattern. (Ula began writing a table. All numbers represent pairs of rabbits.)

Month	Adults	Teenagers	Babies	Total
April	0 pr	0 pr	1 pr	1 pr

TOM: That's great, Ula. Okay, so in May, Patrick and Susan grow to be teenagers. And then in June they become adults and have babies, Thomas and Ursula. And in July they have more babies.

Month	Adults	Teenagers	Babies	Total
April	0 pr	0 pr	1 pr	1 pr
May	0	1	0	1
June	1	0	1	2
July	1	1	1	3
August	2	1	2	5

SUE: Wait, how did you get August?

ULA: Well, in August Thomas and Ursula grew up to be adults, and they had babies. So did Patrick and Susan.

PAT: Who are the teenagers?

TOM: Vida and Wanda. They were babies in July, so they are teenagers in August.

PAT: Oh, I think I see a pattern. The number of adults is the same as the number of babies because each pair of adults has a pair of babies.

SUE: Yeah, and the ones who are babies this month become teenagers next month.

ULA: Right. And the adults are whoever were adults last month plus whoever were teenagers last month.

TOM: I think we can figure this out now.

Month	Adults	Teenagers	Babies	Total
April	0 pr	0 pr	1 pr	1 pr
May	0	1	0	1
June	1	0	1	2
July	1	1	1	3
August	2	1	2	5
September	3	2	3	8
October	5	3	5	13
November	8	5	8	21
December	13	8	13	34
January	21	13	21	55
February	34	21	34	89
March	55	34	55	144
April	89	55	89	233

Example of patterns:
- *Oct., 5 adults →*
  5 babies
- *Oct., 5 adults +*
  3 teens →
  Nov., 8 adults

TOM: I think we're done. Tessa will have 233 pairs of rabbits on April 1, 2012. Great job, guys.

SUE: Wait a second. I see a pattern here. Look down the Total column. You just have to add the two numbers above it to get the next number.

PAT: What?

TOM: I see what she means. Look at the top of the Total column. The first two numbers are 1 and 1. The next number is 2, which is 1 + 1. The next number is 3, which is 1 + 2. The next number is 5, which is 2 + 3. Then look, 8 is 3 + 5. So just add the two numbers above to get the next number.

ULA: Neat. And look, the same thing happens in all the other columns, but the pattern starts later. And the numbers in each column are the same: 1, 1, 2, 3, 5, 8, 13, 21, 34, ...

PAT: Wow, that's cool. I think this sequence has a name. The Liberace sequins, maybe?

TOM: Not sequins. *Sequence.*

ULA: I think it's the Fibonacci sequence. It's named after an Italian.

TOM: I guess when you think you're done, maybe you're not. It's kind of neat to look back and find the pattern. With the pattern in our table, it would be really easy to keep going.

SUE: Yeah, but let's not. But we might see that pattern again someday.

This problem illustrates the effectiveness of finding patterns. The group found a pattern in what was going on as they worked the problem, but they didn't catch the overall pattern until they were done. Tom is right. It's always good to look back at your work to notice things you might have missed and to make sure your answer is reasonable.

The sequence in the rabbit problem is called the Fibonacci sequence, which is 1, 1, 2, 3, 5, 8, 13, 21, 34, ... Each term is the sum of the previous two terms. The next term will be 21 + 34, or 55, the following term will be 34 + 55, or 89, and so on. The Fibonacci sequence is named after the Italian mathematician Leonardo of Pisa (ca 1170–1240), called Fibonacci (son of Bonaccio) by an editor of his works in the nineteenth century.

The sequence that bears his name shows up in some surprising places in nature. For instance, if you count the two sets of spirals on a pinecone, you will always get consecutive **Fibonacci numbers** such as 5, 8 or 8, 13 or 13, 21. For example, a pinecone might have 8 spirals going in one direction and 13 in the other. The same pattern also occurs in sunflowers, pineapples, cacti, and other plants. Read more about Fibonacci in an encyclopedia or a book on the history of mathematics.

Also note that the sequence 1, 3, 4, 7, 11, 18, . . . (solved as sequence G on page 113) is often referred to as the Lucas sequence, named after French mathematician Edouard Anatole Lucas (1842–1891). The "seeds" of the Lucas sequence are 1 and 3, rather than 1 and 1 as in the Fibonacci sequence.

NIGHT OF THE HOWLING DOGS

Shawna liked to jog in the evening. One night she noticed an unusual phenomenon. As she jogged, dogs would hear her and bark. After the first dog had barked for about 15 seconds, two other dogs would join in and bark.

In about another 15 seconds, it seemed that each barking dog would "inspire" two more dogs to start barking. Of course, long after Shawna passed the first dog, it continued to bark, as dogs are inclined to do. After about 3 minutes, how many dogs were barking (as a result of Shawna's passing the first dog)? Work this problem before continuing.

Loc solved this problem:

"This sounds like my neighborhood, 24 hours a day. I started out by trying to draw a diagram of what was going on. It was a mess."

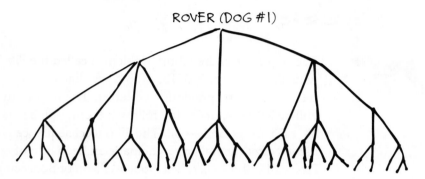

ROVER (DOG #1)

"The biggest problem seemed to be that the first dog doesn't just stop, but continues barking and continues inciting more dogs to stupid barking frenzies. And every other dog that starts barking also incites new dogs to bark.

"I realized I needed to organize the problem into a table. It starts off with one dog barking. Then 15 seconds later, two new dogs start barking. So now there are three dogs barking. Then, 15 seconds later, every dog that is barking gets two more dogs to start barking. So the three barking dogs inspire six new ones to start. I put all of this in a table."

Loc's table is shown on the following page. The Dogs Barking column refers to the number of dogs that *are* barking at the beginning of each round. The New Dogs Barking column refers to the number of dogs that *start* barking during each round. The Total Dogs Barking column shows the number of dogs that are barking at the *end* of each round. The total dogs barking at the end of one round becomes the number of dogs barking for the next round.

Time	Round	Dogs Barking	New Dogs Barking	Total Dogs Barking
0	0	0	1	$1 = 3^0$
:15	1	1	2	$3 = 3^1$
:30	2	3	6	$9 = 3^2$
:45	3	9	18	$27 = 3^3$
1:00	4	27	54	$81 = 3^4$

Loc concluded that at the end of 3 minutes, which would include 12 rounds, there would be 3^{12} dogs barking, or 531,441 dogs.

He finished by saying, "This answer may seem completely unbelievable to you, however, if you've been around my neighborhood, you would probably think the answer was close to dead-on."

MILK LOVERS

Alysia and Melissa and Dante and Melody loved milk. They convinced their older brother, Mark, who did all the shopping, to buy each of them a gallon of milk because they liked it so much. They all put their names on their gallons. One day, they were all really thirsty, and each took ten drinks according to a different system:

Alysia started by drinking half of the milk in her container. Then she drank one-third of what was left. Then she drank one-fourth of what was left, then one-fifth, and so on.

Melissa started by drinking one-eleventh of her milk, then one-tenth of what was left, then one-ninth of what was left, and so on.

Dante started by drinking one-half of his milk, then two-thirds of what was left, then three-fourths of what was left, then four-fifths, and so on.

Melody started by drinking one-half of her milk, then one-half of what was left, then one-half of what was left, and so on.

After each had taken ten drinks, how much milk remained in each container? Work this problem before continuing.

Bimiljit answered the problem in this way:

"This problem seemed a little tough. I examined Melody's usage first because it seemed to be the easiest. I made a chart and looked for a pattern. For example, on the second drink, she drank one-half of the remaining half, which is one-fourth of the whole container. This leaves her with one-fourth of the whole container."

Make a table to
make patterns easily
recognizable.

Drink #	Amount drunk	Amount remaining
1	$\frac{1}{2}$	$1 - \frac{1}{2} = \frac{1}{2}$
2	$\frac{1}{2} \times \frac{1}{2} = \frac{1}{4}$	$\frac{1}{2} - \frac{1}{4} = \frac{1}{4}$
3	$\frac{1}{2} \times \frac{1}{4} = \frac{1}{8}$	$\frac{1}{4} - \frac{1}{8} = \frac{1}{8}$

"The pattern for the amount remaining was pretty obvious: $\frac{1}{2}, \frac{1}{4}, \frac{1}{8}, \frac{1}{16}, \ldots$. So I looked at each denominator as a power of 2, and it turned out that each power matched up exactly with which drink she was taking. For example, $\frac{1}{8}$ is $\frac{1}{2^3}$ after the third drink. So after ten drinks, Melody would have $\frac{1}{2^{10}}$ or $\frac{1}{1024}$ of her milk left. That's not very much.

"I then tried to make the same kind of chart for Alysia because hers seemed to be the next easiest. On her second drink, she drinks $\frac{1}{3}$ of the remaining half. This is $\frac{1}{6}$ of the whole container. Subtracting $\frac{1}{6}$ from $\frac{1}{2}$ gives $\frac{1}{3}$ left in the container after two drinks."

Drink #	Amount drunk	Amount remaining
1	$\frac{1}{2}$	$1 - \frac{1}{2} = \frac{1}{2}$
2	$\frac{1}{3} \times \frac{1}{2} = \frac{1}{6}$	$\frac{1}{2} - \frac{1}{6} = \frac{1}{3}$
3	$\frac{1}{4} \times \frac{1}{3} = \frac{1}{12}$	$\frac{1}{3} - \frac{1}{12} = \frac{1}{4}$
4	$\frac{1}{5} \times \frac{1}{4} = \frac{1}{20}$	$\frac{1}{4} - \frac{1}{20} = \frac{1}{5}$

"I couldn't believe how easy this pattern was once I saw it. The numerator is 1 and the denominator is 1 more than the drink number. The tenth drink would leave $\frac{1}{11}$ of the milk remaining.

"Next I tried doing Melissa's chart. Hers seemed to be a little harder."

Drink #	Amount drunk	Amount remaining
1	$\frac{1}{11}$	$1 - \frac{1}{11} = \frac{10}{11}$
2	$\frac{1}{10} \times \frac{10}{11} = \frac{1}{11}$	$\frac{10}{11} - \frac{1}{11} = \frac{9}{11}$
3	$\frac{1}{9} \times \frac{9}{11} = \frac{1}{11}$	$\frac{9}{11} - \frac{1}{11} = \frac{8}{11}$
.		
.		
.		
10	$\frac{1}{2} \times \frac{2}{11} = \frac{1}{11}$	$\frac{2}{11} - \frac{1}{11} = \frac{1}{11}$

"The pattern wasn't hard to see: The amount of milk remaining keeps going down by $1/11$. But the end was hard to figure out. I figured out that the pattern for the numerator goes 10, 9, 8, 7, 6, 5, 4, 3, 2, 1. So after ten drinks she would have $1/11$ of her milk left. The amazing thing was that this final amount was the same as Alysia's.

"Dante's chart seemed like the hardest. But the chart-and-pattern-finding approach was working so well, I just kept at it."

Drink #	Amount drunk	Amount remaining	Analysis
1	$\frac{1}{2}$	$1 - \frac{1}{2} = \frac{1}{2}$	$\frac{1}{2}$
2	$\frac{2}{3} \times \frac{1}{2} = \frac{1}{3}$	$\frac{1}{2} - \frac{1}{3} = \frac{1}{6}$	$\frac{1}{6} = \frac{1}{2} \times \frac{1}{3}$
3	$\frac{3}{4} \times \frac{1}{6} = \frac{1}{8}$	$\frac{1}{6} - \frac{1}{8} = \frac{1}{24}$	$\frac{1}{24} = \frac{1}{6} \times \frac{1}{4}$
4	$\frac{4}{5} \times \frac{1}{24} = \frac{1}{30}$	$\frac{1}{24} - \frac{1}{30} = \frac{1}{120}$	$\frac{1}{120} = \frac{1}{24} \times \frac{1}{5}$
5	$\frac{5}{6} \times \frac{1}{120} = \frac{1}{144}$	$\frac{1}{120} - \frac{1}{144} = \frac{1}{720}$	$\frac{1}{720} = \frac{1}{120} \times \frac{1}{6}$

"This pattern was much harder to find. I couldn't see any pattern at all in the amount he drank. But the denominators of the remaining amounts looked strangely familiar: 2, 6, 24, 120, 720. Then I remembered where I had seen them. Those are the numbers that show up in **factorials.** You know, like 5! (5 factorial) is $5 \times 4 \times 3 \times 2 \times 1 = 120$. But the factorial in the denominator of each amount remaining was actually the factorial of the next drink number. After three drinks he had $1/4! = 1/24$ of his milk left, and after five drinks he had $1/6! = 1/720$ left. So after ten drinks, he was going to have $1/11! = 1/39,916,800$ of his milk left. Wow! That isn't much at all. Factorials sure get big fast.

"So the answers are Alysia and Melissa each had $1/11$ of their milk left, Dante had $1/39,916,800$ of his left, and Melody had $1/1,024$ of hers left. I can see why Mark bought each of them their own gallon."

Bimiljit was successful with this problem for many reasons, the main reason being that she was unafraid of the problem and persisted in solving it. She found many patterns. They showed up easily because she was organized in her thinking and created her charts so that patterns would show clearly.

Look for a Pattern

The key to finding most patterns is to organize a problem's information in a chart, table, or other systematic list so that the patterns jump out at you.

Try different ways to organize the information so you can see patterns.

- Write down the sequence and extend it if you can.

- Check the differences between the terms.

- A second set of differences—"the differences of differences"—may be useful.

- If seeing the differences doesn't help, try other operations such as multiplication, division, squaring, or use a combination of operations.

- Make a chart that matches each element of the sequence with its position number in the sequence.

- Try to relate the terms to their position numbers.

- Maybe there is a formula that can lead you directly to the answer.

- If nothing else works, be creative—maybe the pattern is really unusual.

- Work on something else for a while and come back with 'fresh eyes.'

- Be persistent, the pattern may be difficult to find.

Patterns turn up in many places. Keep your eyes open for them.

Problem Set A

1. **SEQUENCE PATTERNS**

    Find the next three terms in each sequence and explain your pattern in a sentence.

    a. 2, 5, 10, 17, _____, _____, _____

    b. 64, 32, 16, 8, 4, _____, _____, _____

    c. 4, 12, 10, 30, 28, 84, 82, _____, _____, _____

d. 1, 3, 7, 13, 21, _____, _____, _____

e. 1, 5, 13, 29, 61, 125, _____, _____, _____

f. 1, 5, 13, 26, 45, 71, _____, _____, _____

g. 1, 2, 6, 24, 120, 720, _____, _____, _____

2. MORE SEQUENCE PATTERNS

Find the next four terms in each sequence.

a. 243, 81, 27, 9, 3, _____, _____, _____, _____

b. 4, 9, 8, 13, 12, _____, _____, _____, _____

c. 4, 5, 9, 16, 26, 39, _____, _____, _____, _____

d. 1, 4, 13, 40, 121, 364, _____, _____, _____, _____

e. 3, 1, 4, 5, 9, 14, 23, 37, _____, _____, _____, _____

f. 7, −3, 10, −13, 23, −36, 59, _____, _____, _____, _____

3. AIR SHOW

To keep the spectators out of the line of flight at an air show, the ushers arranged the show's seats in the shape of an inverted triangle. Kevin, who loves airplanes, arrived very early and was seated in the front row, which contained one seat. The second row contained three seats, and those filled very quickly. The third row contained five seats, which were given to the next five people who came. The next row contained seven seats. This seating pattern continued all the way to the last row, with each row containing two more seats than the previous row. All 20 rows were filled. How many people attended the air show?

4. RECTANGULAR DOTS

In this chapter you worked with square and triangular numbers. This problem is about rectangular numbers. Find the pattern in the sequence of diagrams below, and determine how many dots would be in the thirty-fourth diagram.

5. PENTAGONAL NUMBERS

This problem is about pentagonal numbers. Find the pattern in the sequence of diagrams below, and determine how many dots would be in the 17th diagram.

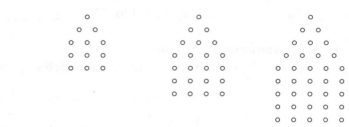

6. BEACH BALL

Kazuko has a beach ball. It is colored with six vertical sections, in order: white, orange, yellow, blue, red, and green. She spins the beach ball, and she notices that the colors whir by very fast. If the first color to go by is white and the ball spins around so that 500 colors go by, what is the 500th color?

7. LAST DIGIT

What is the digit in the ones place of 2^{57}?

8. SQUARES AND TURNS

In the 30th picture, how many squares and how many turns are there?

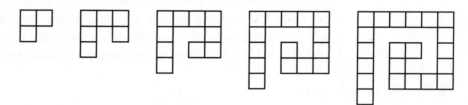

9. DOMINATION

In the card game Domination, there are money cards and triumph cards. The money cards cost 0, 3, 6, and 9 and are worth 1, 2, 3, and 5, respectively. The triumph cards cost 2, 5, 8, and 11 and are worth 1, 3, 6, and 10, respectively. Assume the respective patterns were to continue. What would the next three money cards cost and how much would they be worth? What would the next three triumph cards cost and how much would they be worth?

10. PAIR SEQUENCE

This sequence is formed with pairs of numbers. Fill in the missing two pairs.

(14, 4) , (21, 2) , (26, 12) , (33, 9) , (38, 24) , (___ , ___) , (___ , ___) , (57, 35) , (62, 12) , (69, 54) , (74, 28)

11. ZIG ZAG

In what row and column is the number 5000?

	Col 1	Col 2	Col 3	Col 4	Col 5	Col 6	Col 7
Row 1	1	2	3	4	5	6	
Row 2		12	11	10	9	8	7
Row 3	13	14	15	16	17	18	
Row 4		24	23	22	21	20	19
Row 5	25	26	27	28	29	30	
Row 6		36	35	34	33	32	31
Row 7	37	38	...				

12. JUGGLING

Little Gessop is trying to learn how to juggle. He has scrounged up three balls: two yellow tennis balls and one blue racquetball. He starts his usual juggling procedure by holding the two yellow balls in his left hand and the blue ball in his right hand. Then he throws one yellow ball from his left hand to his right; then he throws the blue ball from right to left just before he catches the first yellow ball; then he throws the second yellow ball from left to right just before he catches the blue ball; then he throws the first yellow ball back to his left hand just before he catches the second yellow ball. (The balls follow a sideways figure-eight pattern.) Gessop considers the first catch to be when he catches the first ball, the second catch to be when he catches the second ball, and so on.

Gessop has been counting how many catches he can make before dropping a ball—or all three. He can't seem to get past 15 catches before he messes up. He's counted his catches several times, and he's begun to notice a pattern, but because he's becoming only more frustrated rather than a better juggler, he gives up. But he's curious to know which hand the blue ball would end up in if he could make it to 100 catches. Gessop is about as good at math as he is at juggling (that is, not very good), and he's made the assumption that there is a 50-50 chance that the ball would end up in his right hand. Where would the blue ball be on the hundredth catch? Is Gessop right about the 50-50 chance?

Student Cory Craig of Sierra College in Rocklin, California, wrote this problem.

13. EMAIL VIRUS

In 1999 a computer macro based on a popular suite of programs was launched through the Internet. This macro came in the form of an email message that replicated itself via the first 50 email addresses in a recipient's address list. For example, person A sends the message to person B, who inadvertently sends a copy of the message to the first 50 people in his email address book.

Assume that there are no duplicate recipients. If the first email is called the first generation, how many total emails will have been sent in the first five generations?

14. BEES

A male bee is born from an unfertilized egg, a female bee from a fertilized one. In other words, a male bee has only a mother, whereas a female bee has a mother and a father. How many total ancestors does a male bee have going back ten generations? (Try drawing a diagram to help organize this problem's information.)

15. PASCAL'S TRIANGLE

This triangle is called Pascal's triangle. Find a pattern that will produce the next row. Then copy the triangle and determine the next four rows.

```
            1
          1   1
        1   2   1
      1   3   3   1
    1   4   6   4   1
  1   5  10  10   5   1
```

16. OTHER PATTERNS IN PASCAL'S TRIANGLE

Look for other patterns in Pascal's triangle. Write down three of them.

17. COIN FLIPS

Pascal's triangle shows up in the solutions of many problems. Consider how Pascal's triangle might help you solve this problem:

One flipped coin can land in two ways: heads (H) or tails (T). Two flipped coins can land in four ways:

HH	HT	TH	TT

Make a list of all the ways that three flipped coins can land. Make a list of all the ways that four flipped coins can land. What does this problem have to do with Pascal's triangle? Could Pascal's triangle help you figure out the number of ways five, six, seven, or more flipped coins can land?

18. FREE INTERNET

A new Internet service has just been announced that gives free Internet access for a year if customers agree to receive an email advertisement every 30 seconds while online. In other words, if you were online for 5 minutes, 10 email ads would show up in your email inbox during that time. Of course, then you'd have to go into your inbox, open the ads, and delete them. While you were opening and deleting them, you'd still be online, so more ads would come in. (To verify that you're living up to your end of the bargain, a return receipt is sent when you open an ad that works only when the computer is still online, so you must remain online while opening and deleting the ads.)

Suppose you've just been online for 1 hour. During that time, 120 emails showed up in your inbox. You know that all of these ads are going to be junk, but you still need to open and delete them. Assume it takes 10 seconds to open and delete each ad. During that time, more ads will come in, and you'll have to open and delete *those* ads. While you're opening and deleting those ads, still *more* ads will come in. You'll finally be able to log off when there are no more ads in your inbox. The ads started coming in 30 seconds after you logged on. Including the original hour you were online, how long will you have been logged on when you've deleted the last ad?

19. REFLECTION

At this time we would like you to reflect on what you have learned so far in this course. Have you learned new problem-solving strategies or become better at strategies you were already familiar with? Have you enjoyed working with other students? What have you liked best about this course so far? What have you liked least? Have you used any of the strategies you've learned so far outside this course?

20. WRITE YOUR OWN

Create your own pattern problem. To start with something easy, write a sequence problem. Then try coming up with a situation to go with it.

CLASSIC PROBLEMS

21. THE HAT THAT DIDN'T SELL

Unable to sell a hat for $20, a haberdasher lowered the price to $8. It still did not sell, so he cut the price again to $3.20, and finally to $1.28. With one more markdown, he will be selling the hat at cost. Assuming that he followed a system in marking his price cuts, can you tell what the next markdown will be?

Adapted from *Mathematical Puzzles of Sam Loyd,* Vol. 2, selected and edited by Martin Gardner.

22. TOWER OF HANOI

The Tower of Hanoi puzzle was invented by the French mathematician Edouard Lucas in 1883 (originally the "Tower of Brahma," in a temple in the Indian city of Benares). He was inspired by a legend that tells of a Hindu temple where the pyramid puzzle might have been used for the mental discipline of young priests. Legend says that at the beginning of time the priests in the temple were given a stack of 64 gold disks, each one a little smaller than the one beneath it. Their assignment was to transfer the 64 disks from the first of three poles to the third, using the second pole as much as they wanted to for transfer purposes. There was one important proviso: A large disk could never be placed on top of a smaller one. The priests worked very efficiently, day and night. When they finished their work, according to the myth, the temple would crumble into dust and the world would vanish. How many moves does it take to do this? Start by only using two disks and figure out how many moves it takes. Then do the puzzle with three disks, then four, and so on. Look for a pattern.

———

Adapted from *The Scientific American Book of Mathematical Puzzles and Diversions* by Martin Gardner.

MORE PRACTICE

1. FIND THE NEXT THREE

Find the next three numbers in each sequence.

a. 17, 23, 32, 44, 59, ___, ___, ___

b. 93, 82, 75, 72, 73, 78, ___, ___, ___

c. 49, 41, 35, 31, 29, ___, ___, ___

d. 1, 8, 27, 64, 125, 216, ___, ___, ___

e. 40, 8, 50, 10, 60, 12, ___, ___, ___

f. 2, 5, 9, 19, 40, 77, 135, ___, ___, ___

g. 1, 4, 5, 9, 14, 23, 37, ___, ___, ___

h. 1, 10, 11, 100, 101, 110, 111, 1,000, 1,001, 1,010, 1,011, ___, ___, ___

2. UNITS DIGIT OF LARGE POWER

Find the units digit in 7^{143}.

3. BORN ON A FRIDAY

Ella was born early in the morning on a Friday. She got married on the 10,000th day of her life. What day of the week was it?

Problem Set B

1. LEGAL EAGLES

There are exactly five parking spaces along the front side of the law offices of Stetson, Neumann, Ostrom, Savidge, and Schoorl. The colors of the lawyers' cars are blue, tan, black, silver, and burgundy. Match the owners with their car colors and their parking spaces using the following clues. Then determine whether Neumann is male or female.

1. Ostrom does not own the silver car.

2. The woman who parks in the fifth space owns the burgundy car.

3. Stetson owns the black car.

4. Schoorl parks her car in the middle space.

5. There are cars parked on both sides of Neumann's car.

6. Ostrom parks his car on one end, but the man who owns the blue car does not.

7. A woman parks in the fourth space.

2. RUDY'S CLOTHES RACK

Rudy examined his rack of clothes—five shirts and four ties—trying to decide what combination to wear. His five shirts included three solids and two patterns. His three solid-color shirts were blue, green, and white. One of his two patterned shirts was a blue-and-green print, and the other was a red-and-white stripe. He needed to match a shirt with a tie. His solid-color ties were white, blue, and yellow, and his patterned tie was a green-and-blue stripe. Rudy has two rules of good taste: You don't wear two solids of the same color; and if you wear a pattern, match it with a solid that is the same color as one of the colors of the pattern. Given those conditions, what is the probability that Rudy will pick out a combination that includes blue?

3. ROO AND TIGGER

Roo and Tigger decided to have a few jumping races. Roo could make three jumps of 2 feet each in the same time that Tigger could make two jumps of 1 yard each. They decided to have three races.

Race 1 was 150 feet up and back (300 feet total).

Race 2 was 100 feet up and back (200 feet total).

Race 3 was 75 feet up and back (150 feet total).

Amazingly, each race had a different outcome. Who won each race and by how much?

4. GOLF MATCH

Clark, Chris, Doug, and Diana are standing on the first tee of their favorite golf course, about to begin a best-ball-of-partners match. (A best-ball match pits two golfers against two other golfers.) They are standing in a square, with two partners standing shoulder to shoulder next to each other on the cart path, directly facing the other two partners standing shoulder to shoulder next to each other on the grass. This standing arrangement is typical of the beginning of a golf match. They shake hands, then throw a tee in the air and let it hit the ground. Whoever it points to will tee off first. Clark is standing diagonally opposite Diana. Chris is facing the person whose name begins with the same letter as that of the name of the person who will tee off first. Partners tee off one after the other. Who will tee off second?

5. COMIC OF THE MONTH

I subscribe to the Comic-of-the-Month Club. Each month I can buy any number of the 48 titles offered by the club. The first month I bought 5 comics for $19.07. The second month I bought 2 comics for $8.72. The next month I bought 6 of the club offerings for $22.52. In May I bought 3 more for $12.17. The club charges a price for each comic and a handling fee for the entire order. How much would it have cost to buy all 48 titles at the same time?

6

Guess and Check

Don't let the informality of the name of this strategy fool you. It is one of the most used and most useful mathematical problem-solving strategies. In the sciences it is called testing hypotheses. A string musician tunes an instrument by checking its pitch, guessing how much to tighten or loosen a string, and then checking the pitch again, constantly refining guesses to zero in on the correct pitch.

Guess-and-check. Even the name sounds bad. It sounds like something your math teachers tried for years to get you to stop doing. When you learn the strategy and start to use it, it may even feel like cheating. But it isn't. Guess-and-check is a powerful tool for solving problems. In the game of golf, when somebody figured out that using a sand wedge is a great way to escape a sand trap on a golf course, the first people to use this strategy may have felt that they were cheating. (Some people use a hand wedge, which is definitely cheating.) But sand wedges are now an accepted part of the game.

The strategy of guess-and-check serves a similar function in mathematics. It isn't cheating, and it really helps! It is tremendously effective in giving you a place to start when you tackle a problem. Guess-and-check helps you understand the problem, and, as you will see in Chapter 13: Convert to Algebra, this strategy gives you a way to set up an algebraic equation.

Why use guess-and-check? Certainly guessing is nothing new to most people. Students have been guessing at answers for years. Some have even gone on to check their answers. The strategy of guess-and-check would seem to arise naturally from most people's educational experience. Unfortunately, this is not the case. Although most people have indeed flirted with the guess-and-check strategy, they probably have not seen its power. The strength of this strategy comes from organizing the problem's information into a useful form. You guess an answer, then evaluate each guess in a systematic way that enables you to advance to more refined guesses.

Guess-and-check is not only a strategy, but also an attitude. When guessing and checking, you must first believe that you can solve a problem, even if you don't understand it well at the outset. Then, through your organization and persistence, you will work toward a solution.

Anybody can guess and anybody can check answers, but not everybody has learned to unite these two activities into a problem-solving tool. The key to guess-and-check is the chart you build to organize the guesses and the operations you perform to check them. The organization of your chart will be the key to writing algebraic equations in Chapter 13: Convert to Algebra.

In this chapter you'll learn how to use the powerful guess-and-check strategy. Plus, you'll see an on-the-job story of an application of the strategy at the end of the chapter.

SATURDAY AT THE FIVE-AND-DIME GARAGE SALE

Cinci held a garage sale, during which she charged a dime for everything but accepted a nickel if the buyer bargained well. At the end of the day she realized that she had sold all 12 items and had a total of 12 nickels and dimes. She had raked in a grand total of 95 cents. How many of each type of coin did Cinci have?

Do this problem before continuing. Even if you think you can write an equation for it, don't. Solve it by guessing a possible answer and checking to see if that answer is correct. Then make another guess, check it, and so on.

The correct answer is seven dimes and five nickels. Let's work through this problem using guess-and-check. The following solution is provided by Kasidra, who gave us a "thought-process" narration of how to solve this problem. As you will see, the method her solution illustrates does not involve just random guesses but rather is a very organized procedure.

"First, make a guess. I usually guess five because five is an easy place to begin. It's a small number, so the calculations are simple.

Process:

1. Make a guess.

"Five what? Five dimes. Or five nickels. It really doesn't matter. I'll make it dimes. Next, set up a chart to keep track of the guesses. This is very important—this is not simply guess-and-guess, it is guess, check, and refine your guess. In order to do this well, you must have an organized chart."

2. Make an organized chart.

Dimes
5

"Not bad for a first guess, but this should be expanded. She has 12 coins in all. So if I guess 5 dimes, she must have 7 nickels to make 12 coins."

Dimes	Nickels
5	7

"Well, that looks a little better, but there must be more to this problem. I've got to figure out a way to see if this guess is correct. I need to know how much these coins are worth, because the problem says she has 95 cents. Expand the chart again."

3. Continue to expand the chart.

Dimes	Nickels	Value of Dimes	Value of Nickels
5	7	$0.50	$0.35

"Looking carefully at the titles here, I see that because the columns of my chart contain numbers, the titles don't have to be so wide. I'll use two lines for the value titles. That way I'll save space but still use descriptive titles. I always want to have one complete guess fit on one line, but I also like titles that contain a lot of information. The easiest way to accomplish this is to draw narrow columns whose titles may take two or three lines."

4. Use narrow columns with descriptive titles.

Dimes	Nickels	Value of Dimes	Value of Nickels	
5	7	$0.50	$0.35	

"Now this chart is beginning to look like it leads somewhere. This slow process is not devised simply to frustrate people. Part of the point of using guess-and-check is that you often don't know where a solution is heading, even if you have a good approach to solving the problem. I'm developing this chart as I go rather than using a predetermined format. If the chart appears to be working, I'll keep it. If I don't seem to be getting anywhere with the chart, I'll get rid of it and start again.

"Here's the next improvement to my chart. I need to know the total value of these coins."

Dimes	Nickels	Value of Dimes	Value of Nickels	Total Value
5	7	$0.50	$0.35	$0.85

"Now get ready for the mega-action on the chart. I add a rating column to tell me how the guess compares to the right answer of 95 cents. This particular guess has a total value of 85 cents, which is less than 95 cents. I need more money so I need to guess more dimes. So I rate my guess as 'low.'"

Dimes	Nickels	Value of Dimes	Value of Nickels	Total Value	Rating
5	7	$0.50	$0.35	$0.85	low

5. When the chart is complete, evaluate your guess for accuracy.

6. Guess again, using prior guesses as a guide.

"Now that I appear to have completed the chart, I evaluate my first guess for accuracy. The guess was wrong, and the guess gave a result that was too low.

"I will continue to guess and will use the previous guess as a guide. I will increase the number of dimes to raise the total value."

Dimes	Nickels	Value of Dimes	Value of Nickels	Total Value	Rating
5	7	$0.50	$0.35	$0.85	low
8	4	$0.80	$0.20	$1.00	high

"The next guess was wrong too, but I did manage to have the results of that guess come out just a little too high. The first guess was too low, and the second was too high. Let's try something in between to see how that comes out. My next guess needs to be between five dimes and eight dimes."

Dimes	Nickels	Value of Dimes	Value of Nickels	Total Value	Rating
5	7	$0.50	$0.35	$0.85	low
8	4	$0.80	$0.20	$1.00	high
7	5	$0.70	$0.25	$0.95	right

"Fortunately, I came up with the right answer on the third guess. The first guess allowed me to make a reasonable second guess, and the first two guesses allowed me to modify my guess and then make what turned out to be a correct guess for the problem.

"Finally I'll check the problem for the question that was asked. The question asked how many dimes and how many nickels Cinci had. I always state my answer in a sentence—I wouldn't want to count on the instructor finding my right answer in the chart. The answer is that Cinci has seven dimes and five nickels.

"In short, here's what I did:

1. I started by making a guess.

2. I followed my guess through to a reasonable conclusion.

3. I evaluated the guess.

4. I modified and guessed again.

5. When I got a correct guess, I checked to see what the question was and answered it in a sentence.

"The chart, it is important to note, was nothing sacred. It was made up as I went and was designed to fit the needs of that particular problem. I also didn't clutter the chart with a lot of unnecessary detail. I did my computations in my mind or on scratch paper."

Incorrect guesses serve as steps toward the solution.

In her solution Kasidra made some excellent points. Another major point is that wrong guesses are important steps on the way to solving a problem. Many people resist guess-and-check at first because they fear being wrong. In some people's school experience, when an answer was wrong, the teacher marked it up with a red pen. When it was right, it was left alone. So many of us became so fearful of the red pen that our primary aim when doing schoolwork was to avoid being wrong. Sometimes the best way to avoid being wrong was to skip problems. Let us suggest here that wrong answers help you find right answers and are an important part of the journey. So guess. It may take you 50 guesses to reach the right answer, but at least you are on your way. It is far better to guess and be wrong than never to have guessed at all.

Arithmetic mistakes prolong the process.

However, we will offer one caution here. Be careful that your arithmetic is correct. Incorrect arithmetic can lead to two difficulties: First, if you make an arithmetic mistake that causes you to rate a high guess as low, or vice versa, you will then be guessing in the wrong direction and will get confused. Second, if you make an arithmetic mistake on a guess that would actually turn out to be the right answer, you will *really* be confused because you won't make that guess again and you might think the problem is not solvable.

Here's another problem to guess and check.

Farmer Jones raises ducks and cows. She tries not to clutter her mind with too many details, but she does think it's important to remember how many animals she has and how many feet those animals have. She thinks she remembers having 54 animals with 122 feet. How many of each type of animal does Farmer Jones have? Do this problem before continuing.

Here's how Vanessa solved this problem:

"I guessed ducks and then subtracted from 54 to get cows. To get duck feet I multiplied ducks by 2 because ducks have 2 feet. To get cow feet I multiplied cows by 4 because cows have 4 feet. Then I added the feet together and compared the total number of feet to 122."

Ducks	Duck Feet	Cows	Cow Feet	Total Feet (122)	Check
20	40	34	136	179	high
10	20	44	176	196	high
40	80	14	56	136	high

"I rated each of these three guesses as high, because I was looking for 122, and all three guesses were above 122. But I was a little confused. I thought my first guess of 20 ducks was too high, because it gave me more than 122 feet. So I guessed a smaller number of ducks the second time. This actually gave me more total feet, so I was guessing in the wrong direction. I thought this was strange, but then I realized that ducks have fewer feet than cows. Decreasing the number of ducks would actually increase the number of cows and add more feet. That's why my guess of 40 ducks got me much closer to the right answer. So I changed the ratings on my first three guesses to low, since the number of ducks was too low.

"I continued guessing. My next guess of 50 ducks was too high, because the number of feet dropped below 122. I correctly rated that guess high because there are too many ducks. So I continued guessing and checking numbers between 40 and 50."

Ducks	Duck Feet	Cows	Cow Feet	Total Feet (122)	Check
20	40	34	136	179	~~high~~ low
10	20	44	176	196	~~high~~ low
40	80	14	56	136	~~high~~ low
50	100	4	16	116	high
45	90	9	36	126	low
47	94	7	28	122	right

Ratings reflect whether the guess is low or high.

State the solution specifically, including units.

"Of course, whenever you work a word problem, you should give your answer in words: There are 47 ducks and 7 cows."

Did you notice the arithmetic mistake in the first line of Vanessa's chart? With her first guess (20 ducks), she should have ended up with 176 in the Total Feet column, not 179. This error didn't affect the solution of the problem. Suppose 20 ducks had been the right answer. Then this error would have led Vanessa to believe that the problem was not solvable.

Do arithmetic carefully—one error can jeopardize your entire solution and create a lot more work later.

Arithmetic errors like this can be very difficult to catch. (In fact, this error wasn't caught in any of the many reprints of this book's first edition!) Our point is that you must do your arithmetic very carefully. If you're having trouble finding a solution to a problem that you think you understand, your arithmetic may be at fault. Check your arithmetic for all your guesses before you proceed further.

Guess-and-check charts can be set up in different ways.

There is more than one way to set up a chart. Vanessa set up her chart like this:

Ducks	Duck Feet	Cows	Cow Feet	Total Feet (122)	Check

The chart could also have been set up like this:

Ducks	Cows	Duck feet	Cow feet	Total feet (122)	Check

The Cows and Duck Feet columns have been reversed, which might help you keep in mind that the ducks and cows have to add up to 54.

Here's still another way to set up the chart:

Ducks	Duck feet	Cow feet	Cows	Sum of animals (54)	Check
20	40				

Let's follow the guess of "20 ducks" through this chart. We know there are 122 feet total, so there must be 82 cow feet because 40 feet are used up on ducks.

Ducks	Duck feet	Cow feet	Cows	Sum of animals (54)	Check
20	40	82			

Because there are 82 cow feet, at 4 feet per cow we get 20.5 cows (dividing by 4). This gives a total of 40.5 ducks and cows, which is low because there are supposed to be 54 animals.

Ducks	Duck feet	Cow feet	Cows	Sum of animals (54)	Check
20	40	82	20.5	40.5	low

Design your chart to suit your needs. Start over if necessary.

Besides, a guess that leads to halving a cow is clearly not correct.

There is nothing sacred about how a chart is set up. Set up your chart in a way that works for you. If your chart doesn't work or seems awkward to use, you have to be willing to scrap it and start again. In any case, using at least *some* kind of chart is very helpful in keeping your work organized and eventually leading you to a solution.

Here's another problem on which to hone your skills.

ALL AROUND THE PLAYING FIELD

The perimeter of a rectangular playing field measures 504 yards. Its length is 6 yards shorter than twice its width. What is its area? Solve this problem before continuing. You might find that a diagram is helpful as well as guess-and-check.

Here is Brad's solution:

"First I drew a picture of a rectangle to represent the field. I figured I had to guess the width and length. It seemed as if it was easier to guess the width, because then I could double it and subtract 6 for the length. I put my guesses on my diagram as well as in the chart I made.

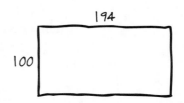

"This problem essentially breaks down into two subproblems: First find the dimensions, and then find the area by using the dimensions. [Breaking down a problem into several subproblems is the subject of Chapter 7: Identify Subproblems.] I decided to worry about the dimensions first, and after I got them right I would get the area. My diagram helped me realize that the perimeter of a rectangle is twice the width plus twice the length. I had to be sure to go around the whole figure."

Width	Length	Twice Width	Twice Length	Perimeter (504)	Rating
100	194	200	388	588	high
60	114	120	228	348	low
80	154	160	308	468	low
90	174	180	348	528	high
85	164	170	328	498	low
87	168	174	336	510	high
86	166	172	332	504	right

"I couldn't believe how many guesses it took me to get the right answer. My diagram was getting really cluttered with all the eraser marks. I actually ended up drawing new rectangles about every third guess to make it easier on my eraser.

"Finally, after all those guesses, all I had to do was multiply the length times the width, 166 yards × 86 yards, to get 14,276 square yards for the area, which answers the question."

Notice in Brad's work that he "bracketed" the right answer. Guessing a width of 100 produced a perimeter that was too high. Guessing a width of 60 produced a perimeter that was too low. He knew that the right answer for the width was between 60 and 100, so he then

Bracket the solution with high and low guesses.

guessed a number between 60 and 100 to further narrow the range in which the answer lay. His next guess, 80 yards, revealed that 80 was too low, and he continued working between 80 and 100 yards.

He also started by guessing the obviously smaller number, the width. Guessing the smaller number generally helps by limiting the range of possible numbers and allows you to multiply and add to get another number rather than having to divide or subtract.

Note that Brad drew a diagram and found it helpful. He was also willing to draw several diagrams when his original became too messy. As mentioned in Chapter 1, many people resist drawing a diagram. However, those who go to the extra trouble (about five seconds' worth) find that it pays off in increased understanding.

CASCADES STATE PARK

Emi and Margit had stopped at the bottom of one of the highest waterfalls in Cascades State Park. As Emi looked up at the waterfall, she said, "Wow, I think the top of that fall is about 20 feet more than three times the height of that young redwood!" Margit, of course, had a different opinion. She said, "No, I think it's about 50 feet less than four times the height of the redwood." If both are approximately right, about how tall is the redwood and how high is the waterfall? Work this problem before continuing.

Jameela, Bart, and Harvey worked together on the problem at the campus coffeehouse. Bart started a guess-and-check chart:

Height of the waterfall	Height of redwood based on Emi's estimate for the waterfall
300	?

Bart muttered, "To find the height of the fall, I would have to subtract 20 and then divide by 3 to get 93.333."

Jameela looked at what Bart was doing and said, "That's getting ugly. Why don't we start with the height of the redwood?"

As he listened to Jameela, Harvey started a new chart and guessed 50 feet for the height of the redwood.

Emi's estimate is 3 times height guess plus 20. Margit's estimate is 4 times height guess minus 50.

Height of the redwood	Emi's estimate for the waterfall	Margit's estimate for the waterfall	Rating
50	170	150	Wrong

Bart agreed, "This is much better. Three times 50 plus 20 is easy to figure, and so is 4 times 50 minus 50. Let's try 30. That's another easy number."

"While you're doing that, I'll try 100," said Jameela. They collected the results in one chart:

Height of the Redwood	Emi's Estimate for the Waterfall	Margit's Estimate for the Waterfall	Rating
50	170	150	Wrong
30	110	70	Worse
100	320	350	Still off, but Margit is higher

Bart noticed, "My guess was off in the same direction as the first guess, but it was farther off."

Jameela added, "This time my results are further apart, but now Margit's estimate is higher. In the first guess Emi's was higher."

Harvey said, "I'm really confused. I don't know which way to guess next. Maybe we should put in another column that would help us figure it out."

"How about a difference column?" Jameela asked. "Would that help?"

Bart said, "I don't understand what you mean."

Harvey said, "I do. If we subtract Emi's estimate minus Margit's estimate, and call that column 'difference Emi – Margit' it will tell us how close we are. We need to get the difference to be zero."

Height of the Redwood	Emi's Estimate for the Waterfall	Margit's Estimate for the Waterfall	Difference Emi – Margit	Rating
50	170	150	20	Wrong
30	110	70	40	Worse
100	320	350	−30	Still off, but Margit is higher

Bart asked, "Why is the last difference negative?"

Jameela answered, "Because Margit's estimate was bigger. We are always subtracting Emi's minus Margit's. $320 - 350 = -30$."

Bart said excitedly, "Oh I see. And the negative answer means we went too far. So the last guess is too high, and the first two guesses were too low. Let's change the ratings in the chart."

Height of the Redwood	Emi's Estimate for the Waterfall	Margit's Estimate for the Waterfall	Difference Emi – Margit	Rating
50	170	150	20	low
30	110	70	40	lower
100	320	350	−30	high

Harvey answered, "Exactly. So we need to guess something between 50 and 100."

Height of the Redwood	Emi's Estimate for the Waterfall	Margit's Estimate for the Waterfall	Difference Emi – Margit	Rating
50	170	150	20	low
30	110	70	40	lower
100	320	350	−30	high
80	260	270	−10	high
70	230	230	0	just right

"The redwood is about 70 feet tall and the waterfall is about 230 feet high."

Let's reconsider some of the things the group did that helped them succeed in solving this problem.

1. They weren't afraid to guess.

2. When things didn't go quite right, they were willing to back up and start again. They continued using the method of guess-and-check but changed how they went about making guesses and organizing the chart. Originally Bart guessed the height of the waterfall first, but that got messy. So Jameela suggested guessing the height of the redwood first.

3. They kept guessing even though at first it wasn't clear whether they should guess higher or lower.

4. Because they were comparing the numbers in two columns, they inserted a difference column. A positive difference indicated

the guess was low and a negative difference indicated the guess was high.

5. They kept working until they found an answer.

Working alone, Mona can paint a room in 4 hours. Working alone, Lisa could paint the same room in 3 hours. About how long should it take them to paint the room if they work together? Show the answer to the nearest tenth of an hour. Work this problem before reading on.

Aimee approached the problem like this: "The key is knowing that each of them gets a certain part of the room painted in each hour.

"Mona can paint one-fourth of the room in 1 hour, whereas Lisa can paint one-third of the room. I used my calculator on this, so I set it up as Mona doing 0.25 rooms each hour and Lisa doing 0.33 rooms each hour. I knew 0.33 is a little bit off, but I knew, if I needed to, I could go back and make it more accurate after I got close to the answer. Knowing the work rate, I could calculate the part of the room painted by multiplying. For example, using Mona's work rate, if she works 5 hours she will paint 1.25 rooms (0.25 times 5).

"To check, I wanted the total painted to equal 1, as that represents one room painted."

Her chart looked like this:

1. Guess times 0.25 gives Mona Amount.

2. Guess times 0.33 gives Lisa Amount.

Guess (Hours)	Amount Mona Paints	Amount Lisa Paints	Total Painted	Rating
5	1.25	1.65	2.90	high
2	0.5	0.66	1.16	high
1	0.25	0.33	0.58	low
1.5	0.375	0.495	0.87	low
1.8	0.45	0.594	1.044	high
1.7	0.425	0.561	0.986	low
1.75	0.4375	0.5775	1.015	high

"We were looking for the answer to the nearest tenth, so I figured I was done because 1.7 was too low and 1.75 was too high. Anything in between those two numbers would still round off to 1.7 anyway, so it wasn't important to know the answer to any more decimal places. Who knows? They could have stopped to get a soda, and that makes your answer completely wrong anyway. So I was done. Mona and Lisa painted for approximately 1.7 hours."

Aimee stated very succinctly the dilemma of producing an exact answer. In a real sense, her answer would probably be wrong anyway because of the inexact nature of the problem as presented.

Carlos also worked the problem, and he took a slightly different approach.

"I worked on this problem for a little bit using fractions. I saw that if they worked for 1 hour, Mona painted $1/4$ of the room and Lisa painted $1/3$ of the room. After a few guesses, I realized that I could just add $1/4$ and $1/3$ together to get how much they painted together in 1 hour. So I added them and got $7/12$. So they did $7/12$ of the room in 1 hour. This would mean they would paint $14/12$ of the room in 2 hours. So it wouldn't take them 2 hours to paint the whole room."

Hours	Mona Work Rate	Lisa Work Rate	Mona Work Done	Lisa Work Done	Total Work Done
1	$1/4$	$1/3$	$1/4$	$1/3$	$7/12$
2	$1/4$	$1/3$	$2/4$	$2/3$	$14/12$ (or $1\,1/6$)

"It suddenly occurred to me that to make $7/12$ of a painted room come out to one whole room, I had to multiply by the reciprocal of $7/12$. This is where I realized that I didn't have to spend the rest of my life adding fractions to find the answer. I knew that when I multiplied through by $12/7$, the Total Work Done column would have to come out to 1."

Hours	Mona Work Rate	Lisa Work Rate	Mona Work Done	Lisa Work Done	Total Work Done
$12/7$	$1/4$	$1/3$	$3/7$	$4/7$	$7/7 = 1$

"I don't know if I ever would have guessed $1\,5/7$, but that had to be the answer: $1\,5/7$ hours—or, as the problem asked for, 1.7 hours."

Guess-and-check leads to algebraic equations.

Some people who recognize the next type of problem probably also remember the gut-wrenching feeling of not being able to set up the algebraic equation. Guess-and-check is a useful tool for developing algebraic equations in situations where an equation is desirable. We will explore this further in Chapter 13: Convert to Algebra.

NEXT TRAIN EAST

A train leaves Roseville heading east at 6:00 a.m. at 40 miles per hour. Another eastbound train leaves on a parallel track at 7:00 a.m. at 50 miles per hour. What time will it be when the two trains are the same distance from Roseville? Do not read on until you've worked this problem.

This is a typical algebra problem. In algebra classes you were probably taught to set up a rate-time-distance chart, then choose a variable and write an equation. This type of problem broke down into three subtypes: same-direction problems, opposite-directions problems, and round-trip problems. You probably had to memorize three different equations for the three different subtypes. Many students get frustrated by these problems because they are unable to master the equations.

Guess-and-check can be a lifesaver in an algebra course. Guess-and-check helps you get started with a problem and then, obviously, helps lead to a solution. As you will see in Chapter 13: Convert to Algebra, guess-and-check also helps you set up an equation in cases where your instructor requires one or when guessing becomes too tedious and slow.

Jaspreet encountered the Next Train East problem in his algebra course. "I hate these rate-time-distance problems. I never could figure out what the equation was, so I just skipped them. When the instructor asked me why, I just said, 'I don't do rate-time-distance problems. Some people don't do windows, I don't do that kind of word problem.' Then one of my friends taught me guess-and-check. What a great method! I'll never fear a word problem again.

"This problem was kind of tough. I knew how fast the trains were going, but I didn't know how long they had been traveling or how far they went. I figured I should guess their times, and then I would be able to figure out how far they went. So I started to set up a chart."

6:00 TRAIN TRAVEL TIME	7:00 TRAIN TRAVEL TIME
10 HOURS	10 HOURS

"I wrote down 10 hours for each train, and then I tried to figure out how to check to see if this was right. After I read the problem again, I realized that the time for the 7:00 train had to be 1 hour less because it left an hour later. So I changed my guess, and then I figured out how fast they each went."

6:00 TRAIN TRAVEL TIME	7:00 TRAIN TRAVEL TIME	6:00 TRAIN SPEED	7:00 TRAIN SPEED
10 HOURS	9 HOURS	40 MPH	50 MPH

"I then wanted to put the distances on my chart, but I was running out of room. My friend taught me that one guess should fit on one line of my paper. If it carried over into two lines, it was too confusing. He said to make the titles smaller and use more lines for them. So I crossed out my first chart and started over."

6:00 TRAIN TIME	7:00 TRAIN TIME	6:00 TRAIN SPEED	7:00 TRAIN SPEED	6:00 TRAIN DIST.	7:00 TRAIN DIST.	RATING
10 HR	9 HR	40 MPH	50 MPH	400 MI	450 MI	?

"To get the distances I multiplied time and speed. So 10 hrs times 40 mph is 400 miles for the 6:00 train. And 9 hrs times 50 mph is 450 miles for the 7:00 train."

"I had no idea whether my guess was high or low. That's another thing my friend taught me: Sometimes you can't tell whether your first guess is high or low. He said to make another guess in one direction, then carefully analyze it and figure out if you are better or worse off than you were before. This is good advice, and I try to follow it. I decided that my guess was too low, so I guessed more hours."

6:00 TRAIN TIME	7:00 TRAIN TIME	6:00 TRAIN SPEED	7:00 TRAIN SPEED	6:00 TRAIN DIST.	7:00 TRAIN DIST.	RATING
10 HR	9 HR	40 MPH	50 MPH	400 MI	450 MI	?
14 HR	13 HR	40 MPH	50 MPH	560 MI	650 MI	WORSE

"I decided that this next guess was worse, because the miles between the two trains got farther apart. Then it occurred to me why. In my first guess, the early train had gone 400 miles and the later train had gone 450 miles. Because the later train had already passed the early train, the travel time must be less than I was guessing. This was what my friend meant when he said to really analyze the guess to see which way it is off. So I made my next guess lower than 10, not just lower than 14, because 10 had been too high in the first place. My next guess turned out too low because the later train hadn't caught up yet. I got the answer on my fourth guess."

6:00 TRAIN TIME	7:00 TRAIN TIME	6:00 TRAIN SPEED	7:00 TRAIN SPEED	6:00 TRAIN DIST.	7:00 TRAIN DIST.	RATING
10 HR	9 HR	40 MPH	50 MPH	400 MI	450 MI	HIGH
14 HR	13 HR	40 MPH	50 MPH	560 MI	650 MI	HIGHER
3 HR	2 HR	40 MPH	50 MPH	120 MI	100 MI	LOW
5 HR	4 HR	40 MPH	50 MPH	200 MI	200 MI	RIGHT

"It will be 11:00 a.m. when the two trains are the same distance from Roseville. That's another thing my friend told me: Make sure you answer the question. Like this question could have been 'How far away are the two trains from Roseville when the later train catches up?' That answer would be 200 miles. My friend told me to watch out for questions like that and make sure I answer them. Boy, I love guess-and-check. It sure has saved my bacon a few times. "And I don't even like bacon."

Note that Jaspreet would benefit from a difference column for the two distances. His first guess would produce a difference of −50 (400 − 450). His third guess would give a difference of 20(120 − 100). So a negative difference is too high and a positive difference is too low.

Liz solved this problem in a different way: "I drew a diagram of the two trains going in the same direction, one leaving an hour later than the other."

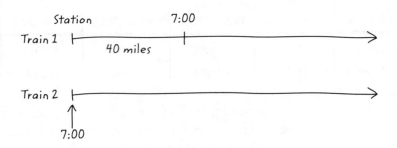

"Then I made a list of possible times and how far each train was from Roseville at each of those times."

Time	Distance of First Train	Distance of Second Train	Difference Between Distances
7:00	40 mi	0 mi	40 mi
8:00	80 mi	50 mi	30 mi
9:00	120 mi	100 mi	20 mi
10:00	160 mi	150 mi	10 mi
11:00	200 mi	200 mi	0 mi

Look for patterns in guess-and-check charts.

"I also noticed a pattern here. The difference between the two trains went down by 10 miles each hour. That's because the second train goes 10 miles per hour faster. 'Eleven o'clock' answers the question."

Applied Problem Solving: Reduce Tax, Increase Bonus

Holly works for a San Francisco law firm as the office manager and bookkeeper, and she is no slouch when it comes to mathematics. She graduated from Cornell University with a major in math. One day, however, all the math she had studied was not enough.

The problem she encountered on that fateful day was a complicated and circular tax-payment problem. The firm's partners used a profit-sharing account in order to reduce their personal taxes and to provide more money and benefits for their deserving employees. Holly's job was to determine the amount of personal income each partner should contribute to the profit-sharing fund.

The fund had been set up so the partners could legally reduce their own personal taxes based on the amount of their contributions to the profit-sharing account. This amount was limited by an Internal Revenue Service equation. The amount could not exceed a certain percentage of the highest possible personal income of the partners, but that percentage could not be determined until the profit-sharing funds had been taken out. Of course, one way to know how much profit-sharing money to take out was to know the personal income of the partners. Which comes first, the income or the profit sharing? This was a tax-law version of the old question about the chicken or the egg.

Holly was trying to lawfully maximize the partners' personal income by reducing their taxes and at the same time increase her own income by maximizing the profit-sharing account. The social security tax was a further complicating factor, because it is a tiered tax on part of your personal income, depending on what plateau that income reaches.

Holly worked on the problem all afternoon. Others in the office helped her work on it, but they were still not able to solve it. She took the problem home and spent several more hours. She originally had three equations in three unknowns, which she had managed to boil down to one very complicated quadratic equation. However, in the course of doing the calculations to get the equation, the limited decimal capacity of her calculator introduced a lot of round-off error. She came up with an estimate that she didn't trust.

Then she called her son and gave him the equation and parameters she had come up with. After about 45 minutes he called back with an answer that he didn't trust. When they considered the difference between their answers, they decided to declare it a "mistrial." At this point, algebra had failed them, and they turned to guess-and-check.

They programmed Holly's computer to guess possible solutions and check them. After about 15 minutes of programming and 45 seconds of run time, they had their answer. By running the check, they knew it was right.

Guess and Check

Using an organized table is the key to success with guess-and-check.

Using traditional algebraic methods requires you to understand the whole problem much earlier in the solution process than you would using guess-and-check. With the guess-and-check strategy, you can develop a fuller understanding of the problem as you work through the process. Guess-and-check is not always taught in algebra courses, although we think it should be. This strategy often works when algebra doesn't, and it helps build algebraic concepts.

The key points to remember about guess-and-check are these:

- Start guessing. As you work through a guess, you'll learn more about the problem.

- Keep your work organized. Guess-and-check helps you organize information. You will defeat this strategy (and yourself) if you don't keep your guesses organized.

- Be ready to start over. As you learn more by working through your guesses, you may discover that your first approach was not productive.

- Start with smaller numbers and build up to bigger numbers. Make your guesses skip around so you can bracket the right answer.

- Remember you are rating the guess. Was the guess too low or too high?

- Sometimes you'll misrate a guess, rating a high guess low or a low guess high. It may take two guesses to determine what is high and what is low. Be patient when this happens, and every time you are about to rate a new guess pay close attention to guesses you made previously.

- In cases where you are comparing two columns, a difference column can help you determine whether a guess is high or low.

- Avoid arithmetic mistakes.

  — If you make a mistake in a rating, you will then make subsequent guesses in the wrong direction.

  — If you make a mistake with the right answer, you may never guess that number again.

- Put a lot of information in your column titles so that you know what each column represents. You may have to use several lines to accommodate the long titles, because you want each guess to fit on one line. Generally the contents of your columns will be numbers, so the columns don't have to be wide, but they do have to be descriptive.

- In problems involving different units – such as mph, hrs, miles – put units into the chart. They can be in the guesses (as in Next from East) or in the column titles.

A guess-and-check chart helps organize information in such a way that you can make that information more useful. Guess-and-check is a powerful strategy. Keep the preceding points in mind, and you will enjoy great success.

Problem Set A

1. DIMES AND QUARTERS

Annette has five more dimes than quarters. The total amount of money she has is $3.30. How many of each coin does she have?

2. MARKDOWN

Dominique bought ski gloves that were marked down 30% to $24.01. What was the price of the gloves before the markdown?

3. TAX

The cost of a basketball was $15.54, including 7.25% sales tax. How much of that cost was the price of the basketball, and how much of it was the tax?

4. REFINANCING

The mortgage payments for Covell's home are about $900 per month. He is going to refinance the loan, which will cost him about $2,500 in fees, and the new payments will be $830 per month. How long will it take him before the new loan starts saving him money?

5. COLLEGE TOWN TRANSIT TICKETS

The College Town Transit System offers two bargain ticket booklets, each good for one month. The Regular Rider discount booklet of transit tickets costs $10.00 plus 25¢ per ride, and the Casual Commuter discount booklet costs only $4.00 but riders have to pay an additional 50¢ per ride. Magoli and Dorián are debating which type of booklet each should buy. Magoli, who rides more often, decides on the Regular Rider booklet, and Dorián chooses the Casual Commuter. For what number of rides will both spend the same amount?

6. CHECKING ACCOUNT

Recently Javier received a letter from his bank concerning his checking account. Under his current plan, each check he writes costs 15¢, and he pays a monthly fee of $1.60. Under the proposed new plan, each check he writes will cost 12¢ and there will be a monthly fee of $2.75. What is the minimum number of checks Javier must write monthly so that the new plan will cost him less than the current plan?

7. BASEBALL CARDS

Giancarlo has two more than three times the number of baseball cards that Antonio has. If Giancarlo gave Antonio 12 of his cards, they would each then have the same number of cards. How many cards does Giancarlo have?

8. A BUNCH OF CHANGE

Plato has 58 coins in nickels, dimes, and quarters. The number of nickels is three less than twice the number of dimes. The total value of the coins is $7.40. How many of each type of coin does Plato have?

9. UNDERGRADS

There are nine men to every ten women at Big State U. There are 26,239 students at the university. How many women are there?

10. STRANGE COINCIDENCE

Ronnie's apartment number plus the square of Harry's apartment number equals 2,240, which happens to be the dorm room number of Harry's girlfriend. Harry's apartment number plus the square of Ronnie's apartment number equals 1,008, which happens to be the dorm room number of Ronnie's boyfriend. What are the apartment numbers of Ronnie and Harry?

11. GOING HOME FROM COLLEGE

Gabby, a college student, headed home for Thanksgiving. He rode his bike at a speed of 18 miles per hour to the town where his two sisters, Ariella and Rachelle, live. They all got in Ariella's car and drove to their parents' home. On the car trip, the siblings averaged 60 miles per hour. The total distance from Gabby's house to their parents' house (via his sisters' house) is 435 miles, and Gabby traveled for 9 hours. How far is it from Gabby's house to his sisters' house?

12. RIDING A HORSE

Neena went riding in the hills. At one point, however, her horse, Dakota, stumbled and was hurt. Neena left Dakota and walked back home to call her vet. Neena figures Dakota walks about twice as fast as she does. If Dakota was hurt about 8 miles into her ride and her whole trip took 4 hours total, how fast did Neena walk?

13. WOMEN'S WORLD CUP SOCCER

Imagine that you traveled back in time to the Fourth of July, 1999. You are at Stanford Stadium, watching the semifinal of the Women's World Cup soccer match. Mia Hamm, star of the United States team, has just received the ball on a breakaway. She is 60 yards from Brazil's goal. Brazilian star Sissi is the closest opponent to Hamm and is 10 yards behind her. Mia takes off for Brazil's goal and, while dribbling the ball, is able to run at a speed of 6.8 yards per second. Sissi takes off at the same time as Hamm and runs at 8.4 yards per second. (She is able to run faster because she isn't dribbling the soccer ball.) Hamm has decided to shoot at the exact moment that Sissi catches up to her, because at that point it will still be difficult for Sissi to block the shot when she is side by side with Hamm. How far away from the goal will Hamm be when she shoots?

14. TELEPHONE SOLICITOR

Keiko is a telephone solicitor. She has been able to convince only 18% of the people she called to donate. If she gets 12 of the next 30 people she calls to donate, she'll barely break 25% for the day. How many calls has she made so far today?

15. EQUAL VOLUME

A box manufacturing company makes rectangular boxes with a square base. The most popular box measures 27 inches wide by 27 inches long by 12 inches high. Two employees are experimenting with increasing the volume of the box. Malcolm is experimenting with increasing the measure of the height and leaving the measure of the square base alone. Rosa is experimenting with increasing the measures of both sides of the square base and leaving the measure of the height the same. They were comparing notes one day when Malcolm said to Rosa, "Wow, what an interesting number. When you increased the measure of the base sides by this number, it gave you a volume increase that was exactly the same as the volume increase I got when I increased the measure of the height by the same number." What was that number?

16. FREE THROWS

Audrey's free-throw percentage so far this season is .875. If she makes only 13 of her next 20 free throws, her percentage will drop to .860. How many free throws has Audrey made this season?

17. CELL PHONE PLANS

Jean Luc was looking for a cell phone. He is considering plans from two companies, both of which have unlimited night and weekend minutes. The Marathon Network plan costs $19.99 per month and comes with 100 free weekday minutes per month. Additional weekday minutes cost 31¢ per minute. The Toucan Communications plan costs $39.99 per month and comes with 250 free weekday minutes per month. Additional weekday minutes cost 47¢ per minute. For what range of weekday minutes is Marathon's plan cheaper? For what range of weekday minutes is Toucan's plan cheaper?

18. WRITE YOUR OWN

Write your own guess-and-check problem. Start with a situation and an answer, then make up the other necessary information.

CLASSIC PROBLEM

19. **THE FIVE NEWSBOYS**

Five clever newsboys formed a partnership and disposed of their papers in the following manner: Tom Smith sold one paper more than one-quarter of the whole lot, Billy Jones disposed of one paper more than one-quarter of the remainder, Ned Smith sold one paper more than one-quarter of what was left, and Charley Jones disposed of one paper more than one-quarter of the remainder. At this stage, the Smith boys together had sold just 100 more papers than the Jones boys had sold. Little Jimmy Jones, the youngest in the bunch, now sold all the papers that were left. The three Jones boys sold more papers than the two Smith boys, but how many more?

Adapted from *Mathematical Puzzles of Sam Loyd,* vol. 2, edited by Martin Gardner.

MORE PRACTICE

1. **STOCK PROFIT**

Warren bought some shares of stock and later sold them for a 15% profit. If he sold the stock for $176.64 per share, how much per share did he buy the stock for?

2. **PAY CUT OR PAY RAISE?**

Barney works in a department store selling suits. He works 30 hours per week, at a certain hourly rate. Recently he was able to work an extra 2 hours every week, but unfortunately his hourly rate was reduced by 5%. His new weekly pay check is $528.96. What was his weekly pay check before when he was working 30 hours at his original hourly rate?

3. **POCKETFUL OF MONEY**

Gates had a pocketful of money in twenties, fifties, and hundreds. The number of twenties was six more than twice the number of hundreds. He had 78 bills, and a total of $3,560. Determine how many of each bill Gates had.

4. POKER

Brandt was dealing a poker game with nickels, dimes, quarters, and dollar bills in the pot. He noticed there were a total of 97 coins in the pot. The number of quarters was 3 less than four times the number of dimes. The number of dollar bills was 8 more than the number of quarters. There was a total of $101.40 in the pot. How many dollar bills were there?

5. DANCING

At Terry's and Jodie's wedding there was a large group of dancers. One group of people was doing country line dancing, and another group of people was doing the chicken dance. The number of line dancers was 7 more than twice the number of chicken dancers. Then suddenly 23 line dancers started doing the chicken dance, which made the two groups equal in size. How many total dancers were there?

6. ROOM KEY

Rich and Mellow were roommates in the freshman dorm. Rich lost his room key so they were sharing Mellow's key until they got another made. One morning, Rich went to an early class without the key. Mellow went to a later class and needed to give Rich the key. At 8:50 exactly, Rich left class and walked toward the dorm. Also at 8:50, Mellow left the dorm and walked towards Rich's class along the same route that Rich would take. Mellow walks 6 yards per minute slower than Rich. They passed each other at 8:54 exactly, and Mellow tossed Rich the key. It is 1,020 yards from Rich's class to the dorm. How fast did each of them walk?

7. CROSS-COUNTRY FLIGHT

Willa flew from Washington, DC, to Los Angeles for a sales conference. When she flew home from Los Angeles, her plane was diverted to New York City because of bad weather in Washington. The flight to the conference was 239 miles shorter than the flight back, but because the prevailing winds blow west to east the flight back east took 36 minutes less time. The speed of the plane from east to west was 445 mph, and the speed of the plane from west to east was 555 mph. How far was the flight from Washington, DC, to LA? (Hint: First figure out how many hours is equal to 36 minutes.)

8. **MORNING COMMUTE**

Danika commutes every day from her home to the hospital where she is a doctor. Most of this drive is on the freeway. One day when she got on the freeway she spent half an hour driving at one speed, but then hit traffic and had to drive 15 miles per hour slower than that for the next 15 minutes until she reached the freeway exit. She drove a total of 36 miles on the freeway that day. How fast was she driving when she first got on the freeway?

Problem Set B

1. **DAILY ROUTINE**

Aji has an argument with his daughter. She says, "You do the same darn thing every day." Aji does go fishing every day but contends that every day is different because he does things in a different order each day. Before he leaves shore in his rowboat, he gets fresh bait, checks the weather, and adjusts his seat cushion. Out in the water, he eats his fruit, puts the meat on his sandwich, drinks his apple juice, and eats his sandwich. Back at shore, after tying his boat to the dock, he picks up the fishing pole in his right hand and the ice chest in his left hand. Then he finally heads back home to have the same argument with his daughter. For how many days could Aji do things in a different order before he has to repeat the order of some prior day?

2. **AFTER THE FOOTBALL GAME**

A group of students went to the pub after the football game on Saturday night. They all ordered from the menu and forgot to tell the server to give them separate checks. The bill totaled $162, including the tip. They decided to split the bill evenly, and they figured out how much each of them owed. But then three people said they had no money. The rest of the people each had to chip in $2.70 extra to cover the tab. How many people were in the group?

3. CATS

I hate cats. It seems as though cats hate me too. I wonder why. My neighbor Elinor loves cats. She seems to attract them in bunches, especially alley cats, tabby cats, and Manx cats. She already had three alleys, five tabbies, and two Manx cats when more of each kind of cat began to show up on her doorstep in March. The alley cats showed up first, on the first of the month. That is, one alley cat showed up on the first, and one new alley cat showed up every day for the rest of the month. The tabby cats began to show up on the fourth. Two showed up on the fourth, and two new ones showed up every day for the rest of the month. Not to be outdone, the Manx cats showed up on the sixth. Four of them came on the sixth, and four new ones came every day for the rest of the month. How many cats did Elinor have at the end of the month?

4. STOCK MARKET

Ms. Edwards and three other students decided to invest part of their college fund in the stock market. They all chose one big-name stock to invest in, and each invested $1,000. After one month, they checked the stock listings to see how they did. It turned out that three of them came out ahead and the fourth lost money. From the clues below, determine the full name of each woman, which stock she invested in, and how much each woman made or lost.

1. Two people made more money than the woman who invested in IBM.

2. Ms. Kortright did not invest in ATT, but the woman who did made the most money.

3. Vickie made $700, which made her the big winner.

4. Ms. McDonald lost $300, and Tina made $200.

5. Nita did not invest in Xerox.

6. The woman who invested in Ford was the only one who lost money.

7. Luann, who is neither Ms. Kortright nor Ms. McElhatton, made $300 less than Vickie.

Seymour, the census taker, came to Larry Longway's house and asked for the ages of the three children living there. Because Larry does not believe in giving information away easily, he gave Seymour the following clues. The clues were given one at a time. After each clue, Seymour really tried to figure it out. If he couldn't figure it out, he then asked for another clue.

Clue 1: The product of their ages is 72.

From this clue, Seymour tried but could not figure out the ages.

Clue 2: The sum of their ages is the same as today's date.

Seymour knew what the date was, but he still could not figure out the ages.

Clue 3: The oldest child loves to eat at Burger Jack.

From this clue, Seymour was able to figure out the ages of the children. What are the children's ages?

7

Identify Subproblems

Breaking down a problem into parts allows you to identify the hard parts and concentrate on solving them. Many complex activities, such as manufacturing automobiles, are broken down into subproblems—for example, attaching a door. These subproblems can be further divided into smaller subproblems that you must solve to achieve the overall goal.

U p to this point, you've learned about two of the three major problem-solving themes of this book, both of which have involved organizing information in some way. When you drew a diagram, you organized information spatially, so diagrams fall under the major theme we call **Spatial Organization.** When you used other strategies, such as making a systematic list or using guess-and-check, you organized information into some sort of table or list. These strategies fall under the major theme we call **Organizing Information.**

The strategy you'll learn about in this chapter is different. The strategy of **subproblems** involves organizing your plan of attack. When you use subproblems, you first move your focus away from the main problem you're working with and instead concentrate on achieving a subgoal. When you've achieved your subgoal, you can then solve the main problem, which is your overall goal. Subproblems fit into the third major problem-solving theme, which we call **Changing Focus.**

When a solution method for a problem is not readily apparent, try using the strategy of subproblems. Here is a simple example that will illustrate the concept. The Scholastic Aptitude Test (SAT) and the Graduate Record Exam (GRE) are full of problems for which you need to use subproblems, such as this one:

If $3x - 1 = 17$, what is $2x - 4$?

To solve this problem, you must first solve for x in the equation $3x - 1 = 17$. You then substitute the value of x into the expression $2x - 4$. Note that you couldn't answer the given question until you'd solved for x first. This "miniproblem" that you solve first is called a subproblem, and it must be solved before you answer the given question.

Listing subproblems will focus your thinking.

Some problems involve many subproblems. To attack these types of problems, you'll find it helpful to list the subproblems before starting on the problem's solution. The list becomes your plan of attack. You can then solve each subproblem and in turn reach a solution to the overall problem. Making a list of subproblems can focus your thinking. If you decide you need help with the overall problem, your list of subproblems will help you determine exactly *where* you need help.

According to our friend and mentor Tom Sallee, a professor at the University of California at Davis, solving a problem by using subproblems is much like crossing a river by using stepping-stones.

If the river is very wide, it isn't possible to jump all the way across it, but by walking through the river on stepping-stones you can make it all the way across. Likewise, for problems that are not possible to solve all at once (jumping across the whole river), you can use subproblems (the stepping-stones) to achieve your goal of solving the overall problem (getting across the river).

The next problem contains some simple subproblems. List the subproblems and solve them before reading on. **Hint:** It is helpful to list each subproblem as a question.)

AT THE STUDENT BOOKSTORE

Shanein is a student at Oregon State. She has four $20 bills. She buys six notebooks at $3.95 each, but they are marked 20% off. She also buys three highlighters at $1.19 each and a textbook for $44.98. There's no sales tax in Oregon. How much change should Shanein receive? Work this problem before continuing.

A typical student-student conversation about this problem might go something like this. (Ghazi has never heard of the strategy called subproblems, although, as you will see, he understands the strategy quite well.)

RICHARD: I'm having some trouble on this problem. Can you help me?

GHAZI: Sure. Let's figure out the parts.

RICHARD: Okay. It's asking about how much change she should receive, so first I have to figure out how much she paid and how much she bought.

GHAZI: I like to organize things by writing down questions and then answering the questions. It seems to work well for me. Since you're doing the learning here, you do the writing.

RICHARD: Sure. Let's both talk, and I'll do the writing.

GHAZI: Okay. Write down the two things you've already identified: the cost and how much money she has.

RICHARD: I'm writing down "How much did Shanein's purchases cost?" and "How much money did Shanein have?"

GHAZI: Actually, the second question should be "How much money did Shanein give the cashier?" Hopefully she has more money stashed somewhere so she can pay her rent and her reg fees.

RICHARD: Okay, I changed it.

GHAZI: So what else is there to figure out?

RICHARD: Well, how much money she gives the cashier is easy: She uses four $20 bills, so that's $80. . . . How much she owes is going to be a pain to figure because I have to work out all the subtotals.

GHAZI: And?

RICHARD: Well, she buys six notebooks for $3.95 each, so that's six times 3.95, which gives me . . .

GHAZI: You know, you're doing everything right, but since you asked me for help I want to show you the way I do stuff, and then you can decide what works best for you. My way of doing these problems will give you one more tool to use. If you don't like it, you can discard it, but if you come across another problem that's tough, you can try this method on that problem and get it on your own.

RICHARD: Okay. I'll do it. You said write down questions, right?

GHAZI: Right. It's like that advice talk show lady says: "What's your question for me?"

RICHARD: Oh, I thought it was a talk show about paranormal occurrences called the Doctor Aura Show, and people ask her "What's your vision for me?"

GHAZI: Very funny.

RICHARD: Okay, Doctor, here are the questions: "How much do six notebooks cost?" "How much is the discount on the notebooks?" "How much is the discounted price on the notebooks?" How am I doing?

GHAZI: Great. What's next?

RICHARD: Umm, I need to find the cost of the highlighters, so the question is "How much do three highlighters cost?" and then

"How much does the $44.98 textbook cost?" which is kind of easy . . .

GHAZI: Yes, but a necessary part of the process.

RICHARD: Now that I have all the parts, I just need to put them together: "How much do all the supplies cost?" and then finally: "How much change should Shanein receive?" This is the question that answers the problem.

GHAZI: Right. It's like your Doctor Aura. She makes people ask questions in order to get them to focus on the issue at hand. By asking math questions, you're focusing on the subproblems that need to be solved in order to solve the big problem.

RICHARD: Okay, I get it.

GHAZI: It's like a car. You have all these subsystems to make the whole thing work: lights, engine, brakes, steering, seating, stereo, cupholder, and so on.

Here are the subproblems Richard wrote and the answer to each one.

How much did Shanein's purchases cost?   Who knows?

How much money did Shanein give the cashier? $80
    4 × $20 = $80

How much do six notebooks cost? $23.70
    6 × $3.95 = $23.70

How much is the discount on the notebooks? $4.74
    .20 × $23.70 = $4.74

How much is the discounted price on the notebooks? $18.96
    $23.70 − $4.74 = $18.96

How much do three highlighters cost? $3.57
    3 × $1.19 = $3.57

How much do all the supplies cost? $67.51
    $18.96 + $3.57 + $44.98 = $67.51

How much change should Shanein receive? $12.49
    $80 − $67.51 = $12.49

Richard concluded: "Shanein got $12.49 back in change."

Notice a couple of things about Richard's work on this problem: The first question he asked was good, but he wasn't ready to answer how much Shanein's purchases totaled. He had to do a significant amount of work on subproblems before he could answer that question. Notice also that the discount can be calculated a few different ways. The first is the way Richard did it, by multiplying the total cost of the items by the discount percentage and then subtracting the amount of the discount from the regular cost. Another approach is to realize that when the discount is for 20% off the price then 80% of the price is still on. Multiplying the price of the six notebooks by 80% gives the discounted price directly. A third approach is to compute the discount on a single notebook first (either by subtracting 20% of the price or by multiplying the price by 80%) and then multiply the discounted price of one notebook by 6.

As you start identifying subproblems, you will find that there is often more than one set of subproblems you can use to solve a problem.

Listing subproblems outlines an action plan.

Problems that appear to be difficult often seem much easier once they're broken into subproblems. Listing the subproblems gives you a plan of action, and your list helps you identify what is known and what you need to figure out.

WATERING THE LAWN

Three quarts of water are needed to water 1 square foot of lawn. How many gallons of water are needed to water a lawn that measures 30 feet by 60 feet? List the subproblems and answer the question before proceeding.

E-Chung wrote the following list of subproblems:

1. How many square feet are in the lawn?

2. How many quarts are needed to water the entire lawn?

3. How many quarts are in a gallon?

4. How many gallons are needed to water the entire lawn?

Note that these subproblems could have been listed in a different order, the only requirements being that question 1 must precede question 2 and question 4 must be last.

By listing the subproblems, E-Chung has clearly laid out the plan. Finding the solution to the problem now seems relatively trivial. In fact, the hardest part of using the subproblems strategy is figuring out what the subproblems are. Once you know what you have to solve, the actual solving is usually fairly easy.

The subproblems don't all have to appear like magic at the same time, well worded and in the right order. In fact, it's likely that when you come up with your subproblems, they'll be in the wrong order and will reflect a couple of different approaches to the problem. Part of solving a problem is arranging the subproblems in the right order. You may find that this task is almost automatic because a subproblem is in the wrong place only if its solution depends on the answer to another subproblem. In that case it should be obvious that you need another answer before you can continue with the subproblem at hand.

This is how E-Chung solved the subproblems he listed:

1. How many square feet are in the lawn?

    Because the lawn measures 30 feet by 60 feet, the area of the lawn is 30 × 60 or 1,800 square feet.

2. How many quarts are needed to water the entire lawn?

    One square foot of lawn requires 3 quarts of water, so 1,800 square feet of lawn requires 3 × 1,800 or 5,400 quarts of water.

3. How many quarts are in a gallon?

    Four quarts are in 1 gallon.

4. How many gallons are needed to water the entire lawn?

    The whole lawn needs 5,400 quarts. There are 4 quarts in a gallon. Dividing 5,400 quarts by 4 gives 1,350 gallons to water the entire lawn.

This problem could also be solved with the set of subproblems that Romina used:

1. How many gallons does it take to water 1 square foot of lawn?

   Three quarts equals ¾ gallon, which is what is needed for 1 square foot of lawn.

2. What is the area of the lawn?

   The area is 1,800 square feet.

3. How many gallons does it take to water the entire lawn?

   Because each square foot of lawn requires ¾ gallon and the area of the lawn is 1,800 square feet, it takes ¾ × 1,800 or 1,350 gallons.

You should have noticed by now that there is often more than one set of subproblems for solving a given problem.

THE CAR BARGAIN

Paul went into the local new-car lot to buy a car. He knew the kind of car he wanted, because his friend Barbara Gain had bought the same car the day before. Barbara received a 30% discount on the car, which listed at $15,000. The salesperson offered Paul the $15,000 car at a 20% discount instead. When Paul protested, the salesperson offered an additional 10% off the 20% discounted price. This offer satisfied Paul and he bought the car, convinced he had paid the same price as Barbara. Had he? Solve this problem before continuing.

Pragnesh wrote this list of subproblems:

1. What is 30% of $15,000?

2. How much did Barbara pay for the car?

3. What is 20% of $15,000?

4. What is the sale price that Paul protested?

5. What is 10% of this new price?

6. What is the final price that Paul paid for the car?

7. Who paid more and by how much?

8. How many subproblems do I have to write before the instructor is satisfied?

Again, listing the subproblems gave Pragnesh a plan. Solving the subproblems does not seem too hard, even though the original problem looked formidable.

1. What is 30% of $15,000?

    Thirty percent of $15,000 is $4,500. This represents the amount of money that Barbara saved.

2. How much did Barbara pay for the car?

    Barbara paid $15,000 less her discount of $4,500, for a net price of $10,500.

3. What is 20% of $15,000?

    The first discount that the salesperson offered Paul was 20%. Twenty percent of $15,000 is $3,000.

4. What is the sale price that Paul protested?

    The original sale price that Paul was offered was $15,000 less $3,000, for a net price of $12,000.

5. What is 10% of this new price?

    The new price was $12,000, and 10% of this price is $1,200.

6. What is the final price that Paul paid for the car?

    Paul paid the discounted $12,000 less $1,200, for a final price of $10,800.

7. Who paid more and by how much?

    Barbara paid $10,500 and Paul paid $10,800, so Paul paid $300 more. He shouldn't have been so happy.

8. How many subproblems do I have to write before the instructor is satisfied?

    Eight is more than enough. You only needed seven.

Percentages and Mixture Problems

Problems about percentages can be very confusing. Listing subproblems for them can help you understand what's going on. Remember that a given percentage of *different* amounts is never the same amount. For example, suppose a baseball player gets a hit 40% of the time (this is a 0.400 batting average). Is this player one of the greatest who ever lived, or is he some fluke? Well, it depends on how many times he has been at bat. Suppose he just came up from the minor leagues and has only 5 at-bats. If he gets a hit 40% of the time, he's made 2 hits. Big deal. But suppose he has played all season and has 500 at-bats. Now, if he gets a hit 40% of the time, he's made 200 hits. That *is* a big deal, and undoubtedly he will be remembered in the Baseball Hall of Fame if he can get that number of hits consistently every season. Forty percent yields very different results in each of these situations. Percentages serve as comparisons, but you have to be careful with them.

Mixture problems can be some of the most confusing problems a person faces in algebra. When you understand the strategy of subproblems, however, you find that these problems are actually quite straightforward. In this chapter you won't use equations to solve mixture problems. You will use subproblems. In Chapter 13: Convert to Algebra, you will solve mixture problems using algebra.

The following two problems, Paint and Chocolate Milk, are mixture problems. Use subproblems to solve them.

PAINT

A mixture is 25% red paint, 30% yellow paint, and 45% water. If 4 quarts of red paint are added to 20 quarts of the mixture, what is the percentage of red paint in the new mixture? List subproblems and solve this problem before reading on.

Were you confused by this problem? Did all the mixture-problem demons come rushing out of the closet of your mind? After you recovered, were you able to write down some subproblems?

Melanie and Kirk worked on this problem:

MELANIE: Arghh, I've never been able to do mixture problems.

KIRK: Me neither. I get all mixed up. Let's try it with the subproblem idea.

MELANIE: Okay, I'm game. What do we need to know?

KIRK: What, er, hmm . . . how about, what is the percentage of red paint in the new mixture?

MELANIE: That's the question. Can we figure that out right now?

KIRK: Well, what do we need to know to figure it out? This problem is making me see red.

MELANIE: We need the amount of red paint in the final mixture and the total amount of paint in the final mixture. Then we can divide and get the percentage.

KIRK: Yello, that's good. Let's start writing these down. (He wrote down the subproblems listed below.)

1. How many quarts of red paint are in the new mixture?

2. How many quarts of paint are in the new mixture?

3. What percentage of the new mixture is red paint?

MELANIE: Okay, let's figure these out. How much red paint is there in the final mixture? How do we figure that out? We don't even know how many quarts of red paint are in the original mixture.

KIRK: You're right. That's another subproblem. I'll add it to the list. (He added it at the top of the list and renumbered the other subproblems. His revised list is shown next.)

1. How many quarts of red paint are in the original mixture?

2. How many quarts of red paint are in the new mixture?

3. How many quarts of paint are in the new mixture?

4. What percentage of the new mixture is red paint?

$0.25 \times 20 = 5$

KIRK: Okay, I think we're in business. The original mixture is 25% red. There are 20 quarts in the original mixture, so 25% of 20 means 5 quarts of red paint in the original mixture.

MELANIE: Great. Okay, so the new mixture contains the original 5 quarts plus the 4 that were added, and that makes 9 quarts of red paint.

KIRK: Now we've got things stirred up. Our next subproblem asks how many quarts of paint are in the new mixture. Well, that's 'red'ily apparent. It's 24. We started with 20, and we just added 4.

MELANIE: Finally, we just have to divide to find the percentage of red paint in the final mixture: $9/24 = 0.375$. That's 37.5% red. We did it. These subproblems made this problem not so bad.

Note that you could also figure the new percentages of yellow paint and water in the same way. If you couldn't solve this problem before reading the solution, figure out the percentages of yellow paint and water by yourself before continuing.

The next problem, Chocolate Milk, may bring up more ghosts from algebra. Solve this problem by using a combination of the subproblems and guess-and-check strategies. (You could use algebra, but try solving this problem without it.)

CHOCOLATE MILK

Augustus is trying to make chocolate milk. He has made a 10% chocolate milk solution (this means that the solution is 10% chocolate and 90% milk). He has also made a 25% chocolate milk solution. Unfortunately, the 10% solution is too weak and the 25% solution is way too chocolaty. He has a whole lot of the 10% solution but only 30 gallons of the 25% solution. How many gallons of the 10% solution should he add to the 25% solution to make a mixture that is 15% chocolate? (Augustus is sure the 15% solution will be absolutely perfect.) Solve this problem before continuing.

Pak worked on this problem. His solution is a combination of identifying subproblems and guess-and-check. He started by listing a subproblem:

1. How much chocolate is in the 30 gallons of 25% solution?

$0.25 \times 30 = 7.5$

Twenty-five percent of 30 is 7.5 gallons of chocolate in the 30 gallons of solution. This also means that there were 22.5 gallons of milk in the 30 gallons of solution.

Then Pak was stuck. He couldn't think of any other subproblems. So he guessed. "I wasn't really sure how guess-and-check was going to help me. But I decided to guess how many gallons of the 10% solution should be added, because that is what the question was asking. I guessed 20 gallons for my first guess. I started to set up my chart, but I wasn't sure what else was going to go in the chart. But my guess of 20 gallons of 10% solution had to be the first column."

Gallons of 10% Solution
20 gal

Pak didn't know what to write next. Then he realized that there was another subproblem lurking. "I thought about what else I could figure out from that guess. I realized that if I had 20 gallons of the 10% solution, then I could figure out how much of those 20 gallons was chocolate. Since it was 10% chocolate, I multiplied 20 times 0.10 to get 2 gallons. So 2 gallons of those 20 gallons are chocolate. That also means that the other 18 gallons are milk, but I didn't think that mattered. Chocolate is the most important thing after all!" So Pak added this information to his chart, along with the answers to the subproblem he had figured out earlier.

Gallons of 10% Solution	Gallons of Choc in 10% Soln	Gallons of 25% Solution	Gallons of Choc in 25% Soln
20 gal	2 gal	30 gal	7.5 gal

Now Pak had to determine how to check his guess. He realized that checking the guess involved more subproblems.

2. How much chocolate is in the new mixture (for that guess)?

3. How many gallons of solution are in the mixture (for that guess)?

4. What percentage of the new mixture is chocolate (for that guess)?

Pak added three more columns to his chart to account for the new subproblems. "I looked carefully at my chart so far, and realized that since I had 20 gallons of 10% and 30 gallons of 25%, that I could add those together and get 50 gallons of mixture." [subproblem 3 above] "I also needed to know how much of that was chocolate, so I added the 2 gallons of chocolate to the 7.5 gallons of chocolate to get 9.5 gallons of chocolate in the mixture." [subproblem 2 above] "Now I needed to know what percentage of the mixture is chocolate." [subproblem 4 above] "I had to think about this for a while. I figured it had something to do with division, but I wasn't sure which way to divide. First I divided 50 by 9.5. That gave me about 5.26, but that didn't make any sense. Since I'm mixing a 10% solution with a 25% solution, the answer should be somewhere between 10% and 25%. It can't get any weaker than 10% and it can't get any stronger than 25%. So dividing 50 by 9.5 is definitely wrong. So then I tried dividing 9.5 by 50. That seemed weird, but I tried it anyway. So $9.5 \div 50 = 0.19$. I had to think about what that meant. Then I realized that I was getting a decimal, which represented a percentage. So 0.19 actually means 19%. That makes sense, because it's in between 10 and 25. Then I saw something else that was kind of cool. 50 gallons is half of 100 gallons. If I had 9.5 gallons of chocolate out of 50 gallons of mixture, then I would have 19 gallons of chocolate out of 100 gallons of mixture, and that would be 19%.

$$\% \text{ choc} = \frac{gal\ choc}{gal\ mix}$$

"Finally I had to figure out how to rate this guess. I wasn't sure, but I thought the guess was high, since I was looking for 15% and 19 is higher than 15. I wrote high, and then I made a lower guess. This time I guessed 5 gallons of the 10% solution. I used all the same calculations to figure out the final percentage of chocolate in the mixture."

Gallons of 10% Solution	Gallons of Choc in 10% Soln	Gallons of 25% Solution	Gallons of Choc in 25% Soln	Total Gallons of Choc	Total Gallons of Mix	% of Choc in Tot Mix	Rating
20 gal	2 gal	30 gal	7.5 gal	9.5 gal	50 gal	0.19 = 19%	high
5 gal	0.5 gal	30 gal	7.5 gal	8 gal	35 gal	0.229 = 22.9%	???

"I was confused. My guess of 20 gallons gave me an answer of 19%. I thought that was high, so I made my next guess lower. But my guess of 5 gallons ended up with an answer of 22.9%, which is much further away from 15%. After thinking about this for a while I figured it out. I'm guessing the gallons of the 10% solution, which is the weaker solution. So if the amount of the 10% solution goes up, the overall strength of the mixture is going to go down. So both of these guesses were actually high. I need more 10% solution to get the final percentage down to 15%. I crossed out high and wrote low for my first guess, and I rated my second guess as lower. So my next guess had to be more than 20. It took me a while, but I got it after four more guesses."

Gallons of 10% Solution	Gallons of Choc in 10% Soln	Gallons of 25% Solution	Gallons of Choc in 25% Soln	Total Gallons of Choc	Total Gallons of Mix	% of Choc in Tot Mix	Rating
20 gal	2 gal	30 gal	7.5 gal	9.5 gal	50 gal	0.19 = 19%	~~high~~ low
5 gal	0.5 gal	30 gal	7.5 gal	8 gal	35 gal	0.229 = 22.9%	~~???~~ lower
30 gal	3 gal	30 gal	7.5 gal	10.5 gal	60 gal	0.175 = 17.5%	low
50 gal	5 gal	30 gal	7.5 gal	12.5 gal	80 gal	0.156 = 15.6%	low
100 gal	10 gal	30 gal	7.5 gal	17.5 gal	130 gal	0.135 = 13.5%	high
60 gal	6 gal	30 gal	7.5 gal	13.5 gal	90 gal	0.15 = 15%	right

Pak finished by answering the question. "Augustus needs to add 60 gallons of the 10% solution and he will have the perfect chocolate milk."

The Chocolate Milk problem is another problem in which strategies overlap. We will revisit this problem in Chapter 13: Convert to Algebra and discuss where the algebra equations come from.

Lesha solved this problem in a completely different way. Her solution involves a different subproblem:

"This would have been really easy if Augustus had wanted a 17.5% solution because that would be halfway between the 25% solution and the 10% solution. That would mean you'd need the same amount of each solution, so you would need 30 gallons of the 10% solution. Of course, it's not 17.5%, it's 15%.

"But that brought to mind a subproblem: What does the ratio have to be between the 10% solution and the 25% solution?

"I noticed that 15% was one-third of the way from 10 to 25, so I thought I only needed to add in a small amount of the 10% mix to create the 15% mix.

"In terms of how close 10% is to 25%, the 15% mix that's asked for is only one of three parts of the way. Better yet, if you break down the difference into 10%–15%, 15%–20%, and 20%–25%, then getting 10% up to 15% is one part close to 10% and two parts close to 25%. Here's a diagram to show what I mean."

Draw a diagram to aid comprehension.

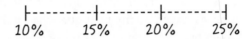

"Fifteen percent is one of three parts of the way toward 25% from 10%, so you have one part on the left and two parts on the right. That means you have to keep this 1:2 ratio when you mix the solutions together. At first I thought it had to be one part of the 10% mixed with two parts of the 25%, but that would actually make the blend closer to the 25%, so I knew I had it backwards.

"Therefore, the right answer is a 2:1 ratio of the weak stuff to the strong stuff—he already has 30 gallons of strong chocolate milk. In a 2:1 ratio, you need to add twice as much as you already have, so twice 30 is 60. He needs 60 gallons of the 10% solution."

Lesha looked at this problem from a different perspective and found that she needed an unusual subproblem. The ratio of the two liquids was the key to her solution. Her diagram helped her find that ratio and solve the problem.

Identify Subproblems

Break the problem into parts and start with the parts you know how to solve.

The strategy of identifying subproblems is very useful for solving complicated problems. Some problems look impossible when you first see them, but after you break them down into their subproblems you can see how to proceed. Listing your subproblems focuses your thinking and helps you more clearly see what you know, what you don't know, and what you can figure out.

Problem Set A

Solve each problem by first listing all the subproblems and then solving them to answer the given question.

1. COFFEE

How many ounces of coffee can be bought for $1.11 if 2 pounds cost $5.92?

2. SHARING EXPENSES

Five housemates held a party. They agreed to share the expenses equally. Leroy spent $20 on soft drinks. Alex spent $5 on paper plates. Kulwinder spent $13 on decorations. Maxx spent $16 on snacks. Bobbi spent $8 on envelopes and paper to send out invitations, and she also spent $7.40 in postage. Who owes money to whom?

3. AIRPLANE SEATS

On an airplane that is two-thirds full, 20% of the passengers are boys, one-fourth of the passengers are women, one-eighth of the passengers are girls, and there are 68 men. How many seats are on the plane?

4. SIX SQUARES

The picture shows six equal squares. The total area is 54 ft². What is the perimeter?

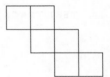

5. SHADED AREA

Find the shaded area in the figure. The large figure is a square, and each arc is one-fourth of a circle.

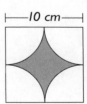

6. SAVINGS PLAN

Harriet saves 10% of her salary every month. Her company has fallen on hard times, so her monthly salary has just decreased from $3,600 to $3,000. She decides to save the same dollar amount each month, even though her salary has been reduced. What percentage of her new salary will her savings be now?

7. TEST AVERAGE

Mr. Howard is a student teacher at the local junior high. His first-period class of 40 students averaged 96% on a recent test. His second-period class of 20 students averaged 90% on the same test. What was the combined average for both classes?

8. CAR TRIP

If Clarence drives 60 miles per hour, it will take him 3 hours to drive to Concordia. How many minutes longer will it take to make the trip if he drives 48 miles per hour?

9. FARGO

Tiffany drove from her home to Fargo, North Dakota, in 2 hours. On the way back home, she drove 54 miles per hour, and it took her 14 minutes longer. At what speed did she drive on the way to Fargo?

10. TEST TRACK

A certain car that is being tested by its manufacturer uses its entire fuel supply in about 38 hours when idling. The same car, when driven at 60 miles per hour on a test track, uses about three-and-a-half times as much fuel per hour as it does when idling. If the engine has been idling for 10 hours and the car is then run at 60 miles per hour, how much longer will the car run before it uses up all of its fuel?

11. SLOW TIME CLOCK

The time clock in a factory is running slow. When an hour passes on the time clock, it has actually been $64\frac{1}{2}$ minutes of real time. A worker that gets paid \$11.80 per hour works an 8-hour shift as measured by this time clock. Beyond an 8-hour day, the worker gets paid time and a half for overtime. How much extra pay is the worker entitled to if he works one week (5 days) in these conditions?

12. RUSH JOB

Thirty men working on a construction job had completed one-third of the job in the past 12 days. The job was behind schedule, and needed to be completed 10 days from now. New workers needed to be hired in order to accomplish that. How many new workers needed to be hired?

13. BOX

The area of the top of a rectangular box is 324 in.^{2}, the area of the front of the box is 135 in.^{2}, and the area of the end is 60 in.^{2}. What is the volume of the box?

14. SWEETENED CEREAL

In her duties as Mom, Libby tries to keep to a minimum the number of empty calories her family consumes. To Libby, empty calories are calories that don't offer any nutritional value, as opposed to the fructose in an apple. Her kids love honey-sweetened cereal. She mixes the honey-sweetened cereal with the unsweetened version of the same cereal. Her kids don't notice until she makes it too bland. From experience, she thinks the mixture is too bland when the amount of the honey-sweetened cereal drops to less than 40% of the mix. Her husband, Labe, has just mixed together a 14-oz package of unsweetened cereal and a 32-oz package of sweetened cereal. How much more unsweetened cereal does Libby need to mix in to make the mix 40% sweetened and 60% unsweetened?

15. STYROFOAM CUP

Find the volume of a Styrofoam cup if the **diameter** of the top is 3 inches, the diameter of the base is 2 inches, and the height is 4 inches. The volume of a cone is given by the formula $V = (\frac{1}{3})\pi r^2 h$.

16. RED ROAD

In right triangle *RED,* angle *R* is the right angle. Point *O* is on segment *ER* and point *A* is on segment *ED*. Segment *OA* is perpendicular to segment *ED*. *EA* = 6, *AD* = 14, and *ER* = 16. Find the area of quadrilateral *ROAD*.

17. WRITE YOUR OWN

Write your own subproblems problem. The easiest types of problems to make up are those like the Coffee problem in this problem set or the Watering the Lawn problem in this chapter.

CLASSIC PROBLEMS

18. NINE COINS

There are nine identical-looking coins. One of the coins is counterfeit and weighs less than the other coins. The only scale available is a balance scale, on which you can weigh any number of coins against each other. Using the scale only twice, figure out a way to find the counterfeit coin.

Source unknown.

19. TWELVE COINS

You have 12 identical-looking coins, one of which is counterfeit. The counterfeit coin is either heavier or lighter than the rest. The only scale available is a simple balance. Using the scale only three times, find the counterfeit coin.

Adapted from *Games for the Superintelligent* by James Fixx.

MORE PRACTICE

1. GRAZING BUSINESS PART I

Cliff runs a grazing business. He will bring his flock of 8 sheep, 20 lambs, 5 goats, and a llama to your field and they will "mow" the weeds by eating them. For one job the lambs ate the weeds on 2/5 of the land. The sheep ate the weeds on 34% of the land. The goats ate the weeds on $\frac{1}{4}$ of the land. And the llama ate the weeds on the remaining 20 square yards of land. How much land was "mown"?

2. GRAZING BUSINESS PART 2

Suppose that in the problem above, the flock was supposed to eat for 10 hours. Further suppose that things did not go quite as planned. Unfortunately 11 of the lambs ran away after 7 hours. Two of the goats took a nap for 2 hours. And the sixth sheep is sick, and only ate half of what she should. How many square yards of land didn't get eaten?

3. RESPOND, PLEASE

When Rob and Danielle got married, they invited 200 people. Eighty percent of the invitees sent back the RSVP card. Ninety percent of the people who sent in an RSVP said they would attend. When the wedding day arrived, 7 people who had responded yes did not show up, but half of the people who had responded no did show up. One-fourth of the people who had not sent in an RSVP showed up. They had chairs for 150 people. Were they able to accommodate everyone?

4. PAYCHECK PART I

Roseanne works in a factory. She makes $13.75 per hour and works 30 hours per week. Taxes (federal, state, and social security) deduct 28% of her gross pay. She then takes 10% of her take-home pay and puts it into a savings account. This leaves her with how much money to spend each week?

5. PAYCHECK PART 2

In the previous problem, Roseanne gets a 4% hourly raise, and at the same time starts working 36 hours per week. Her tax rate goes up to 30%. She decides that she was fairly comfortable living on the amount of money she used to have to spend every week. So she adds just $20 per week to the amount she has available to spend, and puts the rest of her extra pay into her savings account every week. How much is she now saving every week?

6. THE PALMETTO GRAPEFRUIT

Judi purchased a huge organic grapefruit at her favorite health food grocery store where she receives a 15% discount. The grapefruit was marked $1.69 per pound and it weighed 1.5125 pounds. When Judi was peeling the grapefruit she noticed that the peel was extremely thick. As a matter of fact, the diameter of the peeled grapefruit was only 3¾ inches, while the peel itself was ¾ inches thick. She decided to weigh the peel on her postal scale to find out how much she was paying for the peel alone, which she would throw away in her mulch pile. She found that the peel weighed 8.4 ounces while the fruit itself weighed 15.8 ounces. She turned the grapefruit into juice by putting the fruit into her blender and ended up with 1⅞ cups of juice.

a. How much did Judi pay for the peel that she tossed?

b. How much per pound did Judi pay for the part of the grapefruit she could eat?

c. How much would a quart of juice cost at this rate?

d. What was the diameter of the whole grapefruit including the peel when she purchased it?

This problem was written by Judi Caler, a student at Sierra College in Rocklin, California.

Problem Set B

1. CARROT JUICE

Eric, a health enthusiast, enjoys mixing his own beverages. His current favorite is a carrot drink that contains 40% pure carrot juice. He has been to the store and has found some concentrated juice that is 60% pure carrot juice. His wife, Alicia, found some carrot juice on sale at a different health-food store, and she brought home 80 quarts. It contains 12% pure carrot juice. How many quarts of the concentrated 60% carrot juice does Eric need to add to the 80 quarts that Alicia brought home to produce his perfect drink containing 40% carrot juice?

2. WHO WEIGHS WHAT?

Devon, Frank, Fua, Morris, and Pedro belong to the same gym. The gym rules prohibit the staff from giving out personal information about their clients. However, each of the five men said just enough that you can figure out their exact weights. (Do this problem quickly, because after the workout Morris and Devon are going to an all-you-can-eat buffet.)

DEVON: Pedro weighs 18 pounds more than I do. None of us weighs over 200 pounds.

FRANK: My weight is divisible by 7. Morris weighs 12 pounds more than I do.

FUA: Three people are heavier than I am. Pedro's weight is a prime number.

MORRIS: I'm heavier than Pedro. Devon is the lightest.

PEDRO: Morris's weight is divisible by 10. The five of us together weigh a total of exactly 840 pounds.

3. HRUNKLA APARTMENT HOUSES

Most Hrunkla lived in giant, 12-story apartment houses, and their homes were large square rooms bounded on four sides by corridors. Each room had a single door which opened halfway along a corridor. On even-numbered floors, the doors opened onto the east corridor; on odd-numbered floors, the doors opened onto the north corridor. At each intersection of corridors, there was something like an elevator which could be ridden up or down. Half of the corridors had moving belts on the floor, and no self-respecting Hrunkla would walk if he could ride one of these belts. The belts were so arranged that those on floors 1, 5, and 9 ran to the east; those on floors 2, 6, and 10 ran to the south; those on floors 3, 7, and 11 ran to the west; and those on floors 4, 8, and 12 ran to the north. Describe how a Hrunkla who lived on floor 10 could use these moving belts and elevators to visit a friend who lived in the room directly below his.

From *Make It Simpler: A Practical Guide to Problem Solving in Mathematics* by Carol Meyer and Tom Sallee. ©1983 by Addison-Wesley Publishing Company. Reprinted by permission of Pearson Learning. Used by permission.

FAMILY DAY

Incredibly Huge Motors is planning an employee-and-family day at the baseball park. The company has reserved 6,000 seats with the ball club. Each section at the ballpark has 15 seats in each row and is 18 rows deep. How many sections does the ball club need to set aside for the IHM employees and families?

5. **NIGHTMARES**

I want to tell you about Batiscomb and the problem he was having with nightmares. Batiscomb was having some bad recurring nightmares. Every 19 days he was having a nightmare about 19 ghosts scaring him out of his wits by jumping out of a 1919 Hupmobile and saying "boo" 19 times. (Don't ask me to explain this—it's Batty's nightmare.) He was also having a nightmare every 13 days about 13 black cats crossing his path in front of his house, located at 1313 Thirteenth Street. He had a bad week in April, when on the night of Friday, April 5, he had the nightmare about the 19 ghosts saying "boo" 19 times. The very next night, he had the nightmare about the 13 black cats. Batty knows he can handle these nightmares when they occur from time to time, but lately he's been really worried that they might both occur on the same night. Do they ever occur on the same night by the end of the year? If so, on what date?

8

Analyze the Units

Paying close attention to the interrelationships among the units of a problem can help you organize the information and may lead directly to a solution. Chemists and physicists often use unit analysis to check that they've set up their calculations correctly.

C rane recorded the following conversation on a very bad tape. The recorder ate the tape, and Crane had to wind it back into the cassette manually. When he played the tape back, he was able to transcribe the following:

CLERK: Hi! How're you doing?

CUSTOMER: Oh, it feels like I've been going 24-7. Did you hear what happened on 80?

CLERK: Ten-four, buddy! Whew, must be a hundred.

CUSTOMER: I'd bet on it. It must be one-fifty.

CLERK: Just. Is that 24?

CUSTOMER: No, 18. What's new with you?

CLERK: We're goin' down 85 in 3.

CUSTOMER: Hey, last weekend we did 75 in 2.

CLERK: I heard Steve did 285. Impressive.

CUSTOMER: Naw, that was Laresa. She also got a 77 in the afternoon.

CLERK: Wow! Forrest had 17 in 7 the same day.

CUSTOMER: No hits!

Here's a correct interpretation of the same conversation:

CLERK: Hi! How're you doing?

CUSTOMER: Oh, it feels like I've been going 24 [hours per day]-7 [days per week]. Did you hear what happened on Interstate 80?

CLERK: I sure did. Whew, it's at least 100 [degrees] out today.

CUSTOMER: I'd bet on it. It must be about 1:50 [o'clock].

CLERK: Just about. Is that 24 [karat] gold?

CUSTOMER: No, 18 [karat]. What's new with you?

CLERK: We're goin' down Interstate 85 in 3 [days].

CUSTOMER: Hey, last weekend we biked 75 [miles] in 2 [hours].

CLERK: I heard Steve bowled 285 [pins] at Cameranisi's. Impressive.

CUSTOMER: No, that was Laresa. She also shot a 77 [strokes of golf] in the afternoon.

CLERK: Wow! And her brother Forrest had 17 [strikeouts] in 7 [innings] the same day.

CUSTOMER: Gollee!

Although some people might believe that the bad tape made the conversation difficult to understand, it is clear to any mathematician that the lack of units of measure caused more difficulty. In the conversation between the clerk and the customer, all of the following vital units were obscured in the recording: hours per day, days per week, degrees, hours and minutes in a day, karats, days, miles, hours, pins, strokes of golf, strikeouts, and innings. Think about how many of the phrases were difficult to understand, much less the whole conversation, when the conversation omitted the units. Without units, only a few numbers made any sense.

An answer without units is incomplete.

Unit analysis (also called **dimensional analysis**) involves dealing with units of measure very carefully. People often ignore units of measure in everyday computations, because they are working in a context where specific units are assumed. However, in any situation involving measurement the units have to be known in order to make sense of the answer. An answer is not complete without units.

For example, suppose your boss asked you to find out how much fencing you would need to enclose the company parking lot. You did the research and responded by saying you needed 145 to do the job. So your boss bought 145 yards of fence material at a one-time only sale. Unfortunately, you meant you needed 145 *feet* of fence material. You can be sure the boss will not be pleased to have 290 feet of leftover fencing.

By keeping units organized, you will be able to solve problems easily.

Unit analysis falls under the major problem-solving theme of Organizing Information. By keeping the units of measure organized, especially in the numerators and denominators of fractions, you can solve many problems by manipulating the units to get the answer you're looking for.

This chapter contains three different sections, each discussing how to use units of measure in a different way: in ratios, unit conversion, and compound units.

Section 1: Units in Ratios

A **ratio** is a comparison of two numbers. Ratios are often written as fractions, so even the "two-thirds" in the sentence "Bill ate two-thirds of the pizza" represents a ratio ($^2/_3$). Units of measure are often expressed as ratios. A common example is the listing of gas prices in dollars per gallon:

$$\frac{\text{dollars}}{\text{gallon}}$$

A price of $4.20 per gallon can be listed as $4.20/gallon. The ratio also makes sense if it is listed as $42 for 10 gallons. By looking at the ratio in the form of a fraction and dividing the number in the numerator by the number in the denominator, you simplify the ratio, and the units remain.

$$\frac{\$42}{10 \text{ gallons}} = \frac{\$4.20}{1 \text{ gallon}}$$

AT THE GROCERY STORE

A 12-ounce can of Rosario's Refried Beans sells for 59 cents, and the 16-ounce size of the same brand sells for 81 cents. Which can is the better buy? Work this problem before continuing.

To work this problem, Ananda wrote the units for both cans as similar ratios. The smaller can is arranged as 59 cents per 12 ounces, and the larger can is arranged as 81 cents per 16 ounces. Both have been arranged in terms of price over weight.

$$\frac{59 \text{ cents}}{12 \text{ ounces}} \quad \text{and} \quad \frac{81 \text{ cents}}{16 \text{ ounces}}$$

Then, by dividing the numerators by the denominators, she got

$$\frac{4.9 \text{ cents}}{1 \text{ ounce}} \quad \text{and} \quad \frac{5.1 \text{ cents}}{1 \text{ ounce}}$$

Ananda discovered that the smaller can is the better buy, because a shopper pays less per ounce for the smaller can than for the larger can.

Many grocery stores feature shelf tags that list something called **unit pricing.** Before these labels became common, you could see many shoppers with pained expressions trying to do mental calculations to figure out the best buy. Unit-pricing shelf tags resolve the value issue very quickly. In Ananda's case, the shelf tag for the smaller can would show a unit price of 4.9 cents per ounce, and the shelf tag for the larger can would show a unit price of 5.1 cents per ounce. As she found out, the smaller can is clearly the better buy.

Ananda could have compared the prices for the two cans of beans by finding out many ounces she would get for 1 cent. In that case she would have written these two ratios:

$$\frac{12 \ ounces}{59 \ cents} \quad \text{and} \quad \frac{16 \ ounces}{81 \ cents}$$

Dividing the numerators by the denominators would then give

$$\frac{0.203 \ ounce}{1 \ cent} \quad \text{and} \quad \frac{0.198 \ ounce}{1 \ cent}$$

The smaller can gives her more ounces per cent.

TONI'S TRIP

Toni drove 80 miles in 2 hours and used 5 gallons of gas. Notice that there are three different types of units of measure here—miles, hours, and gallons—measuring distance, time, and volume, respectively. From this information, you can calculate six ratios of quantities, considering two units at a time. The six ratios are miles/hour, hours/mile, gallons/hour, hours/gallon, gallons/mile, and miles/gallon. Calculate the value of each of these ratios. Work this problem before continuing.

Using the calculator to simplify gives these answers.

1. Miles/hour: $\quad \dfrac{80 \ \text{mi}}{2 \ \text{hr}} = \dfrac{40 \ \text{mi}}{1 \ \text{hr}} = 40 \ \dfrac{\text{mi}}{\text{hr}}$

2. Hours/mile: $\quad \dfrac{2 \ \text{hr}}{80 \ \text{mi}} = \dfrac{0.025 \ \text{hr}}{1 \ \text{mi}} = 0.025 \ \dfrac{\text{hr}}{\text{mi}}$

3. Gallons/hour: $\dfrac{5 \text{ gal}}{2 \text{ hr}} = \dfrac{2.5 \text{ gal}}{1 \text{ hr}} = 2.5 \dfrac{\text{gal}}{\text{hr}}$

This ratio is probably more useful as a measure of fuel consumption in airplanes.

4. Hours/gallon: $\dfrac{2 \text{ hr}}{5 \text{ gal}} = \dfrac{0.4 \text{ hr}}{1 \text{ gal}} = 0.4 \dfrac{\text{hr}}{\text{gal}}$

This would be relevant if you wanted to know what amount of time you could drive before worrying about filling up the gas tank.

5. Gallons/mile: $\dfrac{5 \text{ gal}}{80 \text{ mi}} = \dfrac{0.0625 \text{ gal}}{1 \text{ mi}} = 0.0625 \dfrac{\text{gal}}{\text{mi}}$

(or one-sixteenth gallon/mile)

6. Miles/gallon: $\dfrac{80 \text{ mi}}{5 \text{ gal}} = \dfrac{16 \text{ mi}}{1 \text{ gal}} = 16 \dfrac{\text{mi}}{\text{gal}}$

This is the familiar **mpg** (miles per gallon) as a relative measure of fuel economy in cars.

You can think of the word **per** as meaning "divide," as you've probably figured out. Writing mph (or mpg) sometimes misleads students to think of mph as a single unit. Instead, you must read **mph** as "miles per hour," which is actually the ratio miles/hour. Even miles/hour is dangerous to write, because the slanted fraction bar doesn't call out the distinction of numerator and denominator. You should get into the habit of visualizing and writing

$\dfrac{\text{miles}}{\text{hour}}$ and $\dfrac{\text{miles}}{\text{gallon}}$

Section 2: Unit Conversion

Convert to units you can conceptualize.

Looking back at the Toni's Trip problem, you can see that the answer to part 2 does not convey much meaning. You probably do not have a sense of how long 0.025 hour is. It would be nice if that number of hours were converted into a quantity with a unit you could conceptualize, such as seconds. The problem then becomes, How do we change 0.025 hour into seconds? or, in this case, How do we change 0.025 hour/mile into seconds/mile? You know that there are 60 minutes in 1 hour, and 60 seconds in 1 minute. This is a problem of **unit conversion,** which can be done by **canceling** units.

Start with the quantity 0.025 hour/1 mile and then multiply by the fractions shown below. Multiplying by 60 minutes/1 hour will convert hours into minutes but will not change the value of the ratio, because 60 minutes is the same as 1 hour. It is like multiplying by 1. You can visualize the ratio as a giant 1:

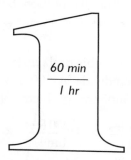

Similarly, multiplying by 60 seconds/1 minute will convert minutes into seconds:

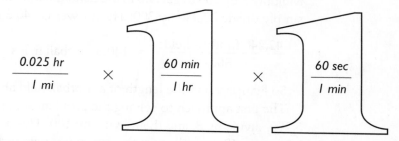

Because you are multiplying, when the same unit of measure (for example, minute and minute, or hour and hour) appears in a numerator and in a denominator, it can be "canceled out." So "hour" in the numerator of the first fraction cancels "hour" in the denominator of the second fraction, and "minutes" in the numerator of the second fraction cancels "minute" in the denominator of the third fraction.

$$\frac{0.025 \text{ hour}}{1 \text{ mile}} \times \frac{60 \text{ minutes}}{1 \text{ hour}} \times \frac{60 \text{ seconds}}{1 \text{ minute}} = \frac{90 \text{ seconds}}{1 \text{ mile}}$$

*Be sure to cancel units in **both** numerator and denominator!*

Unit analysis is a fantastic way to solve any problem involving any kind of unit. The fraction technique is *the* way to do it. You may have some success with easy problems without doing unit analysis, but when you come up against a complex problem you will need this strategy.

RUNNING FOOTBALL FIELDS

Francisco ran 8 miles. Being a football fan, Francisco wondered how many times he had run the equivalent of a full football field (100 yards). How many times had he? Work this problem before continuing. (Note: 1 mile = 5,280 feet)

Look to cancel all other units, leaving the desired unit for your answer.

To solve this problem, you should multiply 8 by 5,280, divide by 3, and then divide by 100. Do you believe it?

Examine this problem solution using the canceling-units approach:

$$\frac{8 \text{ miles}}{1} \times \frac{5,280 \text{ feet}}{1 \text{ mile}} \times \frac{1 \text{ yard}}{3 \text{ feet}} \times \frac{1 \text{ football field}}{100 \text{ yards}}$$

After you cancel units, football field is the unit remaining. To compute the answer, multiply 8 by 5,280 to get 42,240 in the numerator. Multiply 3 by 100 to get 300 in the denominator. To finish the problem, simply divide 42,240 by 300. The answer is 140.8 football fields.

$$\frac{42,240 \text{ football fields}}{300} = 140.8 \text{ football fields}$$

So Francisco ran the length of a football field about 141 times.

The first approach to solving the problem was to multiply 8 by 5,280, divide by 3, and then divide by 100. This approach turns out to be correct. If you still resist the canceling-units technique, ask yourself, "Which of the two solutions can I more readily verify to be correct?" or "Which solution could I easily explain to another person?"

Using Manipulatives for Unit Analysis

Students often ask, "How do you know which way to set up the fractions?" Suppose you have a problem that involves the ratio 65 miles per hour. Should you set up this ratio as 65 miles over 1 hour, or should you use 1 hour over 65 miles?

$$\frac{65 \text{ miles}}{1 \text{ hour}} \qquad \text{or} \qquad \frac{1 \text{ hour}}{65 \text{ miles}}$$

One way to figure out how to set up a fraction is to use **manipulatives.** Manipulatives are objects that can be moved or positioned, in this case small pieces of paper. We will explore the strategy of using manipulatives more fully in Chapter 10: Create a Physical Representation. Here we will use them to solve unit-analysis problems. Consider the next problem.

GAS CONSUMPTION

A car traveling at 65 miles per hour gets 25 miles per gallon and travels for 45 minutes. How many gallons does the car use during that time? Read the solution to this problem, making the manipulatives that the solution suggests and setting up the multiplication needed for the solution.

For each piece of information that the problem gives, make a manipulative. On one side of a small piece of paper, write the fraction given. On the opposite side, write the **reciprocal** of the fraction. For example, for the information about the travel time, 45 minutes, write a fraction showing 45 minutes over 1 trip and, on the back, 1 trip over 45 minutes.

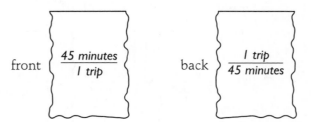

What else do we know here that might be useful? Because the problem involves both hours and minutes, it might be useful to have a manipulative that says 60 minutes equals 1 hour.

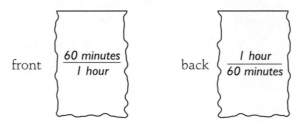

We also know that the car can travel 25 miles per 1 gallon.

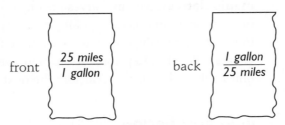

Also, we know that the car has been traveling at 65 miles per 1 hour.

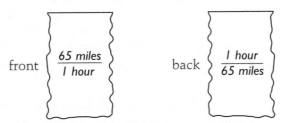

Focus on the desired unit.

Now place the manipulatives side by side. The intention is to get everything to cancel out except gallons. We want to know how many gallons were used during the trip, so the objective is to end up with gallons over trip. Suppose we put the manipulatives together so we start with gallons and end with trip.

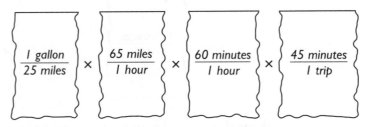

To calculate, multiply numerators and multiply denominators, then divide.

The miles would cancel, the hours would not cancel, and the minutes would not cancel. We could correct this by flipping over the 60 minutes/1 hour fraction. Now the hours cancel, the miles cancel, and the minutes cancel. This leaves us with gallons over trip, which is the unit we want. Now do the calculation.

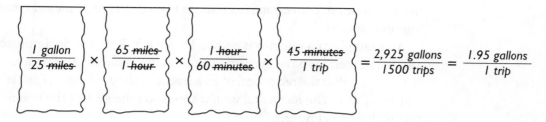

$$\frac{1\ gallon}{25\ \cancel{miles}} \times \frac{65\ \cancel{miles}}{1\ \cancel{hour}} \times \frac{1\ \cancel{hour}}{60\ \cancel{minutes}} \times \frac{45\ \cancel{minutes}}{1\ trip} = \frac{2,925\ gallons}{1500\ trips} = \frac{1.95\ gallons}{1\ trip}$$

Note that it is quite easy to do this problem upside down and still get everything to cancel out. But this method would be wrong because it leads to a peculiar answer.

$$\frac{25\ \cancel{miles}}{1\ gallon} \times \frac{1\ \cancel{hour}}{65\ \cancel{miles}} \times \frac{60\ \cancel{minutes}}{1\ \cancel{hour}} \times \frac{1\ trip}{45\ \cancel{minutes}} = \frac{1\ trip}{1.95\ gallons} = \frac{0.5128\ trip}{1\ gallon}$$

Students often think this answer means that the trip used 0.5128 gallon. But because the gallon unit is on the bottom, it actually means that the car covered 0.5128 of the trip on 1 gallon of gas. Although trips/gallon may be a useful unit, it was not what the problem asked for. So be cautious—make sure that the units in your answer end up where you want them.

Using One-n-oes

The manipulative approach to unit analysis can be very effective. Make manipulatives for all the known information in the problem. These manipulatives will be unique to that particular problem. For example, in the Gas Consumption problem, 25 miles per gallon, 65 miles per hour, and 45 minutes per trip were all quantities unique to that problem.

You will find it useful to have a set of manipulatives for information that will never change, such as the number of minutes in an hour and the number of feet in a mile. We will call these manipulatives one-n-oes because they represent one and look like dominoes. A **one-n-o** (pronounced WON-n-oh) is a fraction that equals 1. We will use one-n-oes to convert measurements from one type of measurement unit into another.[1]

An example of a one-n-o is shown below left. Both the denominator and the numerator have the same value, so the value of this fraction is 1. **Notice** that the reciprocal of this fraction, shown below right, is still equal to 1. The reciprocal would appear on the *back* of the one-n-o that is shown below left.

It's a good idea to make standard one-n-oes from colored paper.

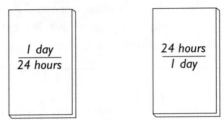

$$\frac{1 \text{ day}}{24 \text{ hours}}$$

$$\frac{24 \text{ hours}}{1 \text{ day}}$$

In your stock of one-n-oes, you should have at least all those that follow, as well as any others you think you will use all the time. Remember that each side of a one-n-o is the reciprocal of the other side.

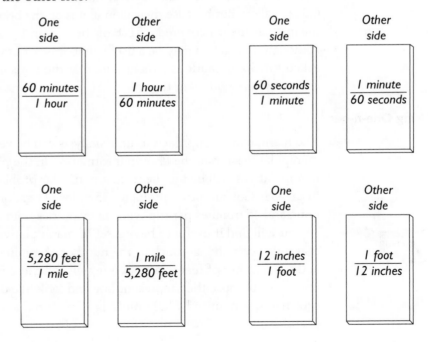

One side	Other side	One side	Other side
$\frac{60 \text{ minutes}}{1 \text{ hour}}$	$\frac{1 \text{ hour}}{60 \text{ minutes}}$	$\frac{60 \text{ seconds}}{1 \text{ minute}}$	$\frac{1 \text{ minute}}{60 \text{ seconds}}$

One side	Other side	One side	Other side
$\frac{5,280 \text{ feet}}{1 \text{ mile}}$	$\frac{1 \text{ mile}}{5,280 \text{ feet}}$	$\frac{12 \text{ inches}}{1 \text{ foot}}$	$\frac{1 \text{ foot}}{12 \text{ inches}}$

[1]Our friend Carolyn Donohoe-Mather says, "A one-n-o is what gets you to what you wanna know."

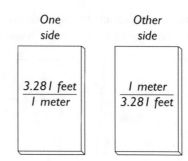

One side

$$\frac{3.281 \text{ feet}}{1 \text{ meter}}$$

Other side

$$\frac{1 \text{ meter}}{3.281 \text{ feet}}$$

When you make your one-n-oes, be careful. Note that it does not make sense to write either

$$\frac{1 \text{ inch}}{12 \text{ feet}} \quad \text{or} \quad \frac{12 \text{ feet}}{1 \text{ inch}}$$

because neither of these fractions has a value of 1. Twelve feet is not the same thing as 1 inch. If it were, a person who is 6 feet tall would be really short.

Look in the appendix for common conversions.

For more of the conversions that you'll use frequently, see the appendix at the back of this book. Also see the appendix for a list of abbreviations for common units of measure and for information about the metric system, which you will need as you work through this chapter.

The trick to using one-n-oes is finding the form (or forms) of 1 that will allow you to solve the problem. Solve the next problem using the standard one-n-oes that you just made. You will also need to make a new manipulative that reflects the information given in the problem.

LEAKY FAUCET, PART I

A leaky faucet drips 1 fluid ounce every 30 seconds. How many gallons of water will leak from this faucet in 1 year? Solve this problem before continuing.

What units are needed? You need the time units of seconds, minutes, days, and years. You also need to know how many fluid ounces are in a gallon, which you can find by using quarts and cups. You'll need a manipulative for the problem information and seven standard one-n-oes. Remember that the reciprocal of the fraction on each one-n-o should be written on the other side of the one-n-o.

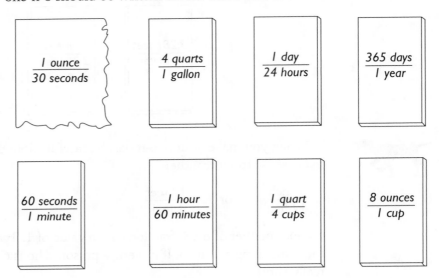

 Focus on the desired end units.

Organize your manipulatives in one long line, making sure that all units cancel except gallons and years. You want to end up with *gallons* on the top and *year* on the bottom. Do this before reading on.

The fractions below work for this problem. **Note** that their order can be rearranged but none of the fractions can be flipped over. Show all the units that cancel.

$$\frac{1\ \cancel{oz}}{30\ \cancel{sec}} \times \frac{60\ \cancel{sec}}{1\ \cancel{min}} \times \frac{60\ \cancel{min}}{1\ \cancel{hr}} \times \frac{24\ \cancel{hr}}{1\ \cancel{day}} \times \frac{365\ \cancel{days}}{1\ yr} \times \frac{1\ \cancel{cup}}{8\ \cancel{oz}} \times \frac{1\ \cancel{qt}}{4\ \cancel{cups}} \times \frac{1\ gal}{4\ \cancel{qt}}$$

Now do the arithmetic. You can use one of two methods to do this. The first method is to multiply all the numbers in the numerators and write down the result. Then multiply all the numbers in the denominators and write down the result. Then divide the resulting numerator by the resulting denominator.

$$\frac{31,536,000\ gal}{3,840\ yr} = \frac{8,212.5\ gal}{1\ yr}$$

With the second method, simply calculate from left to right, multiplying by each number that appears in a numerator and dividing by each number that appears in a denominator. Other than the first 1 appearing in the numerator of the first fraction (1 oz divided by 30 sec doesn't equal 1), all other 1's can be ignored because they don't affect the calculation. The calculator steps for these calculations are as follows:

$$1 \div 30 \times 60 \times 60 \times 24 \times 365 \div 8 \div 4 \div 4 = 8{,}212.5$$

When you calculate using this technique, the result you get is always in the numerator of the fraction. Thus, the answer to the problem is 8,212.5 gal/1 yr. The unit of gallons per year comes from canceling all the other units in the problem.

 Note on rounding: When you use a calculator, don't round any numbers until you finish the problem. Then, and only then, round the answer if you need to.

Metric and English Conversions

Most people in the United States don't currently worry about converting English units of measure to metric units. However, U.S. companies who manufacture goods outside the United States must know how to convert from English to metric units, and scientists in the United States regularly use the metric system. The importance of metric measurement in the United States is illustrated by the $150 million U.S. spacecraft that crashed into Mars in late 1999 because of scientists' failure to convert crucial measurements from English units to metric units. (For more information about metric units of measure, see the appendix at the back of this book.)

CONVERTING WOOD

Raoul had done all the measurements perfectly. He needed a piece of plywood that measured 122 centimeters by 244 centimeters. The problem was, with all of his traveling, he had gotten used to the notion that if it's Tuesday, this must be Belgium. Unfortunately, it was Thursday and he had jetted to the United States the night before. What size piece of wood should Raoul ask for at the local hardware store, where they don't use the metric system? Solve this problem before continuing.

Raoul needs some conversions fast. Raoul, along with most of the rest of the world, is using a metric mind to deal with the English measurement system, which even the English don't use anymore. Of course, Raoul remembered to analyze first: The units of measure in this problem are units of distance. Raoul recalled something from his studies of ancient measurement systems: 1 m (meter) equals approximately 3.281 ft. Let's help him out with his conversions. Change the meters to feet by using 1 m = 3.281 ft.

$$\frac{122 \text{ cm}}{1} \times \frac{1 \text{ m}}{100 \text{ cm}} \times \frac{3.281 \text{ ft}}{1 \text{ m}} = \frac{4.00282 \text{ ft}}{1}$$

Doing the same series of conversions for Raoul's measurement of 244 cm, we get

$$\frac{244 \text{ cm}}{1} \times \frac{1 \text{ m}}{100 \text{ cm}} \times \frac{3.281 \text{ ft}}{1 \text{ m}} = \frac{8.00564 \text{ ft}}{1}$$

As it turns out, Raoul needs a 4-ft-by-8-ft sheet of plywood.

Pipeline

In this section you will have to solve a number of conversion problems, converting one type of distance unit to another type of distance unit. For example, in the Converting Wood problem, you needed to convert centimeters to feet, but the only conversion you could use to go from metric distance to English distance was 1 meter = 3.281 feet. For the rest of the problems in this chapter, you can use a diagram we call a **pipeline** to show you how to get from one type of unit to another.

Using the following conversions, you can make a pipeline diagram that shows the paths used to find those conversions:

1 kilometer = 1,000 meters	1 foot = 12 inches
1 meter = 100 centimeters	1 yard = 3 feet
1 meter = 3.281 feet	1 mile = 5,280 feet

The pipeline diagram represents every conversion in the list exactly once. Each unit of measure is represented in a bubble, and each conversion in the list is represented as a line segment.

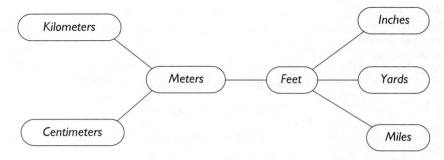

You can use the pipeline to find paths of conversions. For example, to convert from yards to kilometers, start at yards, convert to feet, convert to meters, and then convert to kilometers. This path is represented in the next diagram, with one-n-oes set up to convert 7,000 yards to kilometers.

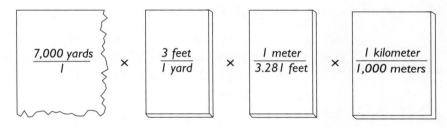

Doing the arithmetic gives an answer of 6.4 kilometers; so 6.4 kilometers is the same as 7,000 yards.

CONVERSION PRACTICE

Work the following problems. You may use any English-to-English conversions and any metric-to-metric conversions you want. However, the only English-to-metric conversion you may use is 1 meter = 3.281 feet.

1. Change 75 kilometers to miles.

2. Change 18 inches to centimeters.

Work these problems before continuing. Round answers to the nearest hundredth.

The strategy of subproblems naturally shows up in these problems, as it does in most unit-analysis problems. Often, you won't be able to go directly from the unit given to the unit asked for, so you must look for intermediate steps along the way. The pipeline diagram provides a visual path for the unit-analysis subproblems.

1. Change 75 kilometers to miles.

   Tim followed the pipeline diagram to take him from kilometers to meters to feet to miles.

   $$\frac{75 \text{ km}}{1} \times \frac{1{,}000 \text{ m}}{1 \text{ km}} \times \frac{3.281 \text{ ft}}{1 \text{ m}} \times \frac{1 \text{ mi}}{5{,}280 \text{ ft}} = 46.6 \text{ mi}$$

   "Then I just multiplied all the numbers on the top and divided by each of the numbers on the bottom. That gave me 46.6 miles."

2. Change 18 inches to centimeters.

   Following the pipeline, we need to go from inches to feet to meters to centimeters. We can use these one-n-oes and a manipulative for 18 inches to solve this problem:

   The inches need to cancel, so the inches in the second fraction need to be in the denominator of the fraction. This, in turn, helps set up all the other one-n-oes.

   There are 45.7 centimeters in 18 inches.

FASTBALL

Aroldis Chapman of the Cincinnati Reds and Stephen Strasburg of the Washington Nationals have been clocked throwing a baseball 100 miles per hour. At that speed, how much time does the batter have to react? (That is, how much time before the ball reaches the plate?) The pitcher's mound is 60 feet 6 inches from home plate. Work this problem before continuing.

Determine the most appropriate units for the final answer.

This problem involves feet, miles/hour, miles, and seconds. The problem is stated in terms of the unit miles/hour, and the answer must be expressed in terms of time. Just decide what unit makes sense. For a pitcher, a fast ball, and a batter's reaction time, it must be seconds or a fraction of a second.

The speed given in the Fastball problem is written as the fraction shown below left. Miles/hour is a measure of distance over time. The answer needs to be given as the number of seconds it takes the ball to traverse 60 feet 6 inches, in other words, as time over distance. Invert the measure given from miles/hour to hours/mile, as in the fraction below right.

$$\frac{100 \text{ mi}}{1 \text{ hr}} \qquad \frac{1 \text{ hr}}{100 \text{ mi}}$$

Now we need to make a series of unit conversions. First, change hours/mile to seconds/mile. The next subproblem is to convert miles to feet, because the plate is 60.5 feet from the pitcher's mound.

$$\frac{1 \text{ hr}}{100 \text{ mi}} \times \frac{60 \text{ min}}{1 \text{ hr}} \times \frac{60 \text{ sec}}{1 \text{ min}} \times \frac{1 \text{ mi}}{5,280 \text{ ft}} = \frac{3,600 \text{ sec}}{528,000 \text{ ft}}$$

Now our fraction is in the form of seconds/feet. All we need to do is multiply by 60.5 ft to find the number of seconds.

$$\frac{3,600 \text{ sec}}{528,000 \text{ ft}} \times \frac{60.5 \text{ ft}}{1} = \frac{217,800 \text{ sec}}{528,000} = 0.4125 \text{ second}$$

At this point, there are no units in the denominator, because we want to give the answer in seconds, not in seconds/feet. The ball reaches home plate in slightly less than half a second. Would you want to bat against Chapman or Strasburg?

Choose Appropriate Units

The answer must make sense.

In the Fastball problem we emphasized which units you would use to best express your answer. When considering how to express the answer for a problem, think about how useful the information given in the answer is. Does it tell you what you need to know in a way that is easy to understand? Can you easily use the information, or would you need to translate it into another form to be able to really use it? For example, which trip description makes more sense, a trip during which you travel 2.1 million feet in 28,000 seconds, or the *same* trip during which you travel 400 miles in 8 hours?

Section 3: Compound Units

The units we have dealt with in this chapter so far are either solitary units (a single unit by itself, such as feet) or units in a ratio (such as price per gallon). Just as putting units into a ratio is a key use of units, we also combine units into what we call **compound units.** Compound units are common in physics and chemistry. The next problem provides an example of a compound unit.

AREA

Find the area of a rectangle with length 3 feet and width 2 feet. Solve this problem before continuing.

The area of a rectangle can be found by using the formula area = length × width.

area = length × width

    = 3 feet × 2 feet

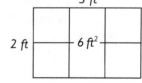

    = 6 square feet (which also can be written feet^2 or ft^2)

The unit in the answer to this problem is a compound unit: square feet. The original unit was a measure of linear distance, and the unit is now a measure of area. The unit square feet is an integral part of the problem and answer. A square foot is the area of a square measuring 1 foot on each side.

If we'd used 36 inches instead of 3 feet and 24 inches instead of 2 feet to find the answer, then the area would have been in the unit square inches: 864 square inches. We can show that this measure is equivalent to 6 square feet (square inches can also be written as inch · inch):

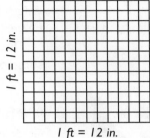

1 ft = 12 in.

$$\frac{864 \cancel{\text{inch}} \cdot \cancel{\text{inch}}}{1} \times \frac{1 \text{ foot}}{12 \cancel{\text{inches}}} \times \frac{1 \text{ foot}}{12 \cancel{\text{inches}}} = 6 \text{ feet}^2$$

Note that to convert square inches to square feet, you can't simply divide by 12 inches. Rather, you must divide by 12 inches squared, or 144 square inches. There are 144 square inches in 1 square foot.

$$\frac{864 \cancel{\text{inches}}^2}{1} \times \frac{1 \text{ foot}^2}{144 \cancel{\text{inches}}^2} = 6 \text{ feet}^2$$

Unit Analysis of Transportation Statistics

Another common type of compound unit—the passenger-mile—shows up in the transportation field, both as a measure of the efficiency of public transportation and in traffic-safety statistics. For example, you might see statistics on deaths per billion passenger-miles, accidents per billion passenger-miles, or even arrests per billion passenger-miles.

Just exactly what is a passenger-mile? It represents 1 passenger traveling 1 mile. For example, a car with 1 passenger traveling 12 miles racks up 12 passenger-miles. A car with 3 passengers traveling 5 miles accounts for 15 passenger-miles. An airplane that travels 2,500 miles with 400 passengers aboard accounts for 1 million passenger-miles.

Cars are quite clearly an inefficient means of transportation because they don't carry very many people and so do not account for very many passenger-miles per car. In contrast, trains, buses, vans, subways, and trolleys carry many more passengers and therefore can count a larger number of passenger-miles per vehicle.

What about the economics of the situation? Occasionally people drive together and share gas expenses. Buses are expensive to buy, and they get only a few miles per gallon in stop-and-go city conditions. A transit district could counter this disadvantage by showing that its buses average over 200 passenger-miles per gallon. When you can pack from 40 to 70 people on a bus, it becomes a very fuel-efficient mode of transportation, despite poor gas mileage. When computing the total costs of running the bus, though, you need to consider the driver's wages and the wages of the support personnel. Again, these costs can be very low when they are considered in terms of passenger-miles.

The next problem—a problem dealing with a car pool—illustrates some of these ideas.

LONG COMMUTE

Gerónimo and three friends regularly drive 208 miles. Their car gets 35 miles per gallon. Gas costs 1.59\frac{9}{10}$ per gallon. They drive at an average speed of 50 miles per hour. Find each of the following:

1. Gallons of gas used

2. Hours the trip took

3. Average feet per second

4. Cents per passenger-mile

Solve this problem before continuing.

Consider the questions one at a time.

1. Gallons of gas used

   To find gallons, consider the given unit, miles per gallon. We want the gallons part, but we need to get rid of the miles part. Miles are also given. Consider multiplying the two quantities together:

   $$\frac{35 \text{ miles}}{1 \text{ gallon}} \times \frac{208 \text{ miles}}{1}$$

This doesn't work because the miles unit will not cancel and the result will be miles²/gallon, which might make sense if you were plowing a field with a tractor but certainly makes no sense here. If the miles-per-gallon fraction were inverted, however, miles would cancel, leaving gallons.

$$\frac{1 \text{ gallon}}{35 \text{ miles}} \times \frac{208 \text{ miles}}{1} = 5.94 \text{ gallons}$$

One student set up the calculation to cancel miles as follows:

$$\frac{35 \text{ miles}}{1 \text{ gallon}} \times \frac{1}{208 \text{ miles}} = \frac{35}{208 \text{ gallons}} = \frac{0.168}{\text{gallons}}$$

Miles would cancel as needed, but does the answer look reasonable? Does it seem possible that a trip of this length could use less than 1 gallon of gas? Take a close look at the units: They're actually 1/gallons instead of gallons. That is, gallons is in the denominator, not in the numerator. That's why this method doesn't work.

2. Hours the trip took

June solved it this way: "I started the problem like this."

$$\frac{50 \text{ miles}}{1 \text{ hour}} \times \frac{208 \text{ miles}}{1}$$

"But then I saw square miles. I didn't want that. I wanted hours. I had to flip the miles per hour to make hours per mile, and then I could multiply and get the miles to cancel, leaving hours."

$$\frac{1 \text{ hour}}{50 \text{ miles}} \times \frac{208 \text{ miles}}{1} = \frac{4.16 \text{ hours}}{1}$$

3. Average feet per second

Terrence said, "This was hard, because I had to convert the miles to feet and then the hours to seconds."

$$\frac{50 \text{ miles}}{1 \text{ hour}} \times \frac{5,280 \text{ feet}}{1 \text{ mile}} \times \frac{1 \text{ hour}}{60 \text{ minutes}} \times \frac{1 \text{ minute}}{60 \text{ seconds}} =$$

$$\frac{264,000 \text{ feet}}{3,600 \text{ seconds}} = 73.3 \text{ feet/second}$$

Make sure your answer seems reasonable.

4. Cents per passenger mile

To find cents per passenger-mile, we need the cost of the trip in cents and the number of passenger-miles. Four people ride 208 miles, so there are 4 × 208 or 832 passenger-miles.

For cost we know that gas costs 1.59^9/_{10}$ per gallon, their car gets 35 miles per gallon, and they travel 208 miles. So far we have

$$\frac{1.599 \text{ dollars}}{1 \text{ gallon}} \qquad \frac{35 \text{ miles}}{1 \text{ gallon}} \qquad \frac{208 \text{ miles}}{1}$$

If we flip the miles/gallon ratio, both the miles and the gallons cancel to leave dollars, which we can convert to cents.

$$\frac{208 \text{ miles}}{1} \times \frac{1.599 \text{ dollars}}{1 \text{ gallon}} \times \frac{1 \text{ gallon}}{35 \text{ miles}} \times \frac{100 \text{ cents}}{1 \text{ dollar}} = 950 \text{ cents}$$

Now we can calculate the cost per passenger-mile:

$$950 \text{ cents} \times \frac{1}{832 \text{ passenger-miles}} = \frac{1.14 \text{ cents}}{1 \text{ passenger-mile}}$$

Note that for Gerónimo and his friends carpooling provided a much more efficient method of transportation than if all four individuals had driven their own cars. In this chapter's Problem Set A, you'll find another problem that illustrates this concept, and you will be asked to compare that situation with this one.

In Geronimo's carpool, the driver is considered a passenger because he is also going to the destination. If the driver is being paid to transport the passengers, such as a taxi driver or a bus driver, then the driver would not be considered a passenger because he or she is not also going to the destination.

Applied Problem Solving: Nerve-Cell Research

Rick, a professor of physiology at a leading university, provided this example of how he uses unit analysis.

"In our research we are very concerned with the timing of our experiments. The computer can sample data in nanoseconds (1 billionth of a second), microseconds (1 millionth of a second), or milliseconds (1 thousandth of a second). We often have to go back and forth between these different computer sampling rates and match them with the frequency responses of the nerve cells, which are in seconds, milliseconds, and microseconds. The computer sampling rate during the experiment must be configured to maximize the efficacy of the nerve activity that is being sampled. Computer storage space is very expensive, so we must be sure that we sample at the right rate. Knowledge of the units involved enables me to record the data I need without wasting memory."

Analyze the Units

Use the manipulatives for the problem information and the one-n-oes to organize the solution.

Many practical problems involve units of measure for things such as distance, time, money, weight, volume, velocity, acceleration, area, and so on. Some measurements involve compound units. Problems involving several units can be daunting. In this chapter we used unit analysis to organize the information in the problems so that canceling the right units would lead to the solution. Here are some key steps in using unit analysis:

- Identify all the units in the problem.

- Consider what units are required in the answer.

- Establish what unit conversions are needed.

- Make manipulatives for the given information and the reciprocals.

- Use one-n-oes for the conversion of units.

- Be flexible in organizing the manipulatives and one-n-oes so that the unwanted units will cancel.

Problem Set A

1. SODA CALORIES

It's been estimated that if you consume an extra 3,500 calories, you will gain 1 pound of fat. One can of soda contains 40 grams of sugar. Sugar has 4 calories per gram. If a person drinks a six-pack of soda every day, how many pounds will he or she gain in 1 year?

2. CHRISTINA'S TRIP

Christina drove 116 miles in 2 hours 15 minutes. She used 4 gallons of gas that cost her $16.52. Find the quantities that are expressed with each of these units (express a quantity with both the number-amount and the unit).

a. miles per hour

b. miles per gallon

c. dollars per gallon

d. feet per second

e. dollars per hour

f. quarts per minute

g. cents per minute

h. cents per mile

i. miles per dollar

j. gallons per hour

3. UNIT CONVERSIONS

Convert from metric to English or from English to metric as indicated. The only metric-to-English conversions you are allowed to use are 1 m = 3.281 ft and 1 gal = 3.79 L. Of course, you may use any English-to-English conversions (such as 1 mi = 5,280 ft) and any metric-to-metric conversions (such as 1 km = 1,000 m).

a. 35 m to feet

b. 170 ft to meters

c. 150 mi to kilometers

d. 47 km to miles

e. 4 ft to centimeters

f. 87 cm to inches

g. 54 in. to millimeters

h. $32 \dfrac{mi}{hr}$ to $\dfrac{meters}{second}$

i. 5 gal to liters

j. 16 L to quarts

4. ANOTHER LONG COMMUTE

Anastoli and two friends drove 87.5 miles and averaged 35 miles per gallon. They drove at an average speed of 50 miles per hour, and the gas cost $3.89⁹/10 per gallon. Find the quantities that are expressed with each of these units:

a. gallons

b. hours

c. dollars/passenger

d. feet/second

e. cents/mile

f. total number of passenger-miles

g. passenger-miles/gallon

h. cents/passenger-mile

Comparing this car with Gerónimo's car from the Long Commute problem, which car do you think is being used most efficiently?

5. IN-HOME SODA MACHINES

At the state fair, Aaron found a booth where in-home soda machines were being sold. The saleswoman told Aaron that if he bought syrup and carbonated it with this machine, it would cost him as little as 25 cents per liter to produce the same soda at home that he'd pay more for in the store. Compare Aaron's cost per liter to the cost per liter of a six-pack (six 12-ounce cans) purchased at your local store.

6. PAINTING CHIPMUNKS

Alvin, Simon, and Theodore went to work helping Dave paint his house. Alvin worked 6 hours, Simon worked $1\frac{1}{4}$ hours fewer than Alvin, and Theodore worked $4\frac{1}{2}$ hours. They were paid a total of $122 for their work. How much did each chipmunk get?

7. FIREPLACE INSERT

Kay wants to buy a new fireplace insert that will burn gas instead of wood. She normally uses two cords of wood in the winter. Wood costs $250/cord. She is interested in knowing how much she will save if she burns gas in the fireplace insert. Gas costs $0.91 per therm. One therm is 100,000 BTU. The fireplace insert burns 33,000 BTU per hour. Kay figures she will burn the fireplace 5 hours per day for 120 days. How much will she save in one winter if she burns gas instead of wood?

8. NEW CAR PURCHASE

Sommer wants to buy a new car. She currently drives a car that gets 24 miles per gallon. She is contemplating buying a car that gets 40 miles per gallon. Sommer drives 15,000 miles per year. Assume that gas costs $3.85 per gallon the entire year. How much will Sommer save on gas in 1 year if she buys the new car?

9. LEAKY FAUCET, PART 2

You realized that more than 8,000 gallons per year was a lot of water to waste with a leaky faucet. You fixed your faucet so that it now takes 11 minutes longer to fill a cup of water than it did when it leaked 1 fluid ounce in 30 seconds. How many gallons will the faucet waste in 1 year now? Refer to the Leaky Faucet problem (Part 1) to solve this problem.

10. CHAINS, FURLONGS, ACRES

An acre is one chain multiplied times one furlong. I know from horse racing that there are 8 furlongs in one mile. I remember that there are 640 acres in one square mile. How many feet are in one chain?

11. FIREFIGHTER

Kevin Stone is a wild-land fire fighter on an engine company. He is responsible for driving and maintaining a water carrier and is assigned to the left flank of a fully involved fire. His engine contains 1 tank of 750 gallons of water and his fire hose allows him to spray 265.3 liters of water per minute. How much time (in minutes and seconds) does he have until he runs out of water? (One gallon is 3.79 liters.)

This problem was written by Sierra College student and firefighter Joe Yates.

12. HOSES IN THE ROAD

Perhaps you have seen the hoses in the road measuring traffic flow and vehicle speed. A car traveling over them will first hit the back hose and then the front hose. A computer measures the time difference between the two. Suppose that the two hoses were 2 feet apart, and the difference between the strikes is 0.044 seconds.

  a. How fast is the car traveling?

  b. If each hose can move independently as much as one inch forward or backward, what is the slowest and what is the fastest that the car could be traveling?

13. SPEEDING UP

You are driving at 50 miles per hour. If you decrease the time it takes you to travel 1 mile by 8 seconds, what is your new speed?

14. READING RATE, PART 1

If you read 15 minutes per day every day and end up reading 12 books of 200 pages each in 1 year, what is your reading rate in pages per minute?

15. READING RATE, PART 2

If you increase your reading speed so that each page takes you 30 seconds less than it did before and you begin reading 20 minutes per day, how many 200-page books can you now read in a year?

16. FLOODS

During the recent floods, Folsom Lake was releasing water at the rate of 50,700 ft³/sec (That's cubic feet per second). The inflow into the lake at the height of the storm was 75,800 ft³/sec. The capacity of the lake is 975,000 acre-feet. One acre-foot is 43,560 cubic feet. Assume that the lake was 70% full at the beginning of this problem. Assume that the inflows and outflows remain the same during the entire problem. How much time will it take for the lake to be filled to capacity?

17. PENALTY KICK

China and the United States are playing each other in the championship match of the 1999 Women's World Cup soccer tournament. Brandi Chastain of the U.S. team is about to take a penalty kick against the Chinese team's goalkeeper, Gao Hong. Chastain will kick the ball from 18 yards away from the goal at 120 miles per hour. How many seconds does the goalkeeper have to react, dive, and reach the ball before it crosses the goal line?

18. SAILING SHIPS

The speed of boats is measured in **nautical miles** per hour, knots for short. The word *knots* is sometimes confusing because it includes the "per hour" part of the phrase "nautical miles per hour." The city of San Miguel on the island of Cozumel, Mexico, has a museum with a room devoted to pirates and sailing ships. One of the displays shows a log chip—a small triangular, weighted device used by fifteenth- and sixteenth-century sailors to measure speed. The log chip was tied to a rope that was wound up on a reel. The sailors tied evenly spaced knots in the rope. They would then throw the rope overboard, and the weight of the log chip would allow the rope to unreel as the boat moved along. The sailors would allow it to unreel for 30 seconds, timed with a sand hourglass. They would then haul in the rope and count the number of knots that had been reeled out. This number was the speed of the boat in nautical miles per hour. If one nautical mile is 6,076.10333 feet (one minute of arc of the earth's circumference), how far apart are the knots? Give your answer in feet.

NURSING

Allyson is a nurse in the intensive care unit of a hospital. She works with units daily, administering drugs through intravenous (IV) tubes. A common dosage for the drug dopamine is 2–5 micrograms per kilogram of body weight per minute. She will put 400 milligrams of dopamine into 250 cubic centimeters of fluid and administer it to the patient through an IV tube. At what rate, in cubic centimeters per hour, should she set the flow to achieve a dosage of 3 micrograms per kilogram per minute for a patient with a mass of 75 kilograms?

20. **DRIVE OR FLY: WHICH IS SAFER?**

Consider these statistics from the National Transportation Safety Board: In 1996, 319 people were killed in U.S. commercial airline accidents. Of course this is bad, but how does it compare to the 42,065 people killed in automobile accidents? Is it fair to compare cars and airplanes?

A way to even out the statistics to get a better feel for the relative safety of these modes of transportation is to use the ratio deaths per passenger-mile. According to the U.S. Department of Transportation, during 1996 the commercial airlines accounted for approximately 445.2 billion passenger-miles (445,200,000,000), whereas automobiles accounted for 3,630 billion passenger-miles. How do the death statistics compare now?

1. Write ratios showing deaths per billion passenger-miles for each form of transportation.

2. How many deaths per year would occur if cars had the same death rate per billion passenger-miles as airplanes? What does this say about the safety of driving versus flying?

21. **WRITE YOUR OWN**

Write your own unit-analysis problem. A unit-conversion problem is relatively easy to come up with. Consider making up your own system of units, possibly involving funny measures of things, such as paper clips or thumbnails.

CLASSIC PROBLEM

22. THE TELEGRAPH POLES

On an automobile trip the other day, I passed a line of telegraph poles that was 3⅝ miles long. With the aid of a stopwatch I discovered that the number of poles that passed per minute, multiplied by 3⅝, equaled the number of miles per hour that I was traveling. Assuming that the poles were equally spaced and that I traveled at a constant speed, what was the distance between two adjacent poles?

Adapted from *Mathematical Puzzles of Sam Loyd,* Vol. 2, edited by Martin Gardner.

MORE PRACTICE

1. RACINE TO CHICAGO

Duane, Jeff, and Glen drove from their home in Racine, Wisconsin, to Chicago to see the Cubs play the Braves at Wrigley Field. The distance was 75 miles, and they averaged 20 miles per gallon. They drove an average speed of 50 miles per hour and spent $14.85 on gas. Find each of the following.   (1 mile = 5,280 feet, and 1 meter = 3.281 feet)

a. gallons

b. hours

c. dollars/gallon

d. feet/second

e. cents/min

f. dollars/passenger

g. kilometers traveled

h. passenger-miles

i. passenger-miles/gallon

j. cents/passenger mile

2. STARBUCKS PART I

Six days every week, Jill goes to Starbucks to get something. She alternates in sequence between a Latte, a Mocha, and a Cappuccino, and always gets the Venti size (20 oz). The Latte has 240 calories. The Mocha has 340 calories, and the Cappuccino has 150 calories. A year is approximately 52 weeks long. Consuming an extra 3,500 calories will cause you to gain one pound of fat. How many pounds of fat per year will Jill gain from feeding her Starbucks habit?

3. **STARBUCKS PART 2**

   Jill decides to only drink Cappuccino, only go 4 times per week, and always order the Tall (12 oz), which only has 90 calories. How many fewer pounds will she gain?

4. **PEDOMETER PART 1**

   I got a new pedometer, which measures the number of steps I take, as well as how many miles I walk. But to measure the miles correctly, you have to enter the length of your stride. I entered 33 inches as the length of my stride, but I wasn't sure it was accurate. Then I took a walk. It showed that I had walked 0.66 miles on my usual neighborhood walk. However, I know that walk is actually 0.7 miles. How many steps did I take on my walk and what is the actual length of my stride?

5. **PEDOMETER PART 2**

   My wife goes for the same walk I took, but her pedometer is set to her correct stride of 30 inches. How many steps does she take on the neighborhood walk of 0.7 miles?

6. **PEDOMETER PART 3**

   I took the neighborhood walk one day, but I accidentally grabbed my wife's pedometer (stride length 30 inches) instead of mine. If I take the same number of steps that I did in Pedometer Part 1, how many miles would my wife's pedometer say I had walked?

7. **PEDOMETER PART 4**

   I walk 2 strides per second. Using my stride length from Pedometer Part 1, what is my speed in miles per hour?

8. **PEDOMETER PART 5**

   How long (in minutes and seconds) does it take for me to do the neighborhood walk?

9. **PEDOMETER PART 6**

   My wife is walking with me and we are both walking at my normal speed. But since her stride is shorter, she has to take more strides per second. Using the answer to Pedometer Part 4 as her speed in miles/hour, how many strides per second does she walk? (Her stride length is 30 inches per stride.)

Problem Set B

1. QUIT WHILE YOU CAN

Smoking has long been connected with serious diseases, such as lung cancer, emphysema, and heart disease. Let's estimate that for every cigarette a person smokes, he or she loses anywhere from 10 to 15 minutes of his or her life. Assuming the worst (15 minutes of life lost for each cigarette), how much shorter would the life span be of someone who smoked two packs a day for 35 years? (Each pack contains 20 cigarettes.) Answer in years, days, hours, and minutes (for example, 42 years 152 days 7 hours and 28 minutes).

2. VOLLEYBALL LEAGUE

Use the league standings for this recreational volleyball league and the schedule for the first three weeks of games to determine which teams won each week. (By the way, the Renegades won their second match.) What happened in each of the three matches that the Buckeyes played?

STANDINGS		
TEAM	WINS	LOSSES
Red Skeletons	3	0
Bombay Bicycle	2	1
Renegades	2	1
Sacto Magazine	2	1
Walleyball	2	1
Buckeyes	1	2
Bill's Thrills	0	3
Whine Sox	0	3

SCHEDULE
WEEK 1
Red Skeletons vs Walleyball
Bombay Bicycle vs Bill's Thrills
Whine Sox vs Renegades
Sacto Magazine vs Buckeyes
WEEK 2
Renegades vs Sacto Magazine
Walleyball vs Whine Sox
Buckeyes vs Bombay Bicycle
Bill's Thrills vs Red Skeletons
WEEK 3
Whine Sox vs Sacto Magazine
Walleyball vs Bill's Thrills
Red Skeletons vs Buckeyes
Bombay Bicycle vs Renegades

3. SPICE ON ICE

The national touring company Spice on Ice visited Seattle, Washington, last year. The producers held a special promotional ticket sale for one hour. During this time, adult tickets sold for $5, junior tickets (ages 9–17) sold for $2, and children's tickets (ages 0–8) sold for the ridiculously low price of 10¢. During this sale, 120 tickets sold for exactly $120. How many of each kind of ticket were sold during the sale?

4. NO CHAIN LETTERS

Dear Friend:

This is a chain letter. Make two copies. Send the two copies and this original to three friends.

Sincerely,
Albin Digas

Albin actually made five originals and sent them out in the first mailing. If the first mailing is considered the first generation and each set of copies constitutes the next generation, how many letters will be in existence after the tenth generation is produced? (Note: Chain letters are illegal.)

5. WHO WAS SNOOZING?

One of the six members of a company's board of directors was suspected of sleeping during a board meeting. It was known that only one board member had actually slept, but no one (except the six members) knew who it was. The company vice president questioned the members, who made the following statements:

DAVIS: The snoozer was either Rawls or Charlton.

RAWLS: Neither Vongy nor I was asleep.

CHARLTON: Both Rawls and Davis are lying.

BOBBINS: Only one of Rawls or Davis is telling the truth.

VONGY: Both Rawls and Davis are telling the truth.

EDWARDS: The snoozer is lying.

When the board chairperson (she was not questioned and is known to tell the truth) was consulted, she said that three of the board members always tell the truth and three of them always lie. She also said that Davis and Edwards can be counted on to do opposite things—when one tells the truth, then the other one lies. Who slept during the meeting?

9

Solve an Easier Related Problem

Sometimes, solving an easier version of a problem will reveal a useful method of solution. In a large-scale painting such as a mural, solving problems of color and composition can seem impossibly hard, so muralists often create smaller, preliminary studies to give them a sense of how to approach the final work.

One of the most famous legends in mathematics involves Carl Friedrich Gauss (1777–1855). The story concerns the young Gauss in fourth grade. One day his teacher was busy and wanted to give the class something to do so that he could get some work done at his desk. He gave the class an assignment: Add up all the numbers from 1 to 100. The teacher believed that this would keep the class occupied for 30 minutes. But after just a short time Gauss walked up to the teacher's desk with the answer written on his slate. The teacher was very impressed and asked him how he had solved the problem. Before we relate the rest of the story, you should try the problem.

FROM ONE TO ONE HUNDRED

What is the sum of the first 100 whole numbers? Work this problem before continuing.

There are many ways to approach this problem. One way, the brute-force problem-solving strategy, is to get out a piece of scratch paper or your calculator and just keep adding. A second way is to use your calculator's constant addition feature and its memory and just press the M+ (memory plus) key 100 times. But Gauss didn't have a calculator, so he must have approached the problem in a different way.

Another approach combines the strategies of subproblems and patterns. Tori used this approach.

Break up big problems into subproblems.

"I decided to break up the problem into subproblems. I split the hundred numbers into groups of ten and computed the sum of each ten. Then I looked for a pattern."

$$1 + 2 + 3 + 4 + 5 + 6 + 7 + 8 + 9 + 10 = 55$$
$$11 + 12 + 13 + 14 + 15 + 16 + 17 + 18 + 19 + 20 = 155$$
$$21 + 22 + 23 + 24 + 25 + 26 + 27 + 28 + 29 + 30 = 255$$

"I quickly saw the pattern, so I added up the sums of the ten groups of ten."

$$55 + 155 + 255 + 355 + 455 + 555 + 655 + 755 + 855 + 955 = 5,050$$

"Subproblems and patterns made this problem easy. Of course, I used a calculator and Carl Gauss couldn't have had one 200 years ago."

So what did Gauss do when he solved the problem? Imagine Gauss, as a boy, telling the story:

"I knew our teacher gave us this problem so we could practice adding, but I didn't need to practice. Besides, I thought there was an easier way. Adding the numbers from 1 to 100 would take a long time, so I thought about an easier problem. What if I add the numbers from 1 to 10? I could just add them in my head..."

$$1 + 2 = 3 \quad 3 + 3 = 6 \quad 6 + 4 = 10 \quad 10 + 5 = 15 \quad 15 + 6 = 21 \quad \text{and so on}$$

"But that wouldn't help with 100 numbers, and it's too easy to make a mistake. I tried rearranging the numbers, but that didn't help. Then I just wrote them out in order and thought, what if I start in the middle. 6 + 5 is 11, and work out from there? Or I could start at the ends, 1 + 10 is 11, and work in."

"I got 11 every time. So the sum of the numbers from 1 to 10 has to be 5 × 11, which is 55. Then I knew what I had to do to solve the problem from 1 to 100."

$$1 + 100 = 101$$
$$2 + 99 = 101$$
$$3 + 98 = 101$$
$$4 + 97 = 101 \quad \text{and so on}$$

"There would be 50 pairs that added up to 101, so the answer had to be 50 × 101, and that's the answer, 5,050!"

This story of what Carl Gauss might have done in the eighteenth century contains a great lesson for today: if a problem seems too big, too hard, or too time consuming, consider an easier problem. Of course, the easier problem has to be related to your original problem. It would not have helped Gauss to solve his problem if he'd just chosen to work on some random multiplication and addition problems. He needed an easier *related* problem.

Common Ways to Create Easier Related Problems

Tom Sallee, a professor at the University of California at Davis, considers the **easier related problem** strategy to be one of the most useful. He finds that there are many ways to make a problem easier and more manageable. Here is his list of the most common ways to make a problem easier:

1. Use a number instead of a variable.

2. Use a smaller or easier number in place of a more difficult one in order to develop the process for solving the problem.

3. Do a set of specific easier examples and look for a pattern.

4. Do a specific easier example and figure out an easier process that will work to solve the problem.

5. Change, fix, or get rid of some conditions.

6. Eliminate unnecessary information.

In this chapter you'll use all of these ways to make a problem easier. This strategy can seem difficult to get the hang of, but **remember,** when a problem seems too hard or too confusing, look for an easier related problem to solve and see if that helps.

A final thought from Sallee: "Easier related problems (abbreviated ERP) are analogous to guess-and-check in a funny way. In guess-and-check you try a number and see if it works. In ERPs you try a process and see if it works."

Solving an easier related problem is another strategy that falls under the major problem-solving theme of Changing Focus. The strategy may involve some elements of organizing information, but the main purpose of the strategy is to move your focus away from a difficult original problem to an easier related problem. After solving your easier problem, you can then decide if you have a plan for solving the original problem. Sometimes it is necessary to do several easier problems before returning to the original problem and solving it.

Use easier numbers.

One way to make an easier related problem is simply to replace the problem's difficult numbers with easier ones. You probably do this all the time. Suppose you're buying lunch and you have only $5.00. Instead of adding $1.59 for a hamburger, $0.89 for a soda, and $1.29 for a small order of french fries, you quickly add easier numbers in your head: $1.50 for the burger, $1.00 for the drink, and $1.25 for the fries. The estimated total is $3.75, so you have enough money. If you had only $4.00, you'd have to consider adding more carefully and determining whether there's tax on the meal. Easier numbers are useful for more than estimating. Sometimes when you replace difficult numbers with easier ones and then experiment with them to explore reasonable answers, you can gain insight into how to actually solve a problem.

SIMPLETOWN ELECTIONS

The clerk of Simpletown has the job of getting materials ready for the next municipal election. There are 29 issues and candidates. In the last election, there were 28,311 registered voters, representing 18,954 households, and they voted at 14 polling places. (In Simpletown all registered voters vote in the election.) The clerk figures she needs a proportionate amount of materials for this election. This time there are 34,892 people registered to vote. How many polling places will be needed? Work this problem before continuing.

Janeen contributed the following comments about this problem: "This problem just totally confused me at first. I had no idea what it was all about. After I reread it a couple of times, I started to think about some way to make the problem easier. I rounded off all of the numbers and made a list."

Use easier numbers.

> Polling Places: 15
> Voters (last election): 30,000
> Households: 20,000
> Issues: 30
> Voters (this election): 35,000

"I looked at these numbers and realized that the polling places and the voters had something in common: Polling places are locations for voters. The numbers must be connected somehow. Because there were 30,000 voters at 15 polling places last election (by my rounded numbers), that would mean there were about 2,000 voters at each polling place.

"The number of households would be important to an elections clerk for deciding the number of information pamphlets to send out, but it probably wouldn't be necessary in calculating the number of polling places needed. Besides, the only information provided for both elections is the number of voters for the two elections. So I assumed that the information about the number of households was not needed.

Eliminate unnecessary information.

"I also figured the number of issues didn't matter. It was probably the same as the last election, but I had no way of knowing because the problem didn't say. It actually would make a difference if there were a whole bunch more issues this time than last time, because it would take someone longer to vote. But I decided to ignore that too.

"Now I needed to know how many polling places were needed for this election. I decided to keep it at 2,000 people per polling place. So I just divided 35,000 by 2,000 and got 17.5. That's either 17 or 18 polling places for this election.

"Now that I had a plan, I went back to the original problem. In my easier related problem, I got that $^{30,000}/_{15}$ polling places is 2,000 voters per polling place. So I divided 28,311 by 14 to find out about how many voters there were per polling place in the original problem. The answer was 2,022.2 (rounded). I then divided 34,892 by 2,022.2 and got 17.25 polling places needed. I figured they would try to set up 17 polling places. This would be 34,892 voters divided by 17 polling places, which gives 2,052.5 voters per location, which is probably no problem. I thought it was interesting that I got basically the same answer with my easier problem."

Janeen's success with this problem can be traced to the following things she did:

1. She organized the information.

2. She simplified the numbers. (Take note of this.)

3. She ignored all the irrelevant information.

It's admirable that, instead of panicking, Janeen just got started on the problem and worked on it. The problem is complicated because it uses large numbers and there's no direct indication that you need to divide to solve it. Janeen also determined the relevant information and ignored the irrelevant information. Most importantly, by simplifying the numbers, she was in a better position to analyze the problem. Janeen used the strategy of solving an easier related problem. For her, the easier related problem was essentially this:

> The clerk of Easierrelatedtown has the job of getting the materials ready for the next municipal election. In the last election, there were about 30,000 registered voters and they voted at 15 polls. The clerk figures she needs a proportionate amount of materials for this election. This time there are about 35,000 people registered to vote. How many polling places will be needed?

This problem is nowhere near as complicated as the original problem, yet it is essentially the same problem. An easier related problem can often help you decide how to proceed with a problem that is difficult. Use the strategy of easier related problems to solve the next problem.

HOW MANY SQUARES?

How many squares are there on a checkerboard? (Hint: It is more than 64.) Work this problem before continuing.

Angie and Isaac used graph paper to work on this problem.

Read the question carefully and understand what it's asking.

ISAAC: This problem is too easy. It's 64—that is, 8-by-8. There are 8 squares on each side, and 8 times 8 is 64.

ANGIE: But that can't be right, because it is too easy. Look, aren't there some other squares on this board? If we divide the board into fourths, we have four more squares that each contain 16 of the little squares. And what about the square that has nine squares in it? (Angie drew the squares shown next.)

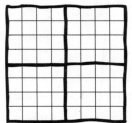 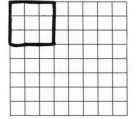

ISAAC: Wow, you're right. This problem is harder than I thought.

ANGIE: Well, let's look at an easier problem. How could we make this one easier?

ISAAC: I know—how about 3 + 5? That's an easier problem, and the answer is 8.

ANGIE: Isaac, the easier problem has to be related to the one we're doing.

Do a specific, easier example.

ISAAC: Okay, okay. I know, we could use a smaller square. How about 7-by-7 instead of 8-by-8?

ANGIE: Good idea, but that still seems too hard.

ISAAC: Well, this is ridiculous. How about 1-by-1? Is that easy enough for you?

ANGIE: Yeah, I think I can handle that. (She drew a 1-by-1 square.) How many squares are here?

ISAAC: I don't know, one?

ANGIE: That was a little too easy. Maybe we should try something a little harder. How about a 4-by-4 square?

Look for a pattern.

ISAAC: Now wait. Maybe if we go to 2-by-2 and then 3-by-3, we might find a pattern.

ANGIE: Okay, let's do 2-by-2. (She drew a 2-by-2 square.) Well, I see four squares.

ISAAC: I see five. There are obviously four little squares, but the whole picture is a square, so that makes five.

ANGIE: Oh yeah. Let's try a 3-by-3 square.

ISAAC: This is getting harder. Let's see, there are nine little squares and one big square, so there are ten.

ANGIE: But wait. There are also some medium-sized squares: squares that contain four squares. They are the same size as the 2-by-2 square we looked at a minute ago.

ISAAC: I see them. They sort of overlap. I think there are four. (He outlined them.)

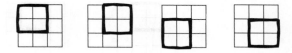

ANGIE: Okay, so there are nine little 1-by-1 squares, four 2-by-2 squares, and one 3-by-3 square. That is a total of 14. I think I'm seeing a pattern here.

ISAAC: Me too. Can we organize this somehow?

ANGIE: I think so. Let's make a chart.

ISAAC: What are we going to put in the chart?

ANGIE: Good question. Well, let's see. So far we have

Organize the information in a chart.

Size of "checkerboard"	Number of squares
1-by-1	1
2-by-2	5
3-by-3	14

ANGIE: Do you see a pattern?

ISAAC: No, I don't. But that last one was interesting. Let's do a 4-by-4.

ANGIE: Okay. (She drew a 4-by-4 square.)

ISAAC: Okay, there are obviously 16 little squares. How many 2-by-2 squares are there?

ANGIE: I don't know, but there is one big square.

ISAAC: Come on, Angie, we've got to do this systematically.

ANGIE: Okay, we're looking for squares that are 2-by-2. I think there are four. We can divide the whole square into fourths, so there are four.

ISAAC: I don't think so. Last time the squares overlapped. I think that happens here too. Let's look at just the top two rows. We can put a square on the left side, one in the middle, and one on the right side. (He outlined them, as shown.)

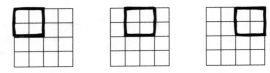

ANGIE: And we can do that in the second and third rows, and in the bottom two rows. So there are nine. I think I see a pattern here.

ISAAC: I do too. I bet there are four 3-by-3 squares.

ANGIE: Yes, there are. (She outlined them.)

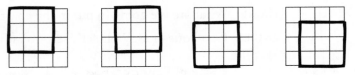

ISAAC: And there is one 4-by-4 square. It seems as though the pattern is the number of squares of each size. We need to organize this.

Size of Board	Number of Squares
1-by-1	1
2-by-2	$4 + 1 = 5$
3-by-3	$9 + 4 + 1 = 14$
4-by-4	$16 + 9 + 4 + 1 = 30$

ISAAC: Hey, look at this. If we reorganize this chart . . .

Size of Board	Size of Squares				Sum
	1 x 1 Squares	2 x 2 Squares	3 x 3 Squares	4 x 4 Squares	
1-by-1	1				1
2-by-2	4	1			5
3-by-3	9	4	1		14
4-by-4	16	9	4	1	30

ISAAC: I see a lot of patterns.

ANGIE: Yeah, check this out. The first column keeps on increasing, and the next row is the next perfect square.

ISAAC: Then it happens again in the second, third, and fourth columns.

ANGIE: Look at the rows: 14 is 1 + 4 + 9, and 30 is 1 + 4 + 9 + 16.

ISAAC: Great, so keep going.

ANGIE: All we have to do is add up all the squares up to the size of the checkerboard. So for the original problem, which is 8-by-8, we will add up $8^2 + 7^2 + 6^2 + 5^2 + 4^2 + 3^2 + 2^2 + 1^2$.

ISAAC: The answer is (using his calculator) 204. Great job! We could do this for any size checkerboard.

ANGIE: Yes, we could. I wonder if there is a formula for this?

ISAAC: I bet there is. Maybe we'll learn it later.

DIVISORS AND RECIPROCALS

The divisors of 360 add up to 1,170. What is the sum of the reciprocals of the divisors of 360? Work this problem before continuing.

Rori approached this problem this way: "I can't even imagine listing all the divisors of 360, let alone trying to find all their reciprocals and adding them up. There has to be an easier way. I'm a little lazy anyway, so it's not hard for me to look for an easier way. I decided to change the number 360 to something smaller. Of course, I wouldn't know what the divisors of my new number added up to, but if I picked a small number, I wouldn't have any problem finding its divisors so I could add them up. I picked the number 24, because it has a fair number of divisors but not so many that my easier problem would be hard to figure out.

"The divisors of 24 are 1, 2, 3, 4, 6, 8, 12, and 24. These added up to 60. The problem asked for the sum of the reciprocals of the divisors. So I needed to add up $\frac{1}{1} + \frac{1}{2} + \frac{1}{3} + \frac{1}{4} + \frac{1}{6} + \frac{1}{8} + \frac{1}{12} + \frac{1}{24}$. Yuck. Fractions. Then I needed a common **denominator.** It looked like the least common denominator was going to be 24. Well, that's interesting. So I changed everything to 24ths."

ORIGINAL		NEW
$\frac{1}{1}$	$=$	$\frac{24}{24}$
$\frac{1}{2}$	$=$	$\frac{12}{24}$
$\frac{1}{3}$	$=$	$\frac{8}{24}$
$\frac{1}{4}$	$=$	$\frac{6}{24}$
$\frac{1}{6}$	$=$	$\frac{4}{24}$
$\frac{1}{8}$	$=$	$\frac{3}{24}$
$\frac{1}{12}$	$=$	$\frac{2}{24}$
$\frac{1}{24}$	$=$	$\frac{1}{24}$

"Then I added up all the new fractions, which amounted to adding up the **numerators:** 24, 12, 8, 6, 4, 3, 2, 1. I'd added these numbers before. I knew the sum of these numbers was 60, because they were just the factors of 24. So the answer to my easier problem was $^{60}/_{24}$. My prediction was that the answer to the original problem was going to be the sum of the factors over the number. I tested this prediction on another example.

Find an easier process.

"I tried 10—I picked a smaller number to test my theory quickly. The factors of 10 are 1, 2, 5, and 10. Adding up these numbers gives 18. My prediction for the sum of the reciprocals was $^{18}/_{10}$. I tried it."

$$\frac{1}{1} + \frac{1}{2} + \frac{1}{5} + \frac{1}{10} = \frac{10}{10} + \frac{5}{10} + \frac{2}{10} + \frac{1}{10} = \frac{18}{10}$$

"I noticed that again I just added the numerators, which were the factors of the number in reverse order: 10, 5, 2, 1. The sum of the factors was 18 and the denominator was 10, so the answer was $^{18}/_{10}$.

"Therefore, the answer to the original problem was the sum of the divisors over the number, $^{1,170}/_{360}$."

Rori did a few specific easier examples, and this led her to an easier process for solving the original problem. She didn't even have to actually work the original problem.

The strategy of solving easier related problems is quite relevant for learning traditional math skills. For example, when students are asked to change $^{12}/_{11}$ into a decimal, they often ask, "Which number do you divide by?"

Consider this dialogue between an instructor and a student.

STUDENT: I have to change this to a decimal. How do I punch it into my calculator?

INSTRUCTOR: Think back to an easier problem. For this particular type of problem, I like to use $^1/_2$. How do you punch that in?

STUDENT: I don't know: $1 \div 2$?

INSTRUCTOR: Try it!

STUDENT: No, it's gotta be $2 \div 1$, because you would've said something.

INSTRUCTOR: I did say something. I said, "Try it." If you think it's $2 \div 1$, try it. If you think it's $1 \div 2$, try it.

STUDENT: But I didn't ask about that problem. I need to know about $^{12}/_{11}$. Anybody knows $^1/_2$ is 0.5.

Use an easier number to develop the process.

INSTRUCTOR: Yes, that's why I want you to do the division for $^1/_2$. If you get 0.5 for your answer, then you will have done it correctly and you can use that as a model for doing $^{12}/_{11}$.

STUDENT: Oh, okay. (A short time later . . .) It works when I punch in $1 \div 2$. So does that mean I punch in $12 \div 11$?

INSTRUCTOR: What do you think?

STUDENT: I think I'm right, because if I'm wrong, you would have told me.

INSTRUCTOR: At some point you'll have to rely on your own reasoning. Remember this when you get stuck: Think of a similar problem related to the one you're trying to solve. Pick a problem that you already know how to do or already know the answer to. Do the easier problem, then apply the process you learned to the problem you're really trying to solve. Remember, the more answers you can figure out for yourself, the less you will have to depend on someone's help. That will make you a better problem solver and a better student.

The strategy of using simpler problems to learn mathematical skills can be very powerful. Knowing some facts, such as $^1/_2 = 0.5$, can allow a student to extend this easily remembered concept to the general process of changing a fraction to a decimal. In the preceding situation, the student was confused about which number to divide by. By using what he already knew, he worked an easier related problem and then applied the method he used to the more difficult problem.

Easier related problems are helpful tools for solving algebra problems involving exponents. Many students have difficulty remembering the product rule and the power rule for exponents and often confuse them.

EXPONENTS

Simplify each expression:

$$m^{1/8} \cdot m^{2/3} \qquad (y^{1/3})^{6/7}$$

Work these problems before continuing.

If you took an algebra course recently, these problems may be very easy for you. However, if your algebra is a bit rusty, problems like these can give you trouble. Consider the following two easier related problems. Try multiplying variables with whole number exponents.

Use an easier number to develop the process.

$$x^2 \cdot x^3 \qquad (x^2)^3$$
$$(xx)(xxx) \qquad (xx)(xx)(xx)$$

Work these problems before continuing.

The answers to these two problems are x^5 and x^6. Obviously you can get these answers by counting the number of x's in each expression. But look carefully at the problems and the answers. Is there another way to get the answers besides counting x's? To get the answer to the first problem, add the exponents: $2 + 3 = 5$. To get the answer to the second problem, multiply the exponents: $2 \times 3 = 6$.

Now look again at the Exponents problem. To solve the first expression, $m^{1/8} \cdot m^{2/3}$, you need to realize that this problem looks just

You can use an easier problem to re-create a forgotton rule.

like the expression $x^2 \cdot x^3$ but with exponents that are more complicated. The process used to solve both is the same: Add the exponents. To add fractions you need a common denominator, which in this case is 24.

$$m^{1/8} \cdot m^{2/3} = m^{3/24} \cdot m^{16/24} = m^{19/24}$$

Similarly, the second problem, $(y^{1/3})^{6/7}$, looks just like $(x^2)^3$ but with exponents that are more complicated. The process used to solve both is the same: Multiply the exponents.

$$(y^{1/3})^{6/7} = y^{6/21} = y^{2/7}$$

Students often forget the exponent rules when they need to solve problems that involve complicated exponents. Instead of looking up a rule, re-create it for yourself by using easy exponents and then applying the process you learned to the difficult exponents.

The process used in the Exponents problem and in the example about changing the fraction to a decimal is similar to the process used to solve the problems in this chapter, but with this chapter's problems you probably don't already know the answer to the easier problem you're posing. However, the fact that you're trying to solve an easier problem means that the answer to the easier problem is within your reach. After you solve the easier problem, look for a way to apply the same process to the more difficult problem.

GOOD LUCK GOATS

In the mythical land of Kantanu, it was considered good luck to own goats. Barsanta owned some goats at the time of her death and willed them to her children. To her first born, she willed one-half of her goats. To her second born, she willed one-third of her goats. And last she gave one-ninth of her goats to her third born.

As it turned out, when Barsanta died she had 17 goats. Barring a Solomonic approach, how should the goats be divided?

Work this famous problem before continuing.

This problem involves working an easier related problem. In this case, an even number would probably work better.

Change a condition.

Nikki approached it like this: "Because 17 is an odd number, I decided to try a simpler number: 2. And 2 worked fine for the first born but didn't work for the second born. The number had to be divisible by 2 and also divisible by 3, so I tried 6. However, 6 didn't work well for the third born. So this time I tried 18, which seemed to work."

$\frac{1}{2}$ of 18 is 9 (1st born)

$\frac{1}{3}$ of 18 is 6 (2nd born)

$\frac{1}{9}$ of 18 is 2 (3rd born)

Total: 17 goats distributed to children; 1 goat left over

"It's weird because there's 1 goat left over, but Barsanta didn't really have 18 goats. She had 17. So there isn't really a goat left over because it doesn't really exist. It's also weird because if you really tried to find half of 17, you'd have to divide a goat. So 17 doesn't work, but 18 does work to make 17 work."

Nikki's solution is an example of changing a condition in the problem to make the problem easier. She changed the number of goats Barsanta had, which enabled her to solve the problem.

This problem is a very strange problem. It contains some "bad" information and an incorrect hidden assumption. The assumption is that all the fractions add up to one whole, as they should because they represent a finite set of objects being partitioned. When you were solving the problem, you may have noticed that it is impossible to divide this group of goats into $\frac{1}{2}$, $\frac{1}{3}$, and $\frac{1}{9}$ without using a meat axe. If you add $\frac{1}{2} + \frac{1}{3} + \frac{1}{9}$, the result is $\frac{17}{18}$. This means that Barsanta wrote her will to divvy up only $\frac{17}{18}$ of her estate, not giving instructions about what was to be done with the remaining $\frac{1}{18}$. That is why the additional goat is needed to make the problem work. The extra goat adds the remaining $\frac{1}{18}$, and then the goat is taken away after the other goats are divided up.

You will get a chance to explore this further in problem 14 on page 260.

Another way to make a problem easier is to use numbers in place of variables. Use this technique on the next problem, which deals with the concept of average. Problems like this are common on standardized tests.

AVERAGES

The average of a group of quiz scores is 31.8. There are k quiz scores in the group. The average of 10 of these quiz scores is 24.3. Find the average of the remaining quiz scores in terms of k. Work this problem before continuing.

This problem is complex for several reasons. The given averages are thorny numbers, and the number of quiz scores in the original group is not known. Right off the bat, you can do two things to make the problem easier. You've done this many times before: Make the numbers 31.8 and 24.3 into easier numbers. Use 30 and 25 instead. Now the problem says:

Use easier numbers.

The average of a group of quiz scores is 30. There are k quiz scores in the group. The average of 10 of these quiz scores is 25. Find the average of the remaining quiz scores in terms of k.

However, you may still have trouble figuring out what to do. If you knew the number of quiz scores, the problem might seem more manageable. So make up an easy number for k. Let's say there are 50 scores. So now the problem says

Use a number in place of a variable.

The average of a group of quiz scores is 30. There are 50 quiz scores in the group. The average of 10 of these quiz scores is 25. Find the average of the remaining quiz scores in terms of 50.

It seems a little easier now, although the last sentence may not make a whole lot of sense. For now, pretend that the last sentence of the problem simply says "Find the average of the remaining quiz scores." This is an example of getting rid of a condition.

Get rid of a condition.

If there are 50 scores with an average of 30, then their sum must be 1500. Why is that? To get an **average,** you take the sum of all the scores and divide by the number of scores.

$$\frac{\text{Sum of scores}}{\text{Number of scores}} = \text{Average}$$

For our easier related problem we have

$$\frac{\text{Sum of scores}}{50} = 30$$

The sum of all the scores must be 30 times 50, which equals 1,500.

The next part of the problem says that the average of 10 of these quiz scores is 25.

$$\frac{\text{Sum of 10 scores}}{10} = 25$$

The sum of this group must be 25 times 10, which equals 250. (Notice that this is a subproblem.) The remaining scores must add up to 1,500 minus 250, or 1,250. How many scores are remaining (more subproblems)? We started with 50 scores and have already considered 10 of them, so there must be 40 scores left that add up to 1,250. The average of these scores is

$$\frac{1{,}250}{40} = 31.25$$

Solving this easier related problem gives us a clue about how to solve the original problem. Look carefully at the numbers 1,250 and 40. Where did they come from? The number 1,250 was just 1,500 minus 250, but where did we get those numbers? We got 1,500 by multiplying 30 (the average of all scores) by 50 (the total number of scores). We got 250 by multiplying 25 (the average of the small group of scores) by 10 (the number of scores in the small group). We got 40 by subtracting 10 (the number of scores in the small group) from 50 (the total number of scores).

$$\frac{1{,}250}{40} = \frac{1{,}500 - 250}{50 - 10} = \frac{30 \times 50 - 25 \times 10}{50 - 10}$$

Now recall the original problem. The average of all scores is really 31.8, not 30. The average of a small group of 10 scores is really 24.3, not 25. And there are really k scores in the whole group, not 50. Substitute these numbers into the previous expression:

$$\frac{31.8k - 24.3 \times 10}{k - 10} \quad \text{or} \quad \frac{31.8k - 243}{k - 10}$$

In this problem we used two types of easier related problems. We replaced difficult numbers with easier numbers, and we replaced a variable with a number to see what was going on. Both of these substitutions made the problem much more manageable and gave us a plan of attack. We also temporarily got rid of this condition to find the average in terms of *k* (which was 50 in our easier related problem). After solving the easier problems, the plan for the original problem became clear, and the problem wasn't hard to solve anymore. It was just a matter of applying the procedure learned from the easier problem to the hard problem.

The next problem appeared in Chapter 6: Guess and Check. This time solve it with an easier related problem by changing some of the conditions in the problem.

NEXT TRAIN EAST

A train leaves Roseville heading east at 6:00 a.m. at 40 miles per hour. Another eastbound train leaves on a parallel track at 7:00 a.m. at 50 miles per hour. What time will it be when the two trains are the same distance away from Roseville? Work this problem before continuing.

Change conditions.

Marla solved this with an easier related problem. "It was obvious that the later train was gaining on the slower train 10 miles every hour. So I decided to pretend that the first train had traveled for an hour and then stopped. The second train then left at 10 miles per hour. I thought of this new problem as being the same as the original problem [an example of changing the conditions in the problem]. So the second train had to make up 40 miles, the distance the first train traveling at 40 miles per hour covered between 6:00 a.m. and 7:00 a.m., at 10 miles per hour. Obviously, that meant it would take the second train 4 hours to catch up. So the answer is 4 hours from 7:00 a.m. [when the second train left], which gives 11:00 a.m. I then checked this. The first train goes from 6:00 a.m. to 11:00 a.m., which is 5 hours, and covers 200 miles at 40 miles per hour. The second train goes from 7:00 a.m. to 11:00 a.m., which is 4 hours, and covers 200 miles at 50 miles per hour. Both trains cover the same distance, so the answer 11:00 a.m. must be right."

The next problem was posed by George Pólya, one of the first teachers of problem solving, in his classic book *How to Solve It,* originally published in 1945. Pólya said, "If you cannot solve the proposed problem, try to solve first some related problem. Could you imagine a more accessible related problem?"

INSCRIBED SQUARE

Given any **triangle**, draw a square inside of it so that all four vertices of the square are on the sides of the triangle. Two of the **vertices** of the square should be on one side of the triangle, and each of the other two sides of the triangle should have one vertex of the square.

Make the problem easier by eliminating one of the conditions. You'll need to draw lots of diagrams. Work this problem before continuing.

Ronaldo and Julie worked on this problem.

RONALDO: This problem seems really tough. How can we make it easier?

JULIE: I'm not sure. Let's experiment by drawing triangles and see if we get lucky. (They drew the following triangles and attempted to put the squares inside them. They were not successful, because the inside figures did not look like squares.)

  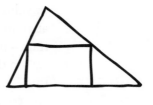

RONALDO: This isn't working at all. We need to make this easier somehow. Can we change the conditions of the problem?

JULIE: Well, we can change the conditions and make it a triangle inside a square. That would be really easy. (She drew the picture below.)

Get rid of a condition.

JULIE: Oops! I don't think that will help. Maybe we can try putting only three of the vertices of the square on the triangle and let the fourth one float around the inside or outside. (She drew the pictures below.)

 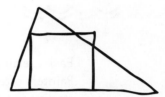

RONALDO: Okay, it's easy to draw just three of the vertices on the triangle. Let's agree to always put the longest side of the triangle on the bottom and to always put two of the vertices on that side.

JULIE: That's a good idea. But wait, I think we should start by picking a point on the left side of the triangle. Then draw a line down to the base. Then measure that distance across the base, and then go up.

RONALDO: That last vertex probably won't be on the triangle, but let's do it anyway. That is relaxing a condition. (They drew the next series of pictures.)

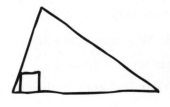 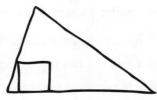 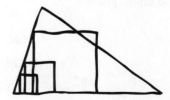

JULIE: Look! I think I see what is going on. All of the fourth vertices form a straight line. It is a visual pattern.

RONALDO: Yes, they do. We can draw that line, and where it intersects the third side of the triangle we have our fourth vertex. Then it's easy to draw the square. (Their final picture is shown below.)

Easier related problems are used in the world in many different places. Painters often paint smaller, simpler versions of their ideas before starting work on the large painting, as described on page 233. Another example is the use of storyboards in the film industry. Walt Disney developed the storyboard for his animated films in the 1930s. A storyboard is essentially a series of pictures that tell the story of a movie when viewed in order. An animator will use the storyboard to visualize what the finished movie will look like. Today storyboards are used in live-action movies as well. They are also used in business to plan ad campaigns. For an extensive and very interesting article about storyboards, see the website Wikipedia at http://en.wikipedia.org/wiki/Storyboard.

Relate an easier problem to explain a difficult problem.

Easier related problems are also used to explain concepts to people who don't understand a particular situation. For example, here's a conversation a father had with his son about traffic. They were driving in rush hour traffic, so cars were moving very slowly.

SON: Why is there so much traffic? Why is everyone driving so slow?

FATHER: Because everyone is going to work right now.

SON: But if everyone just drove 60 miles per hour, then no one would have to slow down.

FATHER: There are too many cars.

SON: I don't understand.

FATHER: Okay, suppose you wanted to leave your classroom at school and go out to the playground. Let's say it wasn't during recess and you just went to the door of the classroom and walked out into the hall. How fast could you run to get to the playground?

SON: Pretty fast.

FATHER: Right. Now suppose the bell had just rung for recess. How fast could you go now?

SON: Not very fast. I'd have a hard time getting out of the room.

FATHER: Why?

SON: Because so many kids would be crowding the door.

FATHER: Right. What would happen when you got out into the hall?

SON: It would be really crowded because there are a bunch of classrooms on that hall. Kids would be coming out of every single classroom and the hall would be packed.

FATHER: So how fast could you go?

SON: Not very fast. If I tried to go fast, I would crash into a bunch of other kids and I'd probably get sent to the office.

FATHER: Right. Well, that's the same thing that happens on the freeway. So many people are going to work that the freeway just can't handle all those cars. So all the cars have to slow down to make sure that they don't crash.

SON: I get it.

The next time you are trying to explain a situation to someone who doesn't understand it, try using an easier related problem.

The Difference Between Subproblems and Easier Related Problems

Students often confuse the strategies of subproblems and easier related problems. The strategy of subproblems involves the process of breaking down a large problem into parts. The whole problem is too large to solve all at once, so you concentrate on solving small, manageable parts of the problem. On the other hand, the strategy of easier related problems doesn't involve solving the original problem at all. You focus instead on a related problem that is easier to solve. You then return to the original problem and solve it by using the method you used when you solved the easier problem.

The first problem in this chapter, "Find the sum of the first 100 whole numbers," provides an example of the use of the two strategies. Remember that Tori split the problem into ten parts or subproblems and looked for a pattern. She used the pattern to write the sum for each subproblem and then added the ten results.

When Tarick saw what Tori was doing, he suggested using two more subproblems to help with the adding. Tori had written the problem as this sum:

$$55 + 155 + 255 + 355 + 455 + 555 + 655 + 755 + 855 + 955 = 5{,}050$$

Tarick said, "There are ten 55's, and then what's left is the hundreds:

$$100 + 200 + 300 + 400 + 500 + 600 + 700 + 800 + 900$$

One of Tarick's subproblems was 10(55).

Another was $100(1 + 2 + 3 + 4 + 5 + 6 + 7 + 8 + 9) = 100(45)$.

The last subproblem was to add the two results: $4{,}500 + 550 = 5{,}050$.

Notice how Gauss's solution was different. He started by finding a method for solving the easier problem by pairing numbers from the outside to the center.

With this easier problem he established a method that he then applied to the original problem.

At first glance, Gauss's easier related problem looks just like Tori's first subproblem. Understanding the difference in strategies lies in seeing that Tori used the first sum $(1 + 2 + 3 + 4 + 5 + 6 + 7 + 8 + 9 + 10)$ as one of several parts that needed to be solved and then combined with other parts to solve the larger problem. Gauss, on the other hand, did not use the answer he got from solving his easier related problem.

Instead he expanded on his method of pairing numbers from the outside in and used it to solve the larger problem.

Here's a final thought on the difference between easier related problems and subproblems. Sierra College student Alexander John Ramm put it this way: "An easier related problem is a different problem that follows the same rules as the original problem. A subproblem *is* the original problem, but just a small part of it."

Solve an Easier Related Problem

The following ways to make a problem easier appear in the list on page 236. The table below names the problem or problems in the text that used each method. Some problems involved more than one technique.

If a problem seems impossible, ask: How can I make this problem easier?

TECHNIQUE	EXAMPLE
1. Use a number instead of a variable.	Averages
2. Use a smaller or easier number in order to develop a process for solving the problem.	Simpletown Elections Exponents Averages
3. Do a set of easier examples and look for a pattern.	How Many Squares?
4. Do an easier example and figure out how to use the same process to solve the harder problem.	Divisors and Reciprocals From One to One Hundred (method 2)
5. Change, fix, or get rid of some conditions.	Good Luck Goats Averages Next Train East Inscribed Square
6. Eliminate unnecessary information.	Simpletown Elections

Using an easier related problem takes a lot of practice. This is a strategy you may have to consciously remind yourself to use.

Problem Set A

1. DIAGONALS

A **diagonal** of a **polygon** is a line segment that connects two nonadjacent vertices of the polygon. A certain convex polygon has 25 sides. How many diagonals can be drawn?

2. SUM OF ODDS

Find the sum of the first 5,000 odd numbers.

3. SWEATER

A clothing store bought a sweater for a certain price and marked it up 70%. The sweater didn't sell, so the store owner took 25% off the original sale price and it sold at that price. What percent profit did the store make?

4. OBSTACLE COURSE

The obstacle course was very challenging. Three out of every 8 contestants fell in the mud at the beginning and gave up. Then 1 out of 5 couldn't climb over the wall. Then 2 out of 9 couldn't climb up the rope and dropped out. Finally, 1 out of 10 couldn't jump over the hurdles and quit. What percent of the contestants made it to the finish line?

5. TV TRUCK

Theotis has to load a truck with television sets. The cargo area of the truck is a rectangular **prism** that measures 8 ft by 21 ft by 11 ft. Each television set measures $1\frac{1}{2}$ ft by $1\frac{2}{3}$ ft by $1\frac{1}{3}$ ft. How many sets can be loaded into the truck?

6. CHINESE NEW YEAR

In China each calendar year is given one of 12 names, which rotate year after year. The year 2000 was the year of the Dragon. The year 2001 was the year of the Snake. The subsequent ten years are, in order, the years of the Horse, Sheep, Monkey, Rooster, Dog, Boar, Rat, Ox, Tiger, and Rabbit. After the year of the Rabbit, the year of the Dragon will occur again, and then the whole cycle will repeat. What will the year 3000 be?

7. SQUARE AND HEXAGON

A square has an area of S^2. A regular **hexagon** has a perimeter of T. If p is the perimeter of the square and h is a side of the hexagon, then find $h + p$ in terms of S and T.

8. ODD AND EVEN

Find the difference between the sum of the first 500 even numbers and the sum of the first 500 odd numbers.

9. POTATOES

To prepare dinner in the army mess hall, Evelyn, who is a member of the Fourth Battalion of the Twenty-Third Regiment, generally uses about 85 pounds of potatoes to feed the 358 people in her unit. She usually assigns 3 soldiers to scrub the potatoes, and it takes them about 20 minutes to complete the job. However, last week she needed to feed about 817 people, beginning at 17:30 hours, because of a special army event. When she arrived at the mess hall tent, she discovered that 131 pounds of potatoes had been sent. She needed to send for the rest right away. How many pounds did Evelyn request, and how many soldiers did she need if she wanted each of them to spend about 20 minutes scrubbing?

10. TWENTY-FIVE-MAN ROSTER

The manager of a baseball team received a strange communication from the team's general manager, who told him to select 25 players for his roster according to this formula:

$1/2$ of the team had to be outfielders and infielders

$1/4$ of the team had to be starting pitchers

$1/6$ of the team had to be relief pitchers

$1/8$ of the team had to be catchers

The manager was taken aback by the innumeracy exhibited by his boss, yet he found a way to comply. How did he do it?

11. LAST DIGIT

What is the last digit in the product $(2^1)(2^2)(2^3)(2^4) \ldots (2^{198})(2^{199})(2^{200})$?

12. REMAINDER

What is the remainder when 2^{1000} is divided by 13?

13. FIFTY-TWO-CARD PICKUP

Naoko found a deck of cards that had been dropped onto the floor. She picked up at least 1 card. She may have picked up 1, all 52, or any number in between. How many possible combinations are there for what she picked up?

14. WRITE A HIDDEN FRACTION PROBLEM

Look back at the Good Luck Goats problem (page 247) and the Twenty-Five-Man Roster problem (page 259).

In each case, these problems were solved by devising an easier related problem. The goats problem was solved by adding a goat and then subtracting a goat. The baseball player problem was solved by subtracting a player and then adding a player.

Here's how those problems work: Each problem contains some bad information and an incorrect hidden assumption. The bad information is that the fractions stated in the problem don't add up to one whole. The incorrect hidden assumption that we make as readers is that the fractions *do* add up to one whole, as they should when we partition a finite set of objects.

In the Good Luck Goats problem, the numbers given are $\frac{1}{2}$, $\frac{1}{3}$, and $\frac{1}{9}$, which have a common denominator of 18: $\frac{9}{18}$, $\frac{6}{18}$, and $\frac{2}{18}$, respectively. The sum of these fractions is $\frac{17}{18}$. This means that Barsanta's will has been written to divvy up $\frac{17}{18}$ of her estate, incorrectly omitting what needs to be done with the remaining $\frac{1}{18}$.

In the Twenty-Five-Man Roster problem, the fractions have a common denominator of 24 and sum to $\frac{25}{24}$.

Now it's your turn: Find more combinations of fractions like these, and write your own Good Luck Goats problem. Is there a pattern for denominators that lend themselves to this type of problem?

For more information about this type of problem, read "The Riddle of the Vanishing Camel" by Ian Stewart in the Mathematical Recreations column of the June 1992 issue of *Scientific American*. (The Mathematical Recreations column was made famous by one of the world's best puzzle makers, Martin Gardner, who wrote for *Scientific American* for many years.)

15. WRITE YOUR OWN

Write your own easier-related-problem problem. Stick with it: These problems can be very difficult to make up. The easiest example to follow may be the Chinese New Year problem (problem 9).

CLASSIC PROBLEMS

16. PAINTING THE LAMPPOSTS

Tim Murphy and Pat Donovan were engaged by the local authorities to paint the lampposts on a certain street. Tim, who was an early riser, arrived first on the job and had painted three lampposts on the south side when Pat turned up and pointed out that Tim's contract was for the north side. So Tim started fresh on the north side, and Pat continued on the south side. When Pat had finished his side, he went across the street and painted six posts for Tim, and then the job was finished. There was an equal number of lampposts on each side of the street. The simple question is, Which man painted more lampposts, and how many more?

Adapted from *Amusements in Mathematics* by Henry Dudeney.

17. BILLIONS

What is the sum of all ten-digit numbers?

From the Problem of the Week website from Ole Miss (University of Mississippi), December 2002. The Problem of the Week contest, run by Dr. David Rock, allows people from all over the world to solve the problem of the week. If you send in a correct solution to the current problem of the week, your name will appear on the list of solvers for that week. To participate, go to http://mathcontest.olemiss.edu

MORE PRACTICE

1. SISTERS AND BROTHERS

I have three more sisters than brothers. Each of my sisters also has three more sisters than brothers. How many more sisters than brothers does my youngest brother have?

Math Teacher Gary Steiger inspired this problem.

2. **SUM OF EVENS**

Find the sum of the first 3,000 even numbers.

3. **BIG DIFFERENCE**

Find the difference between the sum of the first 900 multiples of three and the sum of the first 900 even numbers.

4. **SEQUENCE SUM**

Find the sum of the first 2,500 terms of the sequence 1, 5, 9, 13, 17, 21, 25, ...

5. **HOLEY TRIANGLES**

How many triangles are there in this part of a Sierpinski Triangle?

6. **WINDOW PANES**

Sarah makes special frames for windowpanes, and she has an order to build a frame for a huge 20 × 20 window with 400 small square panes. The outer frame will be wood, but the inner crosspieces between panes will be metal. Sarah wants to cover the internal metal framing with wooden insets. An internal pane will be surrounded by four insets, a pane on the edge of the frame by three, and one on a corner by just two. Sarah needs to know how much material to order for the wooden insets, but she doesn't want to have to count them. How many wooden insets will she need?

Problem Set B

1. COVERING THE GRID

A grid has lines at 90° angles. There are 12 lines in one direction and 9 lines in the other direction. Lines that are parallel are 11 inches apart. What is the minimum number of 12-inch-by-12-inch floor tiles needed to cover all of the line intersections on the grid? The tiles do not have to touch each other. You must keep the tiles intact—don't break or cut them.

2. NINE POINTS

There are nine points on a piece of paper. No three of the points are in the same straight line. How many different triangles can be formed by using three of the nine points as vertices?

3. BASKET PARTIES

Alane sold baskets to people who attended in-home parties. She was frustrated in her attempts to get more people to host sales parties for her. She finally offered to pay $25 to any host who (a) hosted a party for her and (b) arranged for two other friends to host parties during the next month. To her surprise, it worked! She had 100% success: Every host was able to arrange two more parties. She started with five hosts in the first month. If this continues for the entire year, how many parties will there be during the year?

4. ADDING CHLORINE

I have a small circular swimming pool in my backyard for my kids. Last weekend, I set it up and bought chlorine to put into it. The directions on the bottle said to put in 16 fluid ounces of chlorine per 10,000 gallons of water. Of course, our pool holds a lot less water than 10,000 gallons, so I needed to figure out the correct amount of chlorine to put into it. I measured the pool with my tape measure and found it to have a circumference of 27 feet 3 inches and a water height of 21 inches. I knew that the circumference of a circle was $C = 2\pi r$ and the volume of a cylinder was $V = \pi r^2 h$ (r is the radius of the cylinder, and h is the height). I knew that a milliliter is a cubic centimeter. I also knew that there are 3.79 liters in 1 gallon, and 3.281 feet in 1 meter. How many fluid ounces of chlorine did I need to put into the pool? (I have a measuring spoon capable of measuring to the nearest quarter of a fluid ounce.)

5. JOGGING AROUND A TRACK

Dionne can run around a circular track in 120 seconds. Basha, running in the opposite direction as Dionne, meets Dionne every 48 seconds. Sandra, running in the same direction as Basha, passes Basha every 240 seconds. How often does Sandra meet Dionne?

10

Create a Physical Representation

Manipulatives and physical models, as well as acting problems out, allow you to try many possibilities by moving parts or beings quickly. Rescue workers often perform drills to represent problems physically, so that they can be fully prepared in the event of a real emergency.

S ome mathematical concepts were developed to solve real-world problems. Many other mathematical concepts were derived purely from previously developed mathematics and (at least at first) seemed to have no real application. This chapter focuses on bringing problems into the real-world arena. Although few of this chapter's problems are actually "true to life," the strategy for solving them is the same as for solving many real-world problems: create a **physical representation.**

Physical representations fall under the major problem-solving theme of Spatial Organization. Note that a physical representation differs from a diagram in that a physical representation lets you touch the problem, not just represent it with a picture. By using objects or people to solve a problem, you gain a new perspective on it.

Using manipulatives facilitates understanding.

We begin the chapter with a section about the most basic type of physical representation: people **acting out** a problem. In Section 2 you'll explore the use of physical *models* and *manipulatives*. (As you learned in Chapter 8: Analyze the Units, a manipulative is an object that can be moved or positioned.) You'll see that the strategies of acting it out and of using models and manipulatives are closely related. They are different ways of physically representing a problem.

Traditionally, physical representations have been left out of the mathematics curriculum of most schools. This is unfortunate because mathematics has become totally abstract to some people. When problems are represented with objects, the problems become concrete and more easily understood.

Section 1: Act It Out

Of all the strategies presented in this book, the act-it-out strategy is probably the most fun. Acting out a problem gives you a chance to have contact with other people. Other strategies serve to solve the problems in this section, but acting out the problems with a small group of people works very well. No written solutions are given for the problems in this section. The correct solutions should be obvious when you act out the problems.

Three jackals and three coyotes are on a trek across the Mokalani Plateau when they come to a river filled with carnivorous fish. There is a rowboat in sight, and the party decides to use it. (Both species are known for their cleverness.) However, the boat is too small to hold any more than two of the group at a time, so they must traverse the river in successive crossings. There is one hitch, though: The jackals must not outnumber the coyotes at any time, in any place. For example, if two jackals and only one coyote are together on the western side of the river, this problem is reduced to simple subtraction: the jackals will overpower, kill, and eat the coyote. It's okay to have an equal number of each, and it's also okay to have more coyotes than jackals in a given place—neither situation poses a danger to the coyotes, and the coyotes do not pose a threat to the jackals. The trick here is to use the one small rowboat, a lot of sweat, and a little brainpower to ensure the coyotes' safety while both groups cross the river.

Close the book and find some people to act this out with. Also find some object to be physically transported across the room as the boat.

It is usually wise to act out the solution to a problem before actually implementing it. For example, parts of your solution for this problem probably involved some guess-and-check. In real life, you would have to be careful about using guess-and-check, since a wrong guess could lead to an early dinner for the jackals.

While working this problem, you should reach several major understandings. The first is realizing that there can be only two animals in the boat. The second is dealing with the first move: Sending two coyotes over on the first trip means the jackals will eat the third coyote. The next major understanding comes after you get two coyotes and two jackals on the far side. The only thing to do is bring back one of each and then send two more coyotes over. Then, finally, you send the solitary jackal back to begin bringing over the other two jackals.

Sometimes the critical elements of acting out a problem are the physical objects you employ. The next problem reflects this notion.

HORSE TRADER

Once upon a time, there was a horse trader. One morning, the horse trader bought a horse for $60. Just after noon, the horse trader sold that same horse back to the original owner for $70. He then bought it back again just before 5:00 p.m. for $80. By midnight he managed to sell the horse back to the original owner for $90. How much money did the horse trader make or lose on this horse?

Close the book and act out this problem before continuing. You will need something to represent the horse, some play money in different denominations, and at least two people to act out the roles of the horse trader and the original owner.

Acting out this problem allows several people to act as verifiers, who check for mistakes and make sure that all transactions are done correctly. Part of this problem's trickiness is identifying net financial gain and when it occurs. By having a couple of people acting as the trader and the original owner, you can verify in which direction money flowed and how much.

Problem Set A-1

Solve each problem by acting it out.

1. THREE ADULTS AND TWO KIDS

Three adults and two kids want to cross a river by using a small canoe. The canoe can carry two kids or one adult. How many times must the canoe cross the river to get everyone to the other side?

2. THE DOG, THE GOOSE, AND THE CORN

There was once a farmer who, as part of his route to town, used a rickety old boat to cross a wide river. One day he took his dog and went to town just to buy corn. However, in addition to buying corn, he bought a goose that he intended to take home and use to start raising geese. (This farmer was no fool; he already had another goose.) But he also knew that his boat was not reliable: It could handle only himself and one of the other three things he had with him. He feared that if left alone, the dog would eat the goose or the goose would eat the corn. How could he get himself and everything else across the river safely?

3. HOOP GREETING

A group of ten students met on the basketball court in back of the dorm for a pickup game. Everyone there shook hands with everyone else exactly once. How many handshakes took place?

4. SWITCHING JACKALS AND COYOTES

Three jackals are on one side of a river, and three coyotes are on the other side. They have a boat capable of carrying two animals. At no point can the jackals outnumber the coyotes on one side of the river, or the jackals will eat the coyotes. Each group wishes to change sides of the river. Figure out which side of the river the boat must start from and how to manage the groups to get each animal safely to its destination.

5. THE HOTEL BILL

Three sales representatives attending a convention decided to share a room to take advantage of some cheap hotel rates. The clerk at the desk charged them $60, which they paid with cash. A little while later, the clerk discovered that he had made an error: The room should have cost only $55. He dispatched a bellhop with a $5 bill to deliver to the women. The bellhop was dishonest, though, and he kept $2 and returned $1 to each of the three women. So each woman thought that her part of the bill was $19 instead of the original $20.

When Franklin heard this story, he did some calculations. Each woman thought she had paid $19, and 3 times $19 is $57. The bellhop kept $2, and $57 plus $2 is $59. What happened to the missing dollar? Is Franklin's reasoning right or wrong? Determine what happened to the missing dollar and explain your reasoning.

6. PERSIS'S GIFT SHOP

Persis owns a gift shop. One day while she was opening a shipment of figurines she had just received, a stranger walked in and asked the price of a figurine. The figurines had cost Persis $6 each, but she had not decided on a selling price for them. She finally decided on $13 each, and the stranger promptly said he'd take one. Her new clerk accepted his traveler's check for $40 and gave him change. Persis needed more change for the register and signed the traveler's check over to the pharmacist next door, Dr. Drell. An hour later, though, Dr. Drell came back to the gift shop and showed Persis that the traveler's check was actually in French francs, and was worth about $8 at the time. Persis apologized and wrote the pharmacist a check to cover the difference. Who lost how much in this transaction?

Section 2: Manipulatives and Models

In acting out a problem, people get together and assume roles in order to gain insight into the problem. Several brains work on the problem at once, and physical constraints become more and more apparent and more easily respected by using people in the roles. For example, a coyote (played by a person) can be on only one side of a river at any one time. On the other hand, if we work out the Jackals and

Coyotes problem with a diagram, then a particular coyote could appear simultaneously on both sides of the river by accident.

Each problem in Section 1 of this chapter was based on re-creating a problem and acting it out. Some of the problems in Section 1 had people in the roles of animals (such as jackals, coyotes, horses, dogs, and geese), and others had people in the roles of inanimate objects (such as corn).

Although the strategy of acting it out is very effective, it can be impractical (there is no one around when you are working on the problem), too time-consuming (there are people around, but it takes a while to round them up), too embarrassing (very few people will want to be the goose, for example), or too expensive (the only way you can get people to be involved in the Horse Trader problem is to use real money—yours—and they get to keep it; even then you may also have to act out the part of the horse).

Instead of actual people, use objects to represent people.

Let's abstract the act-it-out strategy one step. Instead of using actual people to act out the roles of the characters in the problems, we'll simply represent the characters with manipulatives, objects that can be moved or positioned.

Manipulatives must represent the critical elements of a problem. Different colors help.

Let's look at how we can use manipulatives with the Jackals and Coyotes problem. The advantage of reducing the jackals and the coyotes to little pieces of paper is that you can solve this problem by yourself, without having to hunt up some friends capable of acting like coyotes and jackals. The disadvantage of this method is that you might have only one brain working on the problem. However, using a manipulative to solve a problem is only one step away from actually acting it out with people. Keep in mind that you need to design your manipulatives so that they represent the critical elements of the problem and are in a form that will allow you to work out the problem. Using different colored pieces of paper to represent jackals and coyotes is very helpful.

JACKALS AND COYOTES REVISITED

Solve the Jackals and Coyotes problem again, this time using manipulatives. Choose something to represent coyotes and something to represent jackals. Scraps of paper with *C* written on some and *J* on others work well. Coins, paper clips, or bottle caps also work. Determine the minimum number of river crossings. Work this problem before continuing.

A disadvantage to using pieces of paper rather than actual people is that paper tends to not contribute any creative thinking to a problem. When acting out the Jackals and Coyotes problem, a person playing the part of a coyote would howl if left alone with two jackals. But a little piece of paper labeled *coyote* will probably not have much to say in its own defense.

Despite such limitations, manipulatives can be useful problem-solving tools. You often have to make a conscious effort to look for a way to do the problem using manipulatives. You must be a little inventive. The necessary materials are often right around you. For example, dimes and pennies make very good manipulatives. Not only are they different sizes and different colors, but the back side of each can be distinguished from the front side. If you are in the forest, you can always find twigs, leaves, and pebbles (to represent wolves, bears, and campers, for example). If by chance there is a piece of paper nearby (even a burger wrapper or a soda cup), you can tear it up into little pieces and write labels on the pieces. The manipulatives are there. The trick is to think of using them and then to use them effectively. Sometimes the first manipulative doesn't work very well, and creating a new one may be effective.

Objects that are small and distinct work well for manipulatives.

So far, none of the problems in this chapter have involved any measurement beyond simple counting (such as counting three coyotes). However, many problems involve an element of magnitude, scale, orientation, quantity, positional relationships, correspondence, dynamic relationships, directional movement, or number combinations. The physical representations you use to solve these types of problems must demonstrate these elements, and you must put more thought into creating these representations. We call such representations **models.** A model is a physical representation of a problem that can be used to produce a solution.

A model is a physical representation that can be used to solve a problem.

The strategy of making a model is used extensively to solve real problems, though probably not nearly enough. Models are made for airplanes, cars, freeway bridges, buildings, rockets, filing cabinets, and on and on. Models like these can be used to test designs before moving on to more advanced stages of production. Storyboards, as described in the previous chapter on page 254, can be thought of as a model for a movie. Many of the problems in this chapter require you to build some

First identify the critical elements of the problem. Then make your model.

sort of model. In some problems, relative size is one of the critical elements. In other problems, shape is the critical element. In the following problem, orientation is the key.

FOUR CONTIGUOUS STAMPS

In how many ways can four rectangular stamps be attached together? Be sure to pay attention to the thrust of this chapter. Take care to record each configuration. Work this problem before continuing.

The critical elements in this problem are size and shape, so the four items used as manipulatives in the model of this problem need to be relatively similar in size and shape. It's also important to distinguish the tops of the stamps from the bottoms, because stamps are usually printed all in the same direction. You should have found 19 different configurations.

Orientation is also a feature of the next problem.

LETTER CUBE

Build this cube to see what letter is opposite the letter *T*. Pay attention to the orientation of the letters on the faces of the cube.

To solve this problem, draw the figure shown at right on a piece of paper (graph paper works well), cut it out, and fold it to make a cube. Then write the letters on its faces. Work this problem before continuing.

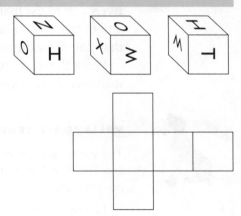

Problems of this type often show up on standardized tests. To solve such a problem, you must visualize the cube in your mind and figure out where each letter is. Some people have well-developed spatial visualization skills and may be able to solve the problem without building the model. Many people aren't very good at these problems

when they first try solving them. Building a model to solve
a problem like the Letter Cube problem can help you develop the ability
to visualize.

After cutting out and folding up your cube, write the letters on its
faces in their correct orientations. The solution for the problem is easily
verified to be the letter *O*.

As you read earlier in this chapter, a model is a physical representation
of a problem that can be used to produce a solution. Examples of
models include architectural models and mock-ups made to scale. These
types of models accurately represent magnitude, scale, and orientation,
which are all critical elements of design and construction. On the other
hand, manipulatives are objects that can be moved around or easily
positioned. Some examples of manipulatives you've worked with so
far in this book are coins and pieces of paper. Manipulatives are found
within a model of a problem. For example, a model of a problem in
which jackals and coyotes cross a river will include the river and the two
riverbanks. It will also include manipulatives that you can move from
riverbank to riverbank to represent the movement of the characters
and of the boat. These manipulatives, as well as the land and water
that you draw on paper, are all part of the model of the problem.

Work the next problem by using a diagram of a volleyball court and
manipulatives, pieces of paper with the players' names written on them.

VOLLEYBALL TEAM

The volleyball team has six players: Virginia, Fannie, Helen, Dermid, Melvin,
and The Prof. Using the following clues, put the players into the starting
positions that will give the best rotation:

1. The players must alternate by gender.

2. Virginia is the team's best server, so she should start in the serving
   position.

3. Melvin and Helen are the team's setters. They must be opposite each
   other at all times.

4. Dermid and Fannie communicate well—it helps to put them next to each other.

5. The Prof is an effective server. He needs to be positioned so he will rotate into the serving position quickly.

(As this diagram shows, half of a volleyball court has six players; the opposing team plays on the other half of the court. From the net, there are three players in the front row and three in the back row. The server is in the right back corner. The players rotate in a clockwise manner. Players are considered to be opposite if they are three positions apart. So server 1 is opposite server 4, server 2 is opposite server 5, and server 3 is opposite server 6.)

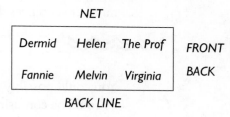

Work this problem before continuing.

To solve this problem, tear up little pieces of paper and write the names of each player on them. You could also solve this problem simply by drawing a diagram, but using manipulatives is faster and requires no erasing. Using manipulatives allows you to try different possibilities quickly, without worrying about whether or not they are correct.

Review the problem to check your solution.

The solution to this problem is shown at right. If you didn't get the correct answer, check two things: (1) Did you use manipulatives, and (2) did you review the clues after you thought you were done to check for compliance?

NET

| Dermid | Helen | The Prof | FRONT |
| Fannie | Melvin | Virginia | BACK |

BACK LINE

Problems like the next one are of great interest to Marcy Cook, a mathematics educator who has created puzzle books and manipulatives used in solving this kind of problem.

Use the digits 0, 1 2, 3, 4, 5, 6, 7, 8, and 9 once each to fill in the blanks of this puzzle:

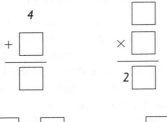

Work this problem by cutting up little pieces of paper and labeling them with the numbers 0 through 9. Then manipulate them by moving them around in the puzzle until you find an arrangement that works. There is more than one possible solution.

There are essentially two different solutions to this problem: One starts with $4 + 1 = 5$, and the other starts with $4 + 5 = 9$. We hope you found both solutions. (Note that you can change both solutions slightly by reversing the numbers in either of the last two number sentences, but those aren't significantly different solutions.)

You could possibly solve the nest problem with matrix logic, but doing so could be confusing. The key to the problem is the position of each person at the table. Using manipulatives is a good way to attack logic problems that involve positioning.

MEXICAN RESTAURANT

Four friends (one is named Janie) went out to dinner at a Mexican restaurant. The hostess seated them in a booth. Each ordered a different meat (pork, mahimahi, beef, or chicken), and each ordered a different kind of Mexican dish (tostada, burrito, fajita, or chimichanga). Use the clues that follow to determine what dish each person ordered, the kind of meat it contained, and where each person was sitting.

1. The person who ordered mahimahi sat next to Ted and across from the person who ordered a burrito.

2. Ken sat diagonally across from the person who ate the fajita and across from the person who ordered beef.

3. The person who ordered a chimichanga sat across from the person who ordered chicken and next to Allyson.

Work this problem before continuing.

The first twist to the Mexican Restaurant problem is determining the names of the people. This information is not stated at the beginning of the problem as usual but instead is hidden in the problem. By reading each clue, you can determine that the names are Janie, Ted, Ken, and Allyson. The meats are pork, mahimahi, beef, and chicken. The dishes are tostada, burrito, fajita, and chimichanga.

We'll use pieces of paper as our manipulatives in this problem, so our first step is to tear a sheet of paper into 12 small pieces and write on them the names of the people, the meats, and the dishes. Then read through the clues and arrange the pieces of paper as the clues suggest. It would be helpful here to make a "booth" manipulative to keep track of position with respect to sides of the table.

Tape the manipulatives together.

The first clue says, "The person who ordered mahimahi sat next to Ted and across from the person who ordered a burrito." So arrange and tape together the Ted, mahimahi, and burrito papers in their relative positions, as shown here:

	burrito
Ted	mahimahi

This clue has another possible interpretation. It is possible for the arrangement to be set up like this:

Ted	mahimahi
	burrito

Flip the manipulatives over and write on the back.

You can accommodate this second possibility by simply flipping over the three pieces of taped paper and writing the information on their other sides.

Tape the other pieces of paper together in their relative positions as suggested by the clues. Now you can make a set of larger manipulatives by combining two or three of the smaller manipulatives. You can then move these larger manipulatives into a very limited number of positions, especially because some of the clues indicate positions on opposite sides of the table. By arranging the large manipulatives on top of each other, you can find many contradictions, such as two people sitting in the same spot or two types of meat in the same position.

For example, start with the Ted-mahimahi-burrito and the Ken-beef-fajita papers. Place the Ted-mahimahi-burrito paper on your desk as your base diagram, and place the Ken-beef-fajita paper on top of it. There are four different ways to do this, three of which require you to turn the paper upside down, turn the paper over to read the back side, or both.

1.

fajita	beef burrito
Ted	Ken mahimahi

This is possible. Beef can be in the burrito and Ken could have ordered mahimahi.

2.

	Ken burrito
fajita Ted	beef mahimahi

This is impossible because beef and mahimahi are matched with the same person. One person cannot order two meats.

3.

Ken	burrito
beef Ted	fajita mahimahi

This is possible. Ted could have ordered beef and the fajita could contain mahimahi.

4.

beef	fajita burrito
Ken Ted	mahimahi

This is impossible, because Ted and Ken are sitting in the same seat.

At this point there are only two possible combinations. Now superimpose the third paper (Allyson-chicken-chimichanga) on top of the two possibilities. Again, arrange the paper in four ways for each possible combination so that you see eight arrangements altogether. If these instructions seem complex, make sure (a) you are using the manipulatives and (b) you understand that this is precisely our point.

Taping the pieces together in their relative positions has made it easier to see which combinations are contradictions and thus don't work. For example, it's easy to see that the Allyson-chicken-chimichanga paper doesn't work in any combination with possibility number 1, but does work with possibility number 3, as shown below.

chicken Ken	burrito
chimichanga beef Ted	Allyson fajita mahimahi

Now you can fill in the remaining information to solve the problem.

Ken chicken tostada	Janie pork burrito
Ted beef chimichanga	Allyson mahimahi fajita

Using manipulatives allows you to try different solutions very quickly. Sometimes taping the manipulatives together helps.

Applied Problem Solving: NASA Uses Tinkertoys

The following article describes a solution to a problem with the Hubble Space Telescope that occurred in outer space.[1]

> The Hubble Space Telescope, all $1.5 billion of it, was put back into working order Monday because a NASA engineer used a

[1]Associated Press, "NASA Fixes Hubble's Antenna," *Sacramento Bee,* May 1, 1990. Article reprinted courtesy of Associated Press.

Tinkertoy, a lamp cord, masking tape and glue to help solve a major problem.

The telescope's No. 2 high-gain antenna, wedged in one position since Friday, was free and sending data through relay satellites.

The National Aeronautics and Space Administration expected calibration and other normal start-up work to begin by tonight and to receive its first pictures from the telescope by this weekend. "The moral of the story is that there is no solution that's too humble," said David Skillman, who built a model of the jammed antenna.

"We were faced with a problem on the telescope that involved quite intricate geometry," he said. "A number of us realized we could benefit greatly from a model. Someone suggested that even a Tinkertoy model could be useful."

He drove to a toy store Sunday afternoon and bought two boxes of the construction toy. He got the other items in a drug store and put the model together in 15 minutes with another engineer, John Decker. . . . What they visualized with the model was matched with computer drawings at Lockheed Missiles and Space Co. in California.

Armed with that knowledge, computer commands were sent to the telescope directing exactly the way the dish should move to back out of its jam.

"The antenna moved beautifully and easily out of its problem and back to normal," said Skillman. "Many times a simple solution is the best solution."

Use a Manipulative

Look for opportunities to use manipulatives.

Because people are not used to solving problems using manipulatives, this unpracticed skill becomes an unused strategy. Very few problems, whether practical or contrived, are presented in a way that tells you to use a manipulative. Rather, you have to look for ways to use manipulatives. The following types of problems lend themselves to the use of manipulatives or models.

Static Arrangements of People

Volleyball rotations, golf foursomes, and starting positions for softball teams are all examples of static arrangements. The question of who is sitting next to whom around a table is also a natural manipulatives problem. It is not necessary to have little figurines of people to solve such problems. Usually all you need to do is tear up little pieces of paper, write the names of the people on them, and draw a representation of the area in which they are to be manipulated.

For example, to arrange the starting positions of a softball team, you will want to at least draw a baseball diamond so that you can set the pieces of paper in their relative positions. This is a good way to arrange players of a team that doesn't have fixed positions. That is, some people are strong in any position they play, others are weak in any position, and some can be weak or strong depending on the position. Having weak players on the field may not be an optimal solution, but it is true-to-life, and an acceptable arrangement can often be found.

Dynamic Arrangements of People

You can also use manipulatives to set up transportation to and from various points for a number of people in a few vehicles. You might have to solve this kind of problem if, say, you're responsible for making sure that prior to a wedding people are picked up from an airport, driven somewhere for shopping, dropped off so they can get their tuxedos fitted, and so on. Many times in such situations there are only a few cars available and many things to be done by various people in various places. Again, by tearing up pieces of paper to represent people, locations, and vehicles and their capacity, you can work out the problem in advance and make sure everyone is taken care of.

SPORTS

Manipulatives are especially useful for solving problems involving movement in sports, as illustrated by Sierra College student John Bingham: "When I arrived at the manipulative part of the book, I found myself drifting back to my youthful past. During my high school years I played water polo. Part of each practice involved manipulatives on a whiteboard. During practice we would participate in specific plays with designated positions and so on. After practice we would have a team meeting to review what we had learned. In the team room was a whiteboard that illustrated a water polo pool. On the board were markers in the shape of water polo hats that had our numbers and names on them. Our coach would re-create a specific situation and expect us to move the appropriate hats to the appropriate place on the model of the pool to show that we had a full understanding of what we had participated in during practice. Explaining and justifying helped us to prepare not only our understanding but also our mind in that I could visualize things I had seen on the board and think ahead to what would possibly happen next."

Problems Involving Spatial Relationships

Rearranging furniture is a great problem to work with manipulatives. By making a scale model of the space and using manipulatives to represent the equipment and furniture to be set up within that space, you can make a number of changes very quickly without risking anybody's back or toes. With manipulatives you can also determine how to landscape a yard by cutting out paper scaled to a plant's expected size at maturity. By moving around trees, bushes, fountains, walkways, or pink flamingos on a scale drawing, you can see where plants may be too crowded, where foot traffic patterns may cause a problem, or where you can set up a visual focal point.

Problems Involving Logical Connections

In Chapter 4: Use Matrix Logic, you solved logic problems. Many of these problems can also be solved with manipulatives. By writing the names of people and their various characteristics on individual pieces of paper and manipulating the pieces, you can make the logical connections. Tape comes in handy in some cases.

Problems Involving Tangible Numbers

Rate-time-distance problems are good examples of problems involving tangible numbers. You can use colored rods or money to represent different speeds, distances, or times.

Abstract Problems Involving Numbers

You can tear up little pieces of paper to represent numbers. This technique proves useful with magic squares and similar problems.

Problems Involving Unit Analysis

Remember that we introduced the use of manipulatives in Chapter 8: Analyze the Units. Some unit analysis problems are virtually impossible to organize without one-n-oes and other reciprocal manipulatives.

Problems Involving Order

Dilyn manages a record store. Every week she receives the list of a music magazine's top 200 albums, which lists the most popular albums in all musical categories in numerical order from 1 to 200. Dilyn wants the list in alphabetical, rather than numerical, order. So she made a copy of the list and gave it to one of her employees, Carrie, who cut up the list into 200 little strips of paper. Carrie alphabetized it by moving the strips around on a large counter and then taped the list back together. Dilyn told me, "We used to have a computer do this. But when the computer did it, the employees wouldn't even notice the new artists that showed up on the list. If a customer came in asking for a new record, my employees probably wouldn't have heard of it. By doing the alphabetizing with the manipulatives every week, Carrie and my other employees stay current with the most popular artists and records. We actually had to move away from technology for the best results."

Problems Involving Organizing Ideas

Alyse Mason, a student at the University of California at Berkeley, says, "You can also use manipulatives to organize abstract ideas. For example, when you write an essay it helps to print it out, cut apart the separate paragraphs or concepts, and physically rearrange them into a logical order. Manipulating abstract ideas in a concrete fashion leads to a better understanding of how ideas relate to one another and allows you to make new connections between them. It is easier to organize complex thoughts if you use physical manipulatives to represent them rather than trying to rationally order them on your computer screen or in your mind."

Using manipulatives improves your ability to visualize.

The strategy of using manipulatives is underused. Despite the obvious applications of manipulatives, many textbooks and instructors favor a more abstract approach. As a result, many students assume they should be able to visualize problems in their heads and become frustrated when they can't. Using manipulatives to create a physical representation not only will help you solve the problem at hand, but also will help you become better at visualizing.

Problem Set A-2

7. TWO JACKALS LOSE THEIR LICENSES

Solve the Jackals and Coyotes problem from page 267 again. All six of them start on the same side of the river, but this time, although all three of the coyotes can operate the boat, only one of the jackals can operate the boat.

8. JACK-QUEEN-DIAMOND

Three playing cards from an ordinary deck of 52 cards lie facedown in a row. There is a queen to the right of a jack. There is a queen to the left of a queen. There is a diamond to the left of a heart. There is a diamond to the right of a diamond. What are the three cards?

9. BASEBALL SEATING

A family of five, consisting of Mom, Dad, and three kids (Alyse, Jeremy, and Kevin), went to a baseball game. They had a little trouble deciding who was to sit where. Alyse would not sit next to either of her brothers. Kevin had to sit next to Dad. Mom wanted to sit on the aisle but not next to any of the children, although she could sit next to her daughter as needed. How was the seating arranged?

10. MAGIC TRIANGLE

Here is a magic triangle. The sum of the digits forming each side of the triangle is 11. Use the digits 1, 2, 3, 4, 5, and 6 (once each), and find the proper location of each.

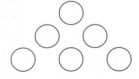

11. MAGIC SQUARE

Here is a magic square. The sum of each row is 12. The sum of each column is also 12, and the sum of each diagonal is 12. Use the digits 0, 1, 2, 3, 4, 5, 6, 7, and 8 (once each), and find the proper location of each. (There is more than one correct solution.)

12. TRUE EQUATIONS

Use the digits 0, 1, 2, 3, 4, 5, 6, 7, 8, and 9 to make true equations of the following. Use each digit once.

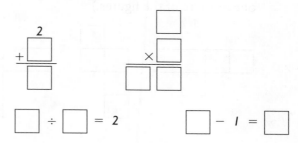

$$\boxed{} \div \boxed{} = 2 \qquad \boxed{} - 1 = \boxed{}$$

13. THREE-ON-THREE BASKETBALL

In a recreational basketball league, each team had only three players: a center, a forward, and a guard. The tallest player on each team was the center; the shortest was the guard. The information in the list below refers to three teams in the league.

1. Leon and Weston are guards.

2. Horace and Ingrid play the same position.

3. Kathryn, the shortest, played her best game against Ingrid's team.

4. Horace and Leon are on the same team.

5. Jerome and Taunia are on the same team. On that team, Jerome is not the shortest, and Taunia is not the tallest.

6. Sasha is shorter than both Weston and Horace.

Who is the forward on Kathryn's team? Who plays guard on Sasha's team? What position does Kedra play, and who else is on her team?

14. CUBIST

Three views of a cube are shown below. The cube has the five vowels A, E, I, O, and U on its faces. One letter appears twice. Fill in the blank space in the last cube with the correct letter in the correct orientation.

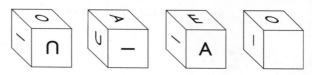

15. FOLDING CUBES

For each figure below, where could you attach a sixth square so that the figure would fold into a cube? (Note: This may be impossible for one or more of the figures.)

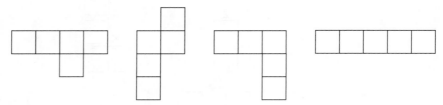

16. CO-ED VOLLEYBALL

This volleyball team has eight players: Allyson, Cheryl, Jan, and Sue are the women; Marty, Ted, Dave, and Harold are the men. The team will have six players on the court and two players off the court at any time. The two players off the court will rotate into the middle position in the back row. The players must alternate by gender. Set up their starting rotation with the following constraints in mind (for help with volleyball rotation, see the Volleyball Team problem in this chapter).

1. Allyson can't be placed next to Dave or Harold because they hog the ball.

2. Sue should be by Harold because she is the best setter and he is the team's best hitter.

3. Dave should probably be next to Sue because he is the team's next best hitter.

4. Allyson prefers to play next to Ted.

5. Harold should get as much time as possible on the front row because he is the team's best front-row player.

6. Dave is the team's best server. He should begin the game serving or be one of the first servers.

7. The weakest players are Jan, Cheryl, and Marty. They should not be placed next to one another.

At Worchestershire Palace, four guards are pacing back and forth performing maneuvers. Two of the guards (Basil and Burton) are wearing blue uniforms, and the other two guards (Roderick and Randall) are wearing red uniforms. The guards are standing in large tile squares on the floor. The tile arrangement with each guard's present location is shown below.

Basil		Burton
Roderick		Randall

One move for a guard consists of the following: He moves one square from his position in one direction (not diagonal), then turns 90° and moves two squares in that direction. (Or he could move first two squares and then one square.) Each move a guard makes is in this L-shape, and he does not stop until he has finished the complete move.

For example, Basil could move one square south and two squares east and end up in the square directly north of Randall's square. No two guards may occupy the same space at the same time. More than one guard may be moving at the same time.

a. How many moves does it take for the blue guards (Basil and Burton) to change places with the red guards (Roderick and Randall)?

b. How many moves does it take for Basil to change places with Roderick (ending with Burton and Randall in their original positions)?

18. **PROBLEM VATS**

In the back room of a scientific supply store, there are two big vats of liquid. The red one contains 10 liters of rubbing alcohol, and the black one contains 10 liters of purified water. Jimmy accidentally pours 3 liters of the alcohol into the water. Then, realizing his mistake and hoping to correct it, he pours 3 liters from the black vat back into the red vat. Each vat again has 10 liters in it. Is there more alcohol in the water, or more water in the alcohol?

19. WRITE YOUR OWN

Write your own act-it-out or manipulatives problem. Maybe you've been on a volleyball team. Perhaps you've had the experience of arranging a car pool for lots of people and different days. There is no limit to the ideas here.

CLASSIC PROBLEMS

20. CROSSING A BRIDGE WITH A FLASHLIGHT

There are four women who want to cross a bridge. They all begin on the same side. You have 17 minutes to get all of them across to the other side. It is night. There is one flashlight. A maximum of two people can cross at one time. Any party who crosses, either one or two people, must have the flashlight with them. The flashlight must be walked back and forth; it cannot be thrown. Each woman walks at a different speed. A pair must walk together at the rate of the slower woman's pace.

Woman 1 takes 1 minute to cross

Woman 2 takes 2 minutes to cross

Woman 3 takes 5 minutes to cross

Woman 4 takes 10 minutes to cross

For example, if Woman 1 and Woman 4 walk across first, 10 minutes will have elapsed when they get to the other side of the bridge. If Woman 4 then returns with the flashlight, a total of 20 minutes will have passed and you will have failed the mission. What is the order required to get all women across in 17 minutes?

Source unknown.

21. STEALING THE CASTLE TREASURE

The ingenious manner in which a box of treasure, consisting principally of jewels and precious stones, was stolen from Gloomhurst Castle has been handed down as a tradition in the De Bourney family. The thieves consisted of a man, a youth, and a small boy, whose only mode of escape with the box of treasure was by means of a high window. Outside the window was fixed a pulley, over which ran a rope with a basket at each end. When one basket was on the ground, the other was at the window. The rope was so disposed that the persons in the basket could neither help themselves by means of it nor receive help from others. In short, the only way the baskets could be used was by placing a heavier weight in one than in the other.

The man weighed 195 pounds, the youth 105 pounds, the boy 90 pounds, and the box of treasure 75 pounds. The weight in the descending basket could not exceed that in the other by more than 15 pounds without causing a descent so rapid as to be most dangerous to a human being, though it would not injure the stolen property. Only two persons, or one person and the treasure, could be placed in the same basket at one time. How did they all manage to escape and take the box of treasure with them?

Adapted from *Amusements in Mathematics* by Henry Dudeney.

MORE PRACTICE

1. CROSSING A RIVER WITH DIFFERENT-SIZE PEOPLE

A group of 14 people come to a river that they need to cross. Five of the people each weigh 100 pounds. Four people each weigh 150 pounds. Three people each weigh 200 pounds, and two people each weigh 300 pounds. To cross the river they are going to use a small boat that will hold at most 300 pounds. What is the fewest number of times that the boat must cross the river in order to get everyone across?

2. SQUARE TABLE

Two men (Kimbuck and Ki) and two women (Kimberly and Kimi) sat around a square table. Their professions are web designer, sales consultant, accountant, and video game designer.

1. Kimi sat across from the video game designer.

2. The accountant sat on Kimberly's left.

3. A woman sat on the sales consultant's right.

4. Ki sat on the web designer's left.

Who is the accountant? Who is the web designer?

3. COLLEGE REUNION

Six college sorority sisters—Summer, Autumn, Wynter, June, April, and January—got together for their twentieth reunion. Since college they had moved to six different states. Two lived out west (Oregon and Washington), two lived back east (New York and New Hampshire), and two lived down south (Oklahoma and Texas). They all sat around a round table.

1. The Texan sat between Autumn and the woman who lives in New Hampshire.

2. Wynter sat across from April.

3. January sat between Summer and the Washingtonian.

4. June and the woman to her left both live out west.

5. April sat between two easterners.

Where does Summer live? Where does January live? Where does Autumn live?

4. WEDDING PICTURE

At a recent wedding, six females—Aura, Aurelia, Ayelet, Jane, Miriam, and Peggy—lined up for a picture. They were the bride, the bride's mother, the bride's sister, the bride's aunt, the bride's cousin, and the bride's niece. Among the six were two pairs of sisters, three mothers, four daughters, and one granddaughter.

1. Peggy stood between her daughter and her granddaughter.

2. The aunt of the bride stood between her daughter and her niece.

3. The bride is Aura's sister.

4. Ayelet is not on either end. Her mother is the maid of honor.

5. Looking at the picture, the flowergirl is on Aura's right.

6. The maid of honor is on one end and the bridesmaid is on the other end.

7. The flowergirl is the bride's niece.

8. Miriam and her mother were in the center of the picture.

9. In the picture, Jane is standing between the bride's cousin and the bride.

List the six people from left to right in the picture. Include their name, their family designation, and their function in the wedding.

5. **WHICH SQUARE STAYS THE SAME?**

The squares below, labeled A–G, each contain a different digit from 1 to 7. No digit is adjacent—either horizontally, vertically, or diagonally—to a consecutive digit. For example, if square D was the digit 6, then squares B, C, E, and G could not be either the digit 5 or the digit 7. There are two ways to arrange the digits 1–7 into the squares. In both arrangements, one square will contain the same digit. Which square is it, and which digit occupies that square?

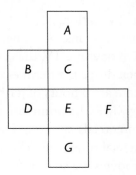

Problem Set B

1. DECREASING NUMBERS

A number is called a decreasing number if it has two or more digits and each digit is less than the digit to its left. For example, 7,421; 964,310; and 52 are decreasing numbers but 3,421; 6,642; 8; and 963,212 are not. How many decreasing numbers are there?

2. WHITE SALE

Debi purchased $91 worth of sheets at a sale on Labor Day weekend, when $1.60 was marked off every article. She returned the sheets on Thursday (she really didn't need sheets anyway), when everything was marked at regular prices. Because Debi is a good customer, the sales clerk gave her credit using regular prices and not sale prices, so Debi got more than $91 worth of credit. In fact, each sheet was now worth a towel and a washcloth. She was ecstatic about this and immediately exchanged her sheets for what she really wanted—towels and washcloths. (The towels, washcloths and sheets were all now marked at regular prices.) Because washcloths cost only $2.70, and she needed more washcloths than towels, she got 6 more washcloths than towels. She came home on Thursday with 16 more articles than she'd bought on Saturday. How many washcloths and towels did Debi go home with on Thursday, and what would she have paid for them if she had bought them on Saturday at the sale prices?

3. KMRCL TV

A local TV station, KMRCL, plans its number of commercials based on what type of show is being broadcast. Each commercial is a minute long. On daytime TV, the station runs 18 commercials every hour and then drops the number down to 16 per hour for the news. (Every day there are 8 hours of daytime programming and 2 hours of news.) During KMRCL's prime time, which is 3 hours per day, the station runs 12 commercials per hour. During the late-night movies, KMRCL sells commercial time really cheap and runs 20 commercials per hour. Late-night movies are shown for 6 hours each day. The station's "other" programming, which is on 5 hours per day, airs with 8 commercials per hour. For the sake of simplicity, assume that for all seven days of the week each type of programming airs for the same number of hours just described. Find how many hours per week KMRCL airs commercials.

4. THUNDER AND LIGHTNING

In a thunder and lightning storm there is a rule of thumb that many people follow. After seeing the lightning, count seconds to yourself. If it takes 5 seconds for the sound of the thunder to reach you, then the lightning bolt was 1 mile away from you. Sound travels at a speed of 331 meters/second. How accurate is the rule of thumb? Express your answer as a percent error.

5. STATE FAIR

Dear Margy,

How's my favorite daughter? How about coming to visit me next week at the state fair? It runs all this week and next, and the way things are going so far, I'm going to be pretty bored.

I'll be at the Hot Spas booth. Let me tell you where it is. But, you know your ol' dad, I can't just tell you straight out. There are seven booths on row 3, and I'm in one of them. On one side of me is Computer Horoscopes—their computers beep all the time. The booth on the other side of me is really quiet. Of course you know your brother, Wayne, works for Encyclopedia Antarctica. He keeps dropping by to visit, as his booth is pretty slow too. He usually comes and goes from my right, so his booth must be in that direction. He told me that the Slice-It-Dice-It-Veggie-Peeler booth is so popular that the vacuum sellers, who were in the booth next to the Slice-It-Dice-It-Veggie-Peeler booth, left the fair because nobody was ever visiting them. So now there are only five booths occupied, because the ladder sellers didn't show up at all. Wayne tells me that the other booth on our row is Foot Massage. He says he went to their booth and it is always really noisy with ticklish people laughing hysterically. The last time Wayne visited, he was in a really grouchy mood. It seems he was promised an end booth, but they are both presently occupied. He also complained about his neighbor, Computer Horoscopes. After he left, someone actually came to see my stuff. She came from my right and wanted to demonstrate her new Slice-It-Dice-It-Veggie-Peeler on my lunch.

That's it. You have enough information to come directly to my booth. See you next week.

Love,
Dad

Which booth is Margy's dad's? Describe the location from the perspective of a visitor facing the row of booths.

Work Backwards

Faced with a problem that appears inextricably complicated from the beginning, you may find it easier to start at the other end and work backwards. Architects and construction firms work backwards from an expected completion date to plan their work schedule.

R ecently, Lucille and her family took a vacation to Hawaii. On the day before they left to come home, they planned their activities for the next day. This is Lucille's tale of the problem they faced:

"Our flight was scheduled to leave at 12:40 p.m., and we needed to return our rental car before we left. We also wanted to spend 30 minutes at the pineapple plantation. It was a 45-minute drive to the airport, with about a 10-minute detour to go to the plantation. We also considered going to the sugarcane museum—it was about 10 minutes out of the way, and we planned to stay there about 1 hour. Based on all this information, we wanted to determine when to leave our condo. We used the strategy of working backwards to solve the problem.

"We started by figuring that we needed to be at the airport 90 minutes before our flight. Since our flight left at 12:40 p.m., we subtracted an hour and a half and decided that we needed to be at the airport at 11:10 a.m. Since we had to return the rental car also, we adjusted that to 11:00 a.m. To figure in the time we needed to visit the plantation, we subtracted the 10-minute drive and the 30 minutes we planned to stay there from 11:00 and got 10:20 a.m. Then we subtracted the 45-minute drive to the airport. This meant we had to leave at 9:35 a.m. If we subtracted the 1 hour and 10 minutes for the sugarcane museum, we would need to leave at 8:25 a.m. Because we had to check out of our condo and we had two little kids, we couldn't leave that early. So we canceled our plans to go to the sugarcane museum, and we left at about 9:30 a.m.

"We visited the plantation without being rushed and we got to the airport in plenty of time."

Working backwards is another strategy that falls under the major theme of Changing Focus. With most other strategies, you work forwards through the information in a problem. To successfully work backwards, you need to change your focus and consider the whole problem in reverse. This is a very useful strategy in certain situations. Much of algebra is based on working backwards, and this strategy is also very useful for planning schedules or agendas, as Lucille's story about her trip to Hawaii demonstrates.

One of the most difficult aspects of working backwards is keeping track of a problem's information and organizing it in a meaningful way. Note that although the solutions in this chapter are organized in various formats, we encourage you to experiment with finding new ways to organize the problems so you can solve them.

The following anecdote provides another example of how working backwards can be organized. In this example, directions and actions when walking from one point to another and back again are reversed.

Amy and Wendy were visiting their sister Kristen in her dorm for a few days. When Amy and Wendy needed to go to the library to do some research, Kristen took them to her window and showed them how to get to the library. She said, "The library's right over there, but you can't get from here to there directly because of all the construction. You have to take a right as soon as you get to the front of the dorm. Walk down Alumni Avenue and then make a left on University. Go past the gym and cut across the quad to the bell tower. Go on the right side of the bell tower and walk about 500 feet forward. You'll see the Café Europa on the corner. That's Regents Drive. Make a left onto Regents. Make a right when you get back to University. It's hard to see the library from in front because of the trees and things, but after you walk about a minute more it'll be there on the left."

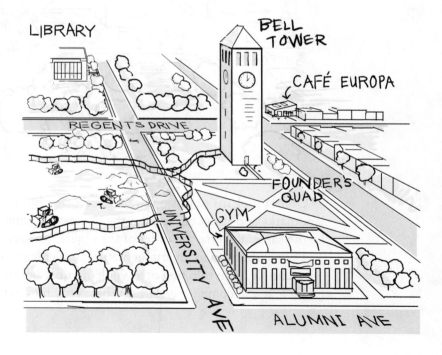

When they needed to return to the dorm, Amy and Wendy didn't want to take a chance on getting lost. They found a student, and he showed them where to go through the library window. He said, "Oh yeah. My dad's company got the construction contract. It's paying for all my tuition and ski trips. Here's how you get to the dorm: Take a right when you get outside the library. Turn left on Regents Drive. When you see Café Europa, make a right. You'll see the bell tower on your right. Go past the bell tower and cut across the quad. When the gym is on your left, make a left. That's University Avenue again. Then make a right on Alumni Avenue. Walk up to the dorm and make a left to get to the security door. You'll be able to call up to your sister's room from there."

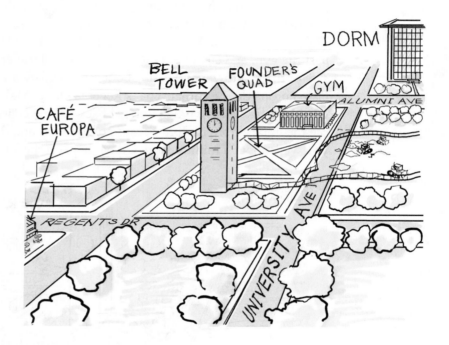

Other problems of this nature will involve objects other than landmarks and street names, but the process for solving them will be the same: Reverse the order of the objects and reverse the actions.

The solutions to many mathematics problems involve this process. For example, I am thinking of a whole number between 1 and 10. I double it and come up with 12. What is my number? Obviously, to reach the answer you must take my "doubled" number, 12, and divide it by 2 to get 6. This is a mathematical example of "reverse the action." Division reverses multiplication and vice versa, just as subtraction reverses addition and vice versa.

Working backwards involves reversing both the order of objects and the actions in a problem.

Consider an unknown number between 1 and 10. Multiply the number by 3. Then add 5 to the result, giving 32. What is the original number? To find the original number, you must reverse the order of the "objects" in the problem and reverse the actions. The first object is the 3 (as in "multiply by"), and the second object is the 5. Thus, the objects you must work with, in reverse order, are the 5 and the 3. Now reverse each action in the problem. Instead of adding 5, subtract 5 ($32 - 5 = 27$). Reverse multiplying by 3 to divide by 3 ($27 \div 3 = 9$). Now you have the original number: 9.

Lo and behold, you got the right answer. To check, work the problem forwards: 9 times 3 is 27, and 27 plus 5 is 32.

Do these three problems:

1. I'm thinking of a number between 1 and 10. If I multiply by 4 and then subtract 3, my answer is 25. What is my number?

2. I'm thinking of a number between 1 and 30. I add 22. I divide by 3. My answer is 12. What is my number?

3. I'm thinking about a number. Divide it by 2, then subtract 1. The answer is $7\frac{1}{2}$. What is my number?

Note that the answers to these problems can be found on pages 7, 14, and 17, respectively. Now on to more serious endeavors.

POOR CHOICES

The night before their debut in Carnegie Hall, the dancers stayed at a hotel on 57th Street. Fourteen members of the ballet company went to an all-night card room to play poker. Half of the remaining dancers went to Madison Square Garden for a special midnight professional wrestling show featuring Buff Bargle. After about an hour, 6 of the dancers who had gone to play poker came back to the hotel broke. The 11 dancers now at the hotel went to bed and got enough sleep, but the rest of the dancers were tired for their debut the next day. How many dancers were in the ballet company? Work this problem before continuing.

Here's an example of how to set up the solution. Follow it from line to line.

A. Fourteen of the
   dancers left.

B. Half of the remaining
   dancers left.

C. Six returned.

D. Eleven were at
   the hotel.

D. Eleven were at
   the hotel.   11

C. Six returned.
   (reverse the action
   to make it 6 less)   5

B. Half of the remaining
   dancers left.
   (reverse by
   multiplying by 2)   10

A. Fourteen of the
   dancers left.
   (reverse by
   adding fourteen.)   24

Check your answers by working forwards from your solution.

So there are 24 dancers in the company. To check, work the problem forwards. Start with 24 dancers. Fourteen of them left to go play poker, leaving 10 dancers. Half of the remaining dancers (which is 5) left to go to the professional wrestling show (leaving 5 dancers still at the hotel). Six dancers returned from the card room (so add 6) which gives the 11 who went to bed on time at the hotel.

DAD'S WALLET

Dad went to the ATM on Wednesday of spring break and withdrew some money. On Thursday morning my brother borrowed half of Dad's money to open a checking account, because he was always short of money. On Friday I needed some money for a date, so I borrowed half of what remained. My sister came along next and borrowed half of the remaining money. Dad then went to gas up the car and used half of the rest of his money, and he wondered why he had only $15 left. How much money did he start with in his wallet? Don't forget to reverse the actions and the order. Don't read on until you've worked this problem.

In order, this is what happened to poor Dad:

A. He got money from the ATM.

B. Brother borrowed half of his money.

C. I borrowed half of what remained.

D. Sister borrowed half of what remained.

E. He spent half of the remainder on gas.

Now it's time to reverse the order and reverse the actions (the reverse actions are shown in orange text).

E. At the end, the wallet has $15 in it.
   Dad spent half on gas.
   He gave back the gas and got back the other half of his money.

D. The wallet now has $30 in it.
   Sister borrowed half of the remainder.
   She gave back what she'd borrowed, which was the same as what she'd left, because she'd borrowed half.

C. The wallet now has $60 in it.
I borrowed half for a date.
I gave back an amount equal to what I left: $60.

B. The wallet now has $120 in it.
Brother borrowed half to start his nest egg.
Brother repaid his loan, half of the money: $120.

A. The wallet now has $240 in it.
So Dad's wallet had $240 in it before all the raids took place.

As you can see, there are different ways to record the information in a problem and use it to solve the problem. You'll see a few other ways to organize information in later examples. Your organization should help, not hinder, the process of working backwards. Be sure to experiment with your own style as you attempt the next problem and the problems that follow.

NUMBER TRICK

Start with a number between 1 and 10.
Multiply the number by 4.
Add 6 to the number you have now.
Divide by 2.
Subtract 5.
Tell me the number you end with, and I'll tell you the number you started with.
Two students, Glenda and Sonia, played this game. Sonia started with a number, did the arithmetic, and told Glenda that she had ended with 12. Glenda then figured out what number Sonia started with. What number did Sonia start with? Work this problem before continuing.

Here's how Glenda found Sonia's number: "I could have tried this problem with guess-and-check, and that definitely would have worked. But I wanted to use working backwards because I thought it would be easier. First I wrote down the given information, and then I worked backwards until I found the original number. I reversed the action each time, so subtracting 5 became adding 5 as I worked backwards, and so on."

Working "up the page" is sometimes the best way to solve a problem.

Organizing information is very important for keeping track of what you are doing. Glenda found, as many solvers have, that the most useful way to solve this problem is to work *up* the page. That is, with the information written in its original order, do your work from the bottom of the page to the top. The finished product looks something like the two columns shown in the following list. Read down the right column for the actions. Then read up the left column for the numbers. (In this example, the reverse actions are not written in.)

	Start with a number between 1 and 10.
7	
	Multiply the number by 4.
28	
	Add 6 to the number you have now.
34	
	Divide by 2.
17	
	Subtract 5.
12	
	The result is 12.

Note that the information and the numbers are staggered every other line to make it easy to read either down or up. In this form the solution is also easy to check. Start with a number (7). Multiply by 4 to get 28. Add 6 to get 34. Divide by 2 to get 17. Subtract 5 to get 12. It checks. Do this problem again, but this time assume that the number Sonia ended up with was 4.

Application: Airline Baggage Tags

The idea of working up a page may seem strange, but this process is very useful in some situations. For example, airline baggage tags list in reverse order the airports you land in. Your original departure point is sometimes shown at the bottom of the tag, but often it is omitted because, of course, the people working at your point of origin know where the airport is located. Each successive airport you land in is shown on the preceding line, along with your flight numbers. Your final destination is shown at the top of the tag. In the tag shown here, a passenger flew from Sacramento, California (not shown at the bottom), to San Francisco (SFO) on Flight 385. Then the passenger flew from San Francisco to Chicago O'Hare (ORD) on Flight 4689, and then from Chicago to New York's John F. Kennedy Airport (JFK) on Flight 1273.

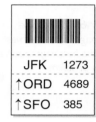

JFK	1273
↑ORD	4689
↑SFO	385

MINTS

After dinner, three friends at a restaurant paid their bill and noticed a bowl of mints on the front counter. Sean took one-third of the mints but returned four because he had a momentary pang of guilt. Faizah then took one-fourth of what was left but returned three for similar reasons. Eugene then took half of the remainder but threw two green ones back into the bowl. Seventeen mints remained after the three friends had each taken same. How many mints were originally in the bowl? Work this problem before continuing.

Write down all the events that occurred, in their proper order.

Start.

Sean took one-third of the mints.

Sean returned four.

Faizah then took one-fourth of what was left.

Faizah returned three.

Eugene then took half of the remainder.

Eugene threw two back into the bowl.

The bowl had 17 mints left.

This problem can be confusing. You may want to act it out with manipulatives, such as pennies, pieces of paper, or actual mints. Or you might imagine a movie of these events and then, starting from the end, run the movie backwards in your head. Keep in mind that to solve this problem you must reverse the order and the action.

Aniko solved this problem as follows: "I began at the end. There were 17 mints left in the bowl when the raid was over."

Start.
Sean took ⅓ of the mints.
Sean returned 4.
Faizah then took ¼ of what was left.
Faizah returned 3.
Eugene then took ½ of the remainder.
Eugene threw 2 back into the bowl.

17

The bowl had 17 mints left.

"Right before that, Eugene threw 2 back into the bowl. This means that there must have been 15 mints in the bowl before Eugene threw 2 back. Right before that, Eugene took half of the mints and left 15 in the bowl. This means that there must have been 30 mints in the bowl before Eugene took his greedy turn."

Start.
Sean took ⅓ of the mints.
Sean returned 4.
Faizah then took ¼ of what was left.
Faizah returned 3.

30

Eugene then took ½ of the remainder.

15

Eugene threw 2 back into the bowl.

17

The bowl had 17 mints left.

"Right before that, Faizah put 3 mints back into the bowl, leaving 30. So just before she put those 3 back, there must have been 27."

At this point, a common mistake is to add 3 and get 33. But think of watching these events in a movie. If there were 33 mints in the bowl before Faizah put 3 back, then there must have been 36 after she returned the 3. But these actions don't match what the problem says. The reverse action of Faizah putting 3 back is for her to take 3 out. So when we work backwards, we take 3 away from 30, which leaves 27. Then, going forwards, she will put 3 back, and 27 plus 3 is 30.

Start.
Sean took ⅓ of the mints.
Sean returned 4.
Faizah then took ¼ of what was left.

27

Faizah returned 3.

30

Eugene then took ½ of the remainder.

15

Eugene threw 2 back into the bowl.

17

The bowl had 17 mints left.

Staggering the lines and the numbers helps organization.

Aniko continued: "Right before that, Faizah took one-fourth of the mints, leaving 27."

A common mistake here is to multiply 27 by 4. However, this doesn't represent Faizah's actions. A diagram can illustrate what happened. (The following type of diagram, invented by student Hao Ngo, is very helpful.)

Aniko explained: "I drew a rectangle to represent all the mints before Faizah took any."

"Faizah would have separated the mints into four parts in order to take one-fourth of them. So separate the rectangle into four parts."

A diagram helps you visualize the fractions.

"Now Faizah takes one of the parts, so cross out one of the parts."

"The three parts that are left total 27 mints. This means, since the parts are all equal, that each part represents 9 mints."

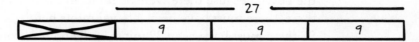

"Therefore, the part that Faizah took, which must be equal to the other three equal parts, must also represent 9 mints."

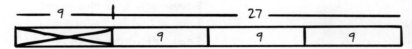

"So with four parts of 9 mints each, there must have been 36 mints in the bowl before Faizah did her dastardly deed.

"Let's see if everything checks at this point. When Faizah arrives at the bowl, she finds 36 mints. She takes one-fourth of them, and ¼ of 36 is 9. Now 36 minus 9 leaves 27 mints in the bowl. Then she puts 3 back—that is, she adds 3—so 27 plus 3 leaves 30 mints in the bowl when Faizah is done. Then Eugene takes half the mints, and half of 30 is 15. Because 30 minus 15 is 15, this leaves 15 mints in the bowl. Then he puts 2 back, so there are 17 left when he is done. Everything checks so far."

Start.
Sean took ⅓ of the mints.
Sean returned 4.

36
Faizah then took ¼ of what was left.

27
Faizah returned 3.

30
Eugene then took ½ of the remainder.

15
Eugene threw 2 back into the bowl.

17
The bowl had 17 mints left.

Let's pick up with Rachel's solution, which shows a different way of handling the fractional parts. "Right before Faizah, Sean returned 4 to the bowl. Because there were 36 mints in the bowl when Faizah got there, there must have been 4 fewer mints before Sean put 4 back. So there must have been 32 mints in the bowl before Sean suffered a pang of guilt."

Start.
Sean took $\frac{1}{3}$ of the mints.

32

Sean returned 4.

36

Faizah then took $\frac{1}{4}$ of what was left.

27

Faizah returned 3.

30

Eugene then took $\frac{1}{2}$ of the remainder.

15

Eugene threw 2 back into the bowl.

17

The bowl had 17 mints left.

"Right before those pangs of guilt, Sean took one-third of the mints. This means he left two-thirds of the mints in the bowl. I used algebra to find how many were in the bowl before that. I used m to represent the number of mints before Sean took one-third of them."

$$(\tfrac{2}{3})m = 32 \qquad \text{Multiply both sides by } \tfrac{3}{2}.$$
$$(\tfrac{3}{2})(\tfrac{2}{3})m = (\tfrac{3}{2})(32)$$
$$m = 48$$

Note that you could also use a diagram like the following to reach the same conclusion.

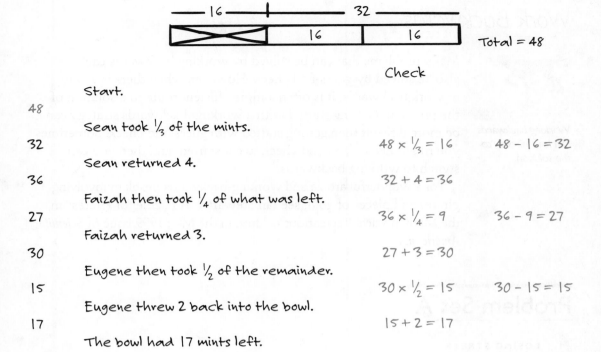

Check

Start.

48

Sean took ⅓ of the mints.

32 $48 \times \frac{1}{3} = 16$ $48 - 16 = 32$

Sean returned 4.

36 $32 + 4 = 36$

Faizah then took ¼ of what was left.

27 $36 \times \frac{1}{4} = 9$ $36 - 9 = 27$

Faizah returned 3.

30 $27 + 3 = 30$

Eugene then took ½ of the remainder.

15 $30 \times \frac{1}{2} = 15$ $30 - 15 = 15$

Eugene threw 2 back into the bowl.

17 $15 + 2 = 17$

The bowl had 17 mints left.

"So there were 48 mints in the beginning."

For this problem, the diagram and the algebraic techniques were both very effective for dealing with the fractional parts that were taken away. In general, however, using algebra is not a very efficient way to solve these types of problems. For instance, if m represents the number of mints the bowl contained at the start, an algebraic equation representing the entire problem would look like this:

$$\left\{\left[m - \left(\frac{1}{3}\right)m + 4\right] - \left(\frac{1}{4}\right)\left[m - \left(\frac{1}{3}\right)m + 4\right] + 3\right\}$$

$$- \left(\frac{1}{2}\right)\left\{\left[m - \left(\frac{1}{3}\right)m + 4\right] - \left(\frac{1}{4}\right)\left[m - \left(\frac{1}{3}\right)m + 4\right] + 3\right\} + 2 = 17$$

If you don't believe working backwards is easier, solve this equation!

Work Backwards

Working backwards can lead directly to the solution.

Many problems that can be solved by working backwards can also be solved by guess-and-check. However, when there is a way to work backwards, it is often a more efficient route to a solution of the problem. Getting started with a working-backwards strategy can be more difficult than getting started with guess-and-check. Sometimes you might need guess-and-check to get started, and then you can switch to working backwards.

For a wonderful article and working-backwards problem involving pirates and pieces of gold, see Ian Stewart's "A Puzzle for Pirates" in the Mathematical Recreations column in the May 1999 issue of *Scientific American*.

Problem Set A

1. LOSING STREAK

Regis was competing on a game show, and he ran into a losing streak. First he bet half of his money on one question and lost it. Then he lost half of his remaining money on another question. Then he lost $300 on another question. Then he lost half of his remaining money on another question. Finally, he got a question right and won $200. At this point, the show ended, and he had $1,200 left. How much did he have before his losing streak began?

2. GENEROSITY

Phil Anthropist likes to give away money. One day he was feeling especially generous, so he went to the park with a wad of money. He gave $100 to a man feeding pigeons. He then gave half of his remaining money to a child licking an ice-cream cone. He then gave $50 to the balloon seller. He then bought a hot dog and paid for it with a $20 bill. "Keep the extra for a tip," he said to the hot-dog seller. Then he gave half of his remaining money to someone giving a sermon on a soapbox. At this point, he had $3 left and stuck it under the collar of a stray cat. How much money did Phil have when he started his good deeds?

3. COOKIES

Barney had a bag of cookies. He ate 35 in the first 10 minutes. He ate one-fourth of the remaining cookies during the next 10 minutes. The next day he ate 20, but then he made 10 more. His wife, Louise, ate 15 cookies while Barney was at work. When Barney got home from work, he ate one-half of the remaining cookies. Louise ate 5 more when he went out into the backyard. Barney then ate 15 more and found that he had only 2 left. How many cookies did he start with? How many did he eat? How much weight did he gain if every 20 cookies translated into 1 pound?

This problem was written by David and Eric from Bob Daniel's class at Centaurus High School in Lafayette, Colorado.

4. THE MALL

My sister loves to go shopping. Yesterday she borrowed a wad of money from Mom and went to the mall. She began her excursion by spending $18 on a new music compact disc. Then she spent half of her remaining money on a new dress. Then she spent $11 taking herself and her friend out to lunch. Then she spent one-third of her remaining money on a book. On her way home she bought gas for $12 and spent one-fourth of her remaining money on a discount tape at the convenience store. Finally, she slipped me $2 when she got home and gave Mom back $10 in change. Mom was furious and demanded an explanation of where the money went. What did Sis tell her? (List the items and the amount spent on each item.)

5. **USED CAR**

A car dealership was trying to sell a used car that no one wanted. First they tried to sell it for 10% off the marked price. Then they tried to sell it for 20% off the first sale price. Finally, they offered it for 25% off the second sale price, and someone bought it for $3,240. What was the original price of the car?

6. **JASMINE'S DANCE TROUPE**

Jasmine was the lead dancer for her dance troupe. She and the troupe's choreographer (also a troupe member) decided that they needed to have one more rehearsal before they performed. She called one-third of the rest of the troupe to let them know about the rehearsal. She then helped prepare dinner and called three more members before sitting down to eat. While her roommates cleaned up the kitchen, Jasmine called two-fifths of the remaining dancers. She did her math assignment, then reached one more dancer on the phone. After completing her chemistry, Jasmine called three-fourths of the remaining troupe members. She edited a rough draft of her essay, then called the last two dancers. How many members are there in Jasmine's dance troupe?

7. **SPEND IT ALL**

Andy wanted to go shopping, but didn't have very much money for what he wanted. He did, however, spend $17 on a shirt, ¼ of the remaining amount on shoes, $741 on a jacket, $2,000 on a horse, ¼ of the remaining amount on a saddle, 4/5 of the rest of his money on horse grain, and $150 for one month's board for the horse. He was left with nothing. How much money did he spend?

This problem was written by Sierra College student Cheyenne Smedley-Hanson.

8. **HOCKEY CARDS**

Jack had quite a few hockey cards, and Jill had some of her own. Jack gave Jill as many hockey cards as she already had. Jill then gave Jack back as many cards as he had left. Finally, Jack gave her back as many cards as she had left. This left poor Jack with no cards and left Jill with 40 cards altogether. How many hockey cards did each of them have just before these exchanges took place? (You might want to try acting this out in conjunction with working backwards.)

9. SHOPPING SPREE

My two sisters pooled their savings and decided to go on a shopping spree at the mall one weekend. They started off at Macy's, each buying a pair of jeans and a T-shirt, spending $50 each. Next they went to the shoe store and my older sister bought a pair of shoes for $66, while my younger sister spent $47 on her shoes. My older sister wanted makeup, so she bought $27 worth of makeup. They were hungry so they spent ¼ of what they had left on a nice lunch. After lunch they went to buy CDs. They bought 2 CDs each at $14 per CD. They then spent half of what they had left on a necklace for my mother. On the way home they bought gas for $20. When they got home with their things, they split the remaining amount of money and each ended up with $6. How much did they spend?

This problem was written by Sierra College student Jeremy Chew.

10. WHAT'S MY NUMBER?

I am thinking of a number. I multiplied my number by 3, subtracted 8, doubled the result, and added 14. Then I added on 50% of what I had and subtracted 11. Then I divided by 5. After all that, I was left with 8. What number did I start with?

11. LOST HIS MARBLES

Livingston is a marble freak and loves to play at school. During the first break, he doubled the number of marbles he had. During the second break, he lost 4 marbles (no big deal). During the third break, he increased his stock of marbles by one-third. During lunch, however, he lost half of his marbles. After school, Livingston played again and tripled what he had. He went home with 72 marbles. How many did he start with?

This problem was written by Sierra College student David Lee.

12. TWO FOR TENNIS

Scott and Jeremy took a basket of balls with them to tennis practice. Scott took out two and threw them at Jeremy. Jeremy chased Scott and accidentally knocked over the basket, dumping out five-eighths of the remaining balls. Scott threw three back in. As they were playing, a hard hit knocked the basket over, and one-fifth of the remaining balls rolled out. Scott put two balls back into the basket, but a couple of minutes later a smash hit knocked the basket over again. Half of the remaining balls rolled out, and then there were seven left in the basket. How many tennis balls did Scott and Jeremy start with in the basket?

13. DONUTS

Four customers came into a bakery. The first one said, "Give me half of all the donuts you have left, plus half a donut more." The second customer said, "Give me half of all the donuts you have left, plus half a donut more." The third customer said, "Give me three donuts." The last customer said, "Give me half of all the donuts you have left, plus half a donut more." This last transaction emptied the display case of donuts. How many donuts were there to start with?

14. GOLF CLUBS

Ken played golf yesterday and shot 107. Considering that he normally shoots in the low 80's or high 70's, this round of golf really frustrated him. It was so frustrating that he decided to buy new golf clubs. But first he had to give his old golf clubs away. He gave half of his golf clubs, plus half a club more, to Daniel. Then he gave half of his remaining golf clubs to Gary. Then he gave half of his remaining golf clubs and half a club more to Will. This left Ken with one club (his putter), which he decided to keep. How many golf clubs did Ken start with before giving them away?

15. WRITE YOUR OWN

Write your own working-backwards problem. Start at the beginning of the problem with your total amount. In other words, start with the number that will be the answer to your problem. Then make up the situation and the steps. By starting with the beginning number (the answer), you will avoid any ugly fractional amounts of things you could get if you started at the end of your problem.

CLASSIC PROBLEMS

16. THE THREE BEGGARS

A charitable lady met a poor man to whom she gave 1¢ more than half of the money she had in her purse. The fellow, who was a member of the United Mendicants' Association, managed, while tendering his thanks, to chalk the organization's sign of "a good thing" on her clothing. As a result, she met many objects of charity on her journey. To the second applicant she gave 2¢ more than half of what she had left. To the third she gave 3¢ more than half of the remainder. She now had one penny left. How much money did she have when she started?

Adapted from *Mathematical Puzzles of Sam Loyd,* Vol. 2, edited by Martin Gardner.

17. X AND O

The game of tic-tac-toe is played in a large square divided into nine small squares.

1. Each of two players in turn places his or her mark—usually X or O—in a small square.

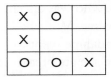

2. The player who first gets three marks in a horizontal, vertical, or diagonal line wins.

3. A player will always place his or her mark in a line that already contains either (a) two of his or her marks or (b) two of his or her opponent's marks, giving (a) priority over (b).

Only the last mark to be placed in the game shown above is not given. Which mark—X or O—wins the game?

Adapted from *The Great Book of Mind Teasers and Mind Puzzlers* by George Summers.

MORE PRACTICE

I. PAINTING NAILS

Marleena went down to the store to find the perfect nail polish color to match her new dress. She was so indecisive that she bought a bunch of bottles and saved her receipt in case she wanted to return some. She got home and realized that the lighting was weird in the store and that she didn't even like half of them. So she took them back. While there she found a new brand that she didn't see before, so she bought one more bottle. When she got home she realized that she already had six of the colors buried in the back of her cabinet, so she went back to the store again and returned the six duplicates. When she got home again, her sister had bought her one more new bottle in still another color. In trade she gave her sister two of her new bottles, which left her with six new bottles. How many bottles of nail polish did she buy originally?

This problem was written by Sierra College student Aletta Gonzales.

2. GUADALUPE'S HOMEWORK

Guadalupe had a ridiculously long homework assignment in her math course. It was a long assignment. Before class was over, she was able to finish 11 problems. When she was in study hall she read her book and then finished one-third of the remaining problems. After soccer practice, she was able to complete one-fourth of her remaining assignment. Her friend, Moua (who was also in her class), came over and they finished 5 more problems. Moua had a long day, so she went home to go to bed. Before Guadalupe went to bed, she finished half of the remaining problems. While she ate breakfast the next morning she finished the remaining 8 problems of homework. How many problems was Guadalupe assigned?

This problem was written by University of Wisconsin – Eau Claire students Ashley Lien, Katie Evenson, Sarah Bianchet, and Alyssa Reindl.

3. BLOCK PARTY

Mrs. Field's class is going to create a city of blocks. To prepare for the lesson, Mrs. Field bought many blocks at the craft store. Unfortunately, she left 4 blocks behind on the counter as she was packing them up to leave the store. Between her home and school she lost 2/5 of the remaining blocks in her car. However, when she got to school she found 8 on the floor of her car and put them back in the bag of blocks. In order to simulate making a city, she decided to have multiple children work together to make a building. The class was split into 6 groups, and 4 blocks were distributed to each group. One group came up and asked for more, so she gave them 5/7 of the remaining blocks. There were 4 blocks not being used. How many blocks did Mrs. Field purchase?

This problem was written by University of Wisconsin – Eau Claire students Matthew Nelson, Katie Fitzpatrick, and Brianne Brickner.

4. BONUS!

Lucky Tiffany—after working so hard at her job she received a bonus paycheck. So after work Tiffany decided to go shopping and reward herself for her hard work. The first thing she did was head over to Raley Field and spend 1/8 of her bonus on some tickets to a Sacramento River Cats baseball game. Then she headed over to the mall where she spent $17 on a DVD movie. She then went over to her favorite department store and bought herself a new shirt for $20 and then spent 1/3 of her remaining money on shoes. Tiffany left the department store and walked over to the music store and bought a CD for $12. After that she went over to the bookstore and bought the new issue of *Math Whiz Magazine* for $5. Tiffany began to think she was being very selfish so she spent 1/3 of her remaining money on lunch with her mother. After lunch she was on her way home and decided to deposit the remaining $50 into the bank for a rainy day. How much was Tiffany's bonus check?

This problem was written by Sierra College student Reina Alvarez.

5. DIEGO AND THE PEANUT TRAIL

Diego, like most other kids, loves anything chocolate. Recently he discovered his mom's chocolate peanut stash in the cupboard. He wasn't too careful, though, and ended up leaving a trail of goodies behind him. This is how I was able to find him as he sneaked from room to room with pockets full of chocolate-covered peanuts. Diego began his quest in the kitchen where two peanuts dropped on the floor. He then went into the living room and left 1/6 of his remaining peanuts on the sofa. He then walked down the hallway and ate three, but then dropped 1/7 of his remaining peanuts in the hallway. Diego went into his room and began to jump on the bed, and ½ of his remaining peanuts fell out of his pockets. Then he went into Uncle Robert's room and gave him 1/3 of the remaining peanuts. He then sees he has very little so he eats six. By the time I catch up to him and he sees me, he drops four and then eats the last two. How many peanuts did Diego start off with?

This problem was written by Sierra College student Robin Shortt.

6. POKÉMON

Curtis loves Pokémon! He went to school on Thursday and traded a bunch of cards to get new ones. He saw Dino and traded 3 of his cards for one of Dino's. Then a girl he liked, Tippi, wanted to trade cards. He was really nice to her because he liked her, so he traded 5 of his cards for 2 of hers. He then put his cards away. When he got home he noticed that 10 of his cards were missing. He was so upset that his mom bought him another pack of 12 cards. He hid half of his cards at home and took the rest to school the next day. He traded ¼ of the cards he brought to school to Dino again and got back 3 of Dino's cards. Curtis now has 9 cards at school. How many cards did he start with? How many cards total does he have now?

This problem was written by Sierra College student Monica Borchard.

Problem Set B

The Camping Trip

1. HOW MUCH DOG FOOD?

The Family family (Mama, Papa, and their three kids, Ed, Lisa, and Judy) were about to go on a camping vacation. They decided to take their five dogs with them, and they needed to know how many cans of dog food to take. They had received a free sample of dog food in the mail, and all the dogs really liked it. The cans of the sample brand of dog food were on sale for $1.24, and the packaging said that three cans of dog food would feed two dogs for one day. The family was planning on going camping for 8 days. How much would the dog food cost them for the 8 days?

2. THE LUGGAGE RACK

The Family family was about ready to leave on their camping trip. They loaded up their van with stuff, but it was getting really full with all kinds of equipment and food, plus five dogs. (Everyone had his or her own dog.) They were trying to decide whether to take the luggage rack for the top of the van to store some of their camping gear or just cram everything inside the van. Papa didn't want to take the luggage rack because he said their gas mileage would be worse because of the increased drag and weight. Mama said she didn't want the family to be crowded inside the van, and she didn't care how much it cost—she still wanted to take the luggage rack. Papa said they got 20 miles per gallon with the luggage rack off and 17 miles per gallon with it on. Their destination was 3 hours away at 55 miles per hour with the rack off, but with the rack on they could travel at only 50 miles per hour. Gas cost $1.89 per gallon. How much more would it cost the Familys to take the luggage rack, and how many extra minutes would it take them to get there?

3. CROSSING THE RIVER WITH DOGS

The five members of the Family family and their five dogs (each family member owned one of the dogs) were hiking when they encountered a river to cross. They rented a boat that could hold three living things: people or dogs. Unfortunately, the dogs were temperamental. Each was comfortable only with its owner and could not be near another person, not even momentarily, unless its owner was present. The dogs could be with other dogs, however, without their owners. The crossing would have been impossible except that Lisa's dog had attended a first-rate obedience school and knew how to operate the boat. No other dogs were that well educated. How was the crossing arranged, and how many trips did it take?

4. DON'T FEED THE ANIMALS

For the camping trip, the Family family had brought many bags of peanuts for snacks. Peanuts were everyone's favorite, so they were well stocked. Unfortunately, the campground was populated by various animals who also enjoyed an occasional peanut. The first night, after the Family family went to bed, a raccoon visited their camp and ate five of the bags of peanuts they had brought. The next day, the Familys ate one-third of the remaining bags. That night, a beaver came to call and ate two more of the bags. The next day, the Familys consumed one-fourth of the remaining bags for breakfast. Then they took the boat trip described in the Crossing the River with Dogs problem, and they had to feed one bag of peanuts to each dog to get the dogs to quiet down. That night, an elephant from a nearby zoo came to the camp and ate four bags of peanuts. The next day, each member of the Family family ate one-ninth of the remaining bags. That night, a spotted owl ate half of the remaining bags. The next day, there were only four bags left. The Familys couldn't decide how to split them among the five of them, so they fed them to the ducks. How many bags of peanuts did the Familys bring with them on the camping trip?

5. LOST IN PURSUIT OF PEANUTS

On the fifth day, after the peanuts were all gone, Ed and Judy took a hike to the store to get some more peanuts. On the way back, they got lost in the middle of a big forest. The forested area they were in measured roughly 13 miles from north to south and 14 miles from east to west. When they first discovered they were lost, they were at a point that was 7 miles from the southern border and 7 miles from the eastern border of the forest. (Of course, they didn't know this or they wouldn't have been lost.) In their indecision as to whether they should walk east (toward the morning sun), west (toward the setting sun), south (the direction from which they came), or north (because they might be wrong about the direction from which they came), they decided to walk in all four directions. They devised a plan whereby they would walk north first for 30 minutes. If they weren't out of the forest yet, they would then turn right and walk east for 60 minutes, then turn right again and walk south for 90 minutes, and so on, adding 30 minutes with each change of direction. (Nobody said this was a good plan.) It turns out that in the forested area they covered about 1 mile every 30 minutes. They also decided to drop a peanut after every 100 yards so that they would be sure they never doubled back on their tracks. They had 50 bags of peanuts with them, and there were 40 peanuts in each bag. On which side of the forest—north, east, south, or west—did they finally emerge, and how long did it take from the moment they were lost until the moment they got out? How many full bags of peanuts did they have left?

12

Draw Venn Diagrams

This specialized strategy is a method of sorting out the elements of overlapping categories. When you recycle, you separate recyclable materials from other trash, then further sort these materials into subsets such as aluminum, glass, and paper. Drawing Venn diagrams is a helpful strategy for solving problems involving objects that can be sorted into different categories.

Mathematics students often use **Venn diagrams** to categorize things because Venn diagrams can clearly show relationships among different categories. As a problem-solving strategy, Venn diagrams fall into the major theme of Spatial Organization. They organize information in a particular way that otherwise could be hard to see. When you draw Venn diagrams, you can use loops, closed circles, or rectangles. Two loops can intersect or be entirely disjoint, or one can be completely inside the other. Each pair of loops represents a different type of relationship between two categories.

In certain types of problems, you will encounter the word *all, some,* or *no.* For example, the word *all* could be used in a statement such as "All roses are flowers." Using a Venn diagram can help you correctly interpret the meaning of these three important words and of other words in a problem.

Several words of caution: When we talk about correctly interpreting the meanings of words in Venn diagram problems, we are talking about the principal, common understanding of the words. For example, you may know a woman named Rose, but the name Rose is not the principal, common understanding of the word *rose.* Don't waste time looking for obscure meanings or interpretations that will render all the information in a problem false.

Use the principal, common understandings of words.

Consider the statement "All roses are flowers." Here is a Venn interpretation of this statement. The diagram shows the word *Flowers.* Anything inside the rectangle is considered to be a flower. Anything outside the rectangle is considered to be "not a flower." Add the word *Roses* to the diagram. Roses is an entirely enclosed subset of the larger set, Flowers. (A **set** is a collection of particular things. Each individual thing in a set is called a **member** or an **element.** A **subset** is a set contained within a set.) Every member of the group called Roses is also a member of the group called Flowers. Inside the rectangle called Roses are all the individual items

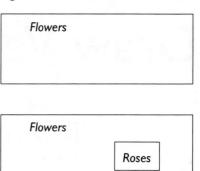

that are classified as roses, such as white roses, red roses, damask roses, and rugosas. Anything outside the Roses rectangle is considered "not a rose."

Think of some examples that are not roses: pink carnations, tiger lilies, apricot blossoms, hamburgers, lawn mowers. Not a single one of these items is considered to be a rose. (Someone might wonder, what if the lawn mower is manufactured by a company called Rose Garden Products, Inc.? Remember, we are using the principal, common understanding of the words in the problem, in which case a rose is a rose is a rose.) Some of these "not rose" items *are* flowers and thus can be placed in the diagram someplace. Let's call the rectangles *regions* and label them A and B, as shown below.

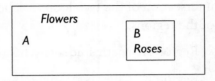

Region A: Flowers but not roses

Region B: Flowers; in particular, roses

In a Venn diagram, all the characteristics of an outer loop apply to everything within that loop, including other loops. In this case, the characteristics of the large rectangle apply to both regions A and B. The characteristics of a **region** are the same as those of its principal loop minus the characteristics of any smaller, wholly enclosed loop. For example, a pink carnation is a flower, but it is not a rose. These are the characteristics described in the statement for region A, so you would place pink carnations in region A because they lack the characteristic necessary to place them in region B. On the other hand, lawn mowers would be placed *outside* both regions, because lawn mowers are not flowers. Figure out where you would place each of the other "not rose" examples listed previously.

Another example of a Venn diagram is shown next. This example shows two loops: a loop that contains students who take mathematics and a loop that contains students who take chemistry. Notice that the word *some* is key in both of these statements:

1. Some math students are also chemistry students.

2. Some chemistry students are also math students.

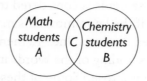

In this diagram, the two loops intersect each other. The math-student loop is labeled A, the chemistry-student loop is labeled B, and the intersection of the two loops is labeled C. What are the characteristics of the regions?

Region A: Math students who are not taking chemistry

Region B: Chemistry students who are not taking math

Region C: Students who are taking both math and chemistry

The **overlapping** section, region C, has the characteristics of both loops and thus of both regions A and B. This was also true in the Venn diagram for the flowers and the roses: The overlap (the inside loop called Roses) had the characteristics of both regions A and B.

With this diagram of math and chemistry students, you can also draw a large rectangle around the two loops to indicate what is called a **universal set.** In this case, the universal set could be assumed to be students. Thus, a student who is taking neither math nor chemistry could be placed in region D, outside both loops. (You could have put a larger rectangle, called Plants, around the diagram for the flowers and roses. This would have indicated the universal set that includes flowers, roses, and plants that are neither flowers nor roses.) The expanded diagram of all students is shown next.

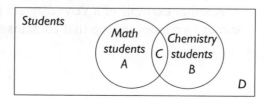

Note that the shapes of the loops are not significant, so shapes may be mixed in a single Venn diagram. In some Venn diagrams, the two regions do not overlap at all. Consider these two statements:

1. No car is a mushroom.

2. No mushroom is a car.

Note that cars and mushrooms are **disjoint** sets, so their diagrams don't overlap. A universal set makes little sense here because cars and mushrooms really couldn't be related with one category, unless we use a category as general as "things." This car-mushroom diagram shows no overlap either conceptually or in the visual representation.

Let's go back to the overlapping student loops and the accompanying statements:

1. Some math students are also chemistry students.

2. Some chemistry students are also math students.

The key in both of these statements, and in many other statements, is the word *some*. However, the word does not automatically indicate overlapping loops. Consider the statement "Some flowers are roses." This statement can be illustrated with the same two loops we showed for the statement "All roses are flowers." Recall that with those two loops, one loop was entirely enclosed within the other.

You need to look carefully at statements that include the word *some*. A Venn diagram may have overlapping loops or may have one loop inside another. If you are having trouble deciding how to draw the Venn diagram of a statement that includes the word *some,* remember that it's safer to draw the loops as **intersecting** and realize that one or more regions may be **empty** (contain no elements). You may not know which regions are empty until the problem is solved.

Your understanding of creating accurate Venn diagrams is critical to understanding the problems in the rest of this chapter. Draw Venn diagrams that represent each of the following statements:

1. Some birds are pets.

2. No dogs are sheep.

3. All poodles are dogs.

4. Some dogs are poodles.

Draw these diagrams before continuing.

The correct answers to the problems are shown below. If you haven't drawn the Venn diagrams yet, draw them before reading on.

1. Some birds are pets.

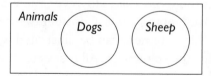

2. No dogs are sheep.

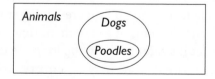

3. All poodles are dogs.

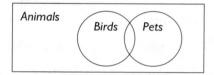

Note that the large rectangle could have been dogs, with only one circle inside it to represent poodles. The large rectangle represents the universal set, and it is up to the person drawing the diagram to decide what that universal set should be.

4. Some dogs are poodles.

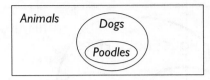

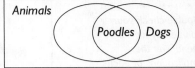

This problem has been illustrated in two ways. In the left-hand diagram, the Poodles circle is shown inside the Dogs circle; in the right-hand diagram, the two circles overlap. In the right-hand diagram, the region of the Poodles circle that is outside the Dogs circle is empty because there are no poodles that are not dogs.

It is perfectly okay for Venn diagrams to contain empty regions, especially if you're not sure how the members of the different sets are related. For example, if you didn't know what poodles and dogs were, you would have to draw the diagram of the statement "Some dogs are poodles" as two overlapping circles.

Empty regions are okay for Venn diagrams.

CATEGORIES

For each list of categories, draw a Venn diagram that shows the relationships among the categories. One of the categories describes the universal set, and the others describe the various loops inside the universal set. The universal set is not necessarily listed first.

1. Household pets, dogs, animals, cats

2. Living things, lizards, apes, chimpanzees, reptiles, dogs, mammals, terriers, dachshunds

3. Place of birth, USA, Canada, Miami, Florida, Orlando, Montreal, Missouri

4. Universities, private universities, public universities, Yale, University of Texas at El Paso (UTEP), Notre Dame

5. Hamburgers, with cheese, double, homemade

Work this problem before continuing.

Takashi drew the Venn diagrams that follow. Are your Venn diagrams equivalent to his? If not, how does your interpretation of the relationships differ?

1. Household pets, dogs, animals, cats

Double-check your solution by logically identifying the implications of each region.

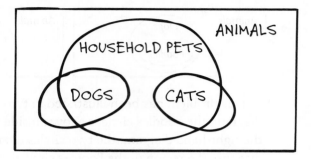

2. Living things, lizards, apes, chimpanzees, reptiles, dogs, mammals, terriers, dachshunds

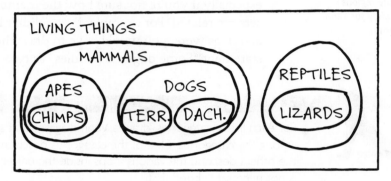

3. Place of birth, USA, Canada, Miami, Florida, Orlando, Montreal, Missouri

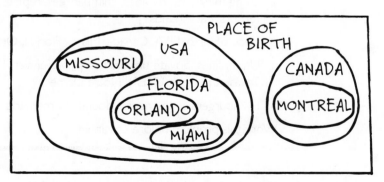

4. Universities, private universities, public universities, Yale, UTEP, Notre Dame

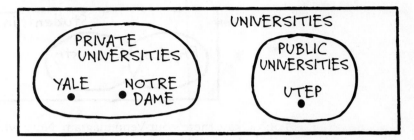

Note that Yale, Notre Dame, and UTEP are shown as points in the diagram, not within circles that represent subsets. These schools are not subsets because each is a *single* university, not a set of universities.

5. Hamburgers, with cheese, double, homemade

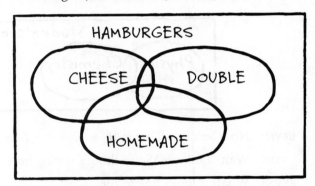

Now you are ready to solve some problems by using Venn diagrams. Draw the diagrams, then fill in the numbers from the problems as they fit.

SCIENCE COURSES

In a group of students, 12 are taking chemistry, 10 are taking physics, 3 are taking both chemistry and physics, and 5 are taking neither chemistry nor physics. How many students are in the group? Work this problem before continuing.

Bryndyn and Toi worked together on this problem. They drew the following diagram.

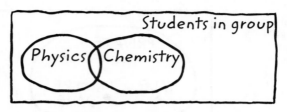

TOI: Okay, there's our Venn diagram. Now what do we do with the numbers?

BRYN: Well, let's see. Let's put the 12 chemistry students in the chemistry-student loop and the 10 physics students in the physics-student loop.

TOI: And put in the 3 students who are in both classes and the 5 that aren't taking either class.

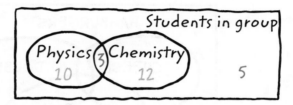

BRYN: Now let's see. That is 10 + 12 + 3 + 5 or 30 students altogether.

TOI: Wait, a minute. Something's wrong here.

BRYN: What? It looks fine to me.

TOI: Aren't we counting the people who are taking both classes twice? From the looks of this diagram, there are 13 students taking physics and 15 taking chemistry.

BRYN: How do you figure that?

TOI: Well, look at the physics loop. Those 3 students in the intersection are taking physics, right?

BRYN: Yeah, so what?

TOI: Well, that means we have 3 students in the intersection taking physics and chemistry, and 10 students in the outer portion of the physics circle taking physics and not taking chemistry. That means we have 13 students taking physics.

BRYN: Oh, I see what you mean. And by the same logic, we have 15 students taking chemistry.

TOI: So what can we do about this?

BRYN: Maybe we'd better start over. I think our original diagram is fine, so let's work with the numbers again.

TOI: Let's start by putting the 3 students who are taking both courses in the intersection of the two loops.

Start with the intersection of the two loops.

BRYN: That sounds like the working-backwards method we already learned.

TOI: Yeah, it does. Okay, of the 12 students taking chemistry, 3 of them are taking both chemistry and physics. This leaves 9 students taking chemistry but not physics.

Subtract to find chemistry only.

BRYN: So let's fill in 9 in the section of the chemistry loop that is outside the physics loop.

TOI: Yeah, that looks great. Now let's do the same thing for the physics loop.

10 − 3 = 7 physics only

BRYN: Of the 10 students taking physics, 3 of them are taking both chemistry and physics. This leaves 7 students taking physics but not chemistry. So fill in 7 in the section of the physics loop that represents physics only.

TOI: Great. Now put the 5 students taking neither course outside the two loops.

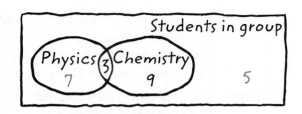

Students in group

Physics 7 3 Chemistry 9 5

Each section of the diagram is unique.

BRYN: Now let's answer the question. The question asks for the number of students in the group. The diagram shows 7 + 3 + 9 + 5 or 24 students. Each number represents a different group of students.

TOI: So, before, we were counting those students who took both classes three times: once in the physics category, once in the chemistry category, and once in the both category.

BRYN: You're right. It did seem too easy the other way. Venn diagrams really help.

MUSIC SURVEY

In a poll of 46 students, 23 liked rap music, 24 liked rock music, and 19 liked country music. Of all the students, 12 liked rap and country, 13 liked rap and rock, and 14 liked country and rock. Of those students, 9 liked all three types of music. How many students did not like any of these types? Work this problem before continuing.

There are three categories in this problem, so you need to draw three loops. Because it is possible to like two or three types of music at once, the three loops must intersect.

Tony worked on this problem. He succeeded in drawing the three loops, but after that point he was stuck. Maria joined him.

TONY: How do you solve a problem like . . . Maria, how do you . . . ?

MARIA: Well, let's figure out what each section represents. Let's label each section with a letter.

You may not need to label sections, or regions, with letters when you are solving problems, but if you do, be sure you know what each section represents. Before reading on, figure out what type or types of music each section in the diagram represents.

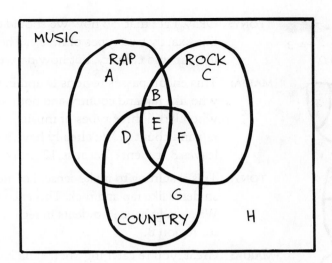

MARIA: The sections represent students who like the following types of music:

> A = Rap only, not rock or country
> B = Rap and rock, not country
> C = Rock only, not rap or country
> D = Rap and country, not rock
> E = Rap, country, and rock
> F = Rock and country, not rap
> G = Country only, not rap or rock
> H = None of the three types of music

If you haven't been able to solve this problem yet, draw the diagram on a piece of paper and fill in the numbers as Maria and Tony go through their explanation. The completed diagram appears at the end of the solution.

TONY: Okay, I get that. Now what do we do? How do we put in the numbers?

MARIA: Remember the strategy in the last chapter?

TONY: You mean working backwards?

MARIA: Yes. Let's work backwards, starting with the last clue. There are 9 students who like all three types of music. This is region E in our Venn diagram, so enter 9 in region E.

*Subtract 12 − 9 = 3
to get region D.*

TONY: Okay, I get that. So now we should consider the students who like exactly two types of music. There are 12 students who like rap and country, but how do we put that in the diagram?

MARIA: This encompasses regions D and E, since region D is students who like rap and country and not rock, and region E is students who like all three types of music, which of course includes rap and country. We already have 9 students in region E. This leaves 3 students (because 12 − 9 = 3) in region D.

TONY: This is starting to make sense. Let me do the next one. Thirteen students like rap and rock. This encompasses regions B and E. We already have 9 students in region E, so this leaves 4 students for region B.

MARIA: Great, you're catching on. (Maria had filled in region F with 5, because 14 − 9 = 5.) But I'm not really sure what to do next. This is as far as I got before I came over to work with you.

TONY: Well, let's keep on working backwards. Now we have to consider the students who like only one type of music. Region G is the students who like country only. We were told that there were 19 students who like country. So far we have 17 students who like country and something else. The country circle encompasses regions D, E, F, and G. Region D has 3 students, region E has 9, and region F has 5. This sums to 17. There are 19 students who like country, so we have 2 students left who like country but none of the other types. Therefore, put 2 in region G.

*19 − (3 + 9 + 5) = 2
so region G = 2.*

MARIA: Okay, I see. Let me do one. (Maria got 6 for region C, and then Tony figured 7 for region A.)

*24 − (4 + 9 + 5) = 6
24 − (4 + 9 + 3) = 7
so 6 rock only
and 7 rap only*

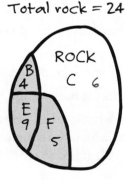

TONY: Great. We have the regions filled. We're done.

MARIA: No we aren't. We haven't answered the question. How many students didn't like any of the three types of music?

TONY: Maria, Maria, Maria, how are we going to answer that?

MARIA: Let's add up all of the numbers in the regions we have so far: $7 + 4 + 6 + 3 + 9 + 5 + 2$ equals 36 students.

TONY: But wait. There are supposed to be 46 students, because that's what the problem said. We're off by 10. We made a mistake.

MARIA: No we didn't. This just leaves 10 students $(46 - 36)$ who don't like any of these three types of music. So put 10 in region H.

TONY: Okay, I see. Now we can answer the question. So, there are 10 students who do not like any of these types of music. Maybe they just like musicals.

Here is the final diagram:

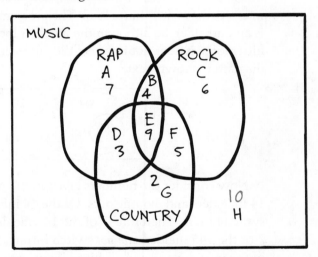

A manager of a baseball team looked over his roster at the beginning of spring training. He noticed the following facts: Every outfielder is a switch hitter. Half of all infielders are switch hitters. Half of all switch hitters are outfielders. There are 14 infielders and 8 outfielders. No infielder is an outfielder.

How many switch hitters are neither outfielders nor infielders? Work this problem before continuing.

This problem is quite different from those you've been working on so far. What the Venn diagram looks like is not obvious from reading the problem.

Start with the outfielders and the switch hitters. The problem states that all outfielders are switch hitters. This means that the outfielder loop must be completely contained within the switch-hitter loop. So draw two loops, with the outfielder loop inside the switch-hitter loop as shown in the diagram below. In the diagram, abbreviate switch hitters as SH, outfielders as OF, and infielders as IF.

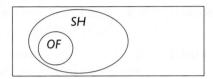

Now consider the infielders. The problem states that half of all infielders are switch hitters. This means that the infielder loop must intersect the switch-hitter loop. The problem also states that no infielder is an outfielder. So although the infielder loop intersects the switch-hitter loop, it does not intersect the outfielder loop. The expanded diagram is shown next.

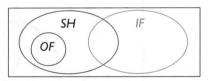

Now consider the numbers given in the problem. It says there are 14 infielders and 8 outfielders. Of the 14 infielders, half of them are switch hitters. Therefore, 7 of the 14 infielders *are* switch hitters, and 7 of the 14 infielders *are not* switch hitters. So put 7 in the intersection of the infielder and switch-hitter loops, and put 7 in the infielders-only part of the infielder loop. Also put 8 in the outfielder loop, because the problem states that there are 8 outfielders.

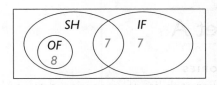

Finally, consider the statement "Half of all switch hitters are outfielders." Because the 8 outfielders represent half of all the switch hitters, there must be 16 switch hitters. So far, the diagram shows 8 outfielders (all of whom are switch hitters) and 7 infielders who are switch hitters. These total 15 switch hitters, so there must be one more switch hitter. This switch hitter is neither an outfielder nor an infielder (maybe he is a catcher), so he must go outside the outfielder and infielder loops but inside the switch-hitter loop. Here is the final diagram:

$$\frac{1}{2}(SH) = OF$$

$$\frac{1}{2}(SH) = 8$$

$$SH = 16$$

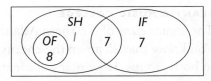

Now answer the question posed by the problem. How many switch hitters are neither outfielders nor infielders? A quick look at the diagram shows the answer to be 1.

Draw a Venn Diagram

Venn diagrams deal with overlapping sets.

You can't use Venn diagrams on very many problems, but they are effective to use with problems that involve overlapping sets. The diagram organizes information spatially and enables you to consider many different pieces of information at the same time. Venn diagrams are usually quick and easy to use and can provide the organization needed to solve problems that are very confusing when attempted with another strategy.

Problem Set A

1. MORE CATEGORIES

For each list of categories, draw a Venn diagram that shows the relationships among the categories. One of the categories describes the universal set, and the others describe the various loops inside the universal set. The universal set is not necessarily listed first.

1. Trumpets, pianos, musical instruments, clarinets, violins, trombones, brass, woodwinds

2. Water vessels, submarines, war boats, sailboats, battleships, ferries

3. Old cows, dairy cows, cows, cows wearing bells

2. NORTHERN ATLANTIC MUSIC COLLEGE

The senior class of Northern Atlantic Music College has 100 students. You know that 25 are in the chorus, 40 are in the orchestra, and 10 are in both groups. How many students are not members of either group?

3. HAMBURGERS AND HOT DOGS

Of 900 people interviewed, 254 said they liked only hamburgers, 461 said they liked only hot dogs, and 140 people said they liked both hamburgers and hot dogs. How many of the people interviewed didn't like either?

4. ROCK BAND

In a third-rate rock band, three members play guitar, four sing, and two do both. Six have no talent for singing or guitar, so they do something else. How many members are in the rock band?

5. TEACHER CANDIDATES

At State U, 43 students are working toward K–8 teaching credentials in the new undergraduate subject-area-emphasis program. Students can choose to emphasize one or two subjects: 18 students are taking the math emphasis courses, 15 are emphasizing science, and 6 are taking the math emphasis but not the science emphasis. How many are emphasizing neither math nor science?

6. EATING VEGETABLES

The staff at Tiny Little Cherubs (abbreviated TLC) day-care center observed the eating habits of their 64 students during several lunches. They observed that 59 children ate green beans, 56 ate cauliflower, 60 ate broccoli, 55 ate green beans and cauliflower, 54 ate cauliflower and broccoli, 56 ate green beans and broccoli, and 53 ate all three. How many children did not eat any of these three types of vegetables? How many children ate green beans but not cauliflower? How many children did not eat broccoli? How many children ate only cauliflower? How many children ate exactly two of these three vegetables?

7. THE FIELD TRIPS

The Silversnake class, the Jellyfish class, and the Radical Dog-Star class (kindergartners and first- and second-graders) were talking about their favorite field trips during the school year. One of the teachers, Ms. Burke, turned the discussion into a math lesson, and the students conducted a survey. Each child wrote down on paper which trip was his or her favorite. (Note that many children named more than one field trip as their favorite.) The survey revealed that 52 wrote down the trip to the river, 50 indicated the trip to the police station, and 44 included the trip to the hardware store. The police station and the river were chosen on 19 papers, 32 papers included both the river and the hardware store, and 25 children wrote down the police station and the hardware store on their papers. Ms. Burke counted 17 papers that included all three and one that did not list any of the three trips. How many children were surveyed? How many children wrote down the river but not the police station? How many children did not list the hardware store?

8. NO PETS ALLOWED

The new owner of a student apartment building near campus was considering establishing a No Pets rule. He decided to survey the current residents to see how many owned pets and what kind of pets they owned. The building had 58 units. The previous owner had not allowed cats or dogs or other mammals, but the residents reported a variety of other pets: 30 owned birds; 38 owned fish; and 30 owned frogs, lizards, or snakes. Of those who had fish, 19 also had a frog, lizard, or snake. The owner found that 18 had both fish and birds and 15 kept a bird and a frog, lizard, or snake. Lastly, he determined that 10 owned fish, a bird, and a frog, lizard, or snake. How many residents could have reported no pets? How many residents owned a bird but not a fish? How many residents do not own a bird?

9. SPORTSMEN

Most of the men of Delta Omicron Gamma play team sports, mostly on intramural teams but several on intercollegiate teams. Of all the members, 18 play basketball, 20 play baseball, and 23 play football. Only 3 play all three sports; 9 play baseball and basketball, 8 play baseball and football, and 10 play basketball and football. The other 14 members are involved in track, tennis, or swimming. How many men are in the DOG house? How many men play baseball only? How many men play football and baseball but do not play basketball?

10. GAMING SYSTEMS

A group of people were surveyed to find out what kind of gaming system they had. Forty-four people had an Xbox. Twenty-eight had a PS3. Thirty-one had a Wii. Eleven had both an Xbox and a PS3. Five had a Wii and a PS3. Eight had an Xbox and a Wii. And 2 people had all three gaming systems. Seven people did not have any of the three gaming systems. How many people were surveyed? How many people didn't have an Xbox? How many people had a PS3 but didn't have a Wii? How many people had an Xbox and a Wii, but didn't have a PS3? How many people had a PS3 and nothing else?

11. FAMILY REUNION

At one family reunion, every niece was a cousin. Half of all aunts were cousins. Half of all cousins were nieces. There were 50 aunts and 30 nieces. No aunt was a niece. How many cousins were neither nieces nor aunts?

12. JUST WHAT ARE THESE THINGS, ANYWAY?

All DERFs are ENAJs. One-third of all ENAJs are DERFs. Half of all SIVADs are ENAJs. One SIVAD is a DERF. Eight SIVADs are ENAJs. There are 90 ENAJs. Draw the Venn diagram. How many ENAJs are neither DERFs nor SIVADs?

This problem was written by Luther Burbank High school student James Davis.

13. MUSIC TASTES

A group of students were surveyed about their taste in music. They were asked how they liked three types of music: rock, country western, and jazz. The results are summarized below.

No one dislikes all three types of music. Six like all three types of music. Eleven don't like country western. Sixteen like jazz. Three like country western and jazz, but not rock. Six don't like rock. Eight like rock and country western. Seven don't like jazz. How many students were surveyed?

14. VIDEO GAMES AND SWIMMING

In a group of 14 boys and 14 girls, the number of boys who like video games but don't like swimming is the same as the number of girls who like swimming but don't like video games. The number of girls who like video games and swimming is the same as the number of boys who like swimming but don't like video games. The number of girls who don't like either video games or swimming is one less than the number of boys who don't like either video games or swimming. Twice as many girls as boys like video games but not swimming. The number of boys who like video games is one more than the number of girls who like video games. Five boys like video games and swimming. How many girls don't like either video games or swimming?

Influenza is a serious disease causing thousands of deaths each year, mostly among the elderly. Influenza symptoms include fever, cough, sore throat, headache, and muscle aches. The Centers for Disease Control and Prevention (CDC) recommends that people in the following groups receive an annual flu shot:

- Seniors, defined by the CDC as everyone 50 years of age or older

- Anyone with a chronic disease or long-term health problem, for example, asthma, heart disease, kidney disease, or diabetes

- People considered "high risk" because of their professions, such as doctors, health-care workers, students, or teachers

Hospitals and other health agencies sponsor annual flu vaccination clinics. One agency reported the following data for all the people vaccinated during one year:

- 98 people who were vaccinated were seniors.

- 80 people reported a chronic disease or long-term health problem, such as heart or kidney disease.

- 6 of the seniors who reported a chronic disease were still working in their high-risk professions.

- Of those who reported having a chronic disease, 52 were seniors.

- Of the 60 people who worked in high-risk professions, 18 were older than 50.

- 8 of the people vaccinated were younger than 50 and did not work in high-risk professions but reported having a chronic disease.

a. How many people, total, received flu shots?

b. How many people in high-risk professions also had a chronic medical condition?

c. What percentage of seniors did not report a chronic medical condition?

This problem was written by Sierra College student and health-care worker Leah Hall.

16. BLOOD LINES

Human blood is classified O, A, B, or AB, depending on whether the blood contains no antigen, an A antigen, a B antigen, or both the A and B antigens. A third antigen, called the Rh antigen, is important to human reproduction. Blood is said to be Rh-positive if it contains this Rh antigen; otherwise, the blood is Rh-negative. For instance, blood is type A+ if it contains the A antigen and the Rh antigen. Blood is type AB− if it contains the A and B antigens but does not contain the Rh antigen.

In a hospital the following data were recorded:

- 22 people were either type A or type AB, 16 of whom had the Rh antigen.

- 27 were either type B or type AB, 18 of whom had the Rh antigen.

- 8 were type AB, 2 of whom were AB−.

- 35 were type O, 5 of whom were O−.

How many patients are listed here? How many patients have type B+ blood? How many patients have type A− blood? How many have exactly two antigens?

17. CAMPUS LIFE

A group of 120 university students was surveyed about whether they had a job, whether they were carrying a full load of courses, and whether they lived on campus. These were the results:

- 49 had a job.

- 52 lived off campus.

- 44 lived on campus and carried a full load but did not have a job.

- 17 had a job and carried a full load.

- 23 did not carry a full load and did not have a job.

- 29 lived off campus, had a job, and did not carry a full load.

- 11 lived off campus, didn't have a job, and didn't carry a full load.

How many students lived on campus, had a job, and carried a full load? How many students lived off campus, did not have a job, and carried a full load?

18. WRITE YOUR OWN

Write your own Venn diagram problem. It's easiest to create Venn diagrams that contain just two circles, but if you're feeling ambitious, try a diagram with three circles. Draw the diagram first and fill in all the numbers. Then create the clues.

Here's an example:

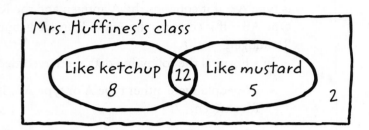

This diagram will serve as the basis for the problem Ben is writing. He notes the following clues:

1. There are 27 students in Mrs. Huffines's class.

2. Eight students like ketchup but do not like mustard.

3. Five students like mustard but do not like ketchup.

4. Twelve students like both ketchup and mustard.

5. Two students like neither ketchup nor mustard.

At this point, Ben can finish the problem by erasing one of the clues. Ben chose to erase clue 4. He then wrote a question based on the clue he crossed out: "How many students like both ketchup and mustard?"

You don't need to follow Ben's example for creating a Venn diagram problem, but if you're struggling with creating your own problem, give his example some consideration.

MORE PRACTICE

1. COLOR MY WORLD

Miss Barker surveyed her fifty preschoolers at Color My World Academy to find out their favorite colors. She asked that each student raise their hand for their favorite color. She found that some preschoolers cannot make up their mind and voted more than once. Fourteen children raised their hand for blue, 20 for yellow, and 25 for green. Five children raised their hand for blue and for yellow, and 3 of those five also voted for green. Five chose yellow and green but not blue, and nobody liked blue and green and not yellow. How many students liked only yellow as their favorite color? How many did not like any of the given colors?

This problem was written by University of Wisconsin – Eau Claire students Alyssa Haugen, Haley Haus, and Erica Johnson.

2. BEAUTY SALON

Anna runs a beauty salon. One day 27 people got haircuts, 21 people got facials, and 18 people got manicures. Eight people got haircuts and facials, 9 people got facials and manicures, and 11 people got haircuts and manicures. Six people got haircuts, facials, and manicures, and 8 people had something else done other than a haircut, facial, or manicure. How many people were in the salon that day? How many people did not get a haircut? How many people got a manicure and a haircut, but did not get a facial? How many people didn't get either a facial or a haircut?

3. MAGAZINE SURVEY

A group of 120 people were surveyed about what magazines they subscribed to. All but a few of them subscribed to Entertainment Weekly, People, and/or Sports Illustrated. It was discovered that 53 people subscribed to *Entertainment Weekly*, 65 to *People*, 35 to *Sports Illustrated*, 23 to *Entertainment Weekly* and *People*, 11 to *People* and *Sports Illustrated*, 16 to both *Entertainment Weekly* and *Sports Illustrated*, and 9 subscribed to all three magazines. How many people subscribed to none of the three magazines? How many subscribed to *Entertainment Weekly* but not *People*? How many subscribed to *People* and at least one of the other two magazines? How many people subscribed to only one of the three magazines?

4. SCHOOL'S OUT

At the end of the semester all of the students in the math class decided to go out to a club to celebrate. They found dancing, pool tables, and a snack bar at the club. Nineteen people danced, 17 went to the snack bar, and 18 played pool. Twelve danced and went to the snack bar, 11 danced and played pool, 10 went to the snack bar and played pool, and 7 did all three things. There were 2 people who just sat at a table and pouted because they didn't pass the class. How many people were in the math class? How many people played pool but didn't dance? How many people danced and went to the snack bar, but didn't play pool? How many people didn't go to the snack bar? How many people only danced and nothing else?

This problem was written by Sierra College student Reina Alvarez.

5. CUP OF JOE

A group of college professors were surveyed as to what they drank in the morning. A total of 23 people drink coffee, 8 drink milk, and 17 drink orange juice. Eleven drink coffee and OJ, 5 drink coffee and milk, and 4 drink milk and OJ. Three drink all 3 drinks, and 4 refused to say what they drink but did admit that they did not drink coffee or milk or OJ. How many people were surveyed? How many don't drink coffee? How many drink milk and coffee but not OJ?

This problem was written by Sierra College student Josh Bridges.

Problem Set B

A Family Holiday

1. HOLIDAY PASTRIES

Every year, as part of the Family family holiday tradition, the whole family bakes something to give to their friends and relatives. This year they decided to make some special pastries. Mama Family was very careful to make sure they baked plenty so there would be plenty to give away. If there was anything the family liked better than baking, it was eating. So during the night, Papa got up and ate one-third of the pastries. He noticed his dog looking very hungry, so he gave 1 pastry to his dog. Ed was the next to get up. He planned to eat one-fourth of the pastries, but when he separated them into four piles, he found there were 3 pastries left over. He fed those 3 to his dog and then ate one-fourth of the remaining pastries. Lisa then got up and ate one-third of the remaining pastries. As she was heading back to bed, her dog whined, so she gave it 2 pastries. Judy was the last to get up, and she consumed 8 pastries but then gave one-eighth of those remaining to her dog. When Mama went into the kitchen the next morning, she found crumbs, ants, and only 14 pastries. Mama sighed, realizing that they would have no pastries to give away again this year. She ate 2 pastries, gave 2 to her dog, and then split the remaining pastries evenly for her whole human family, including herself, to have for breakfast. How many pastries did the family bake, and how many pastries did each person consume?

PRESENTS! OH BOY!

The Family family had two of the kids' cousins, Gail and Keith, over for holiday gift-giving. The Family kids are, of course, named Ed, Lisa, and Judy. The Family parents bought presents for all five kids, but they forgot to buy wrapping paper. Consequently, the gifts were wrapped in strange ways, although every gift was wrapped in something different. One of the gifts was wrapped in a towel.

Interestingly, the kids' heights were exactly 3 inches apart—their heights were 59, 62, 65, 68, and 71 inches.

From the following clues, determine the height of each child, the gift received, and the strange way in which it was wrapped.

1. The child who received a set of Legos is 3 inches shorter than the child who received a gift wrapped in a pillowcase.

2. The child who received shoes is not 59 inches tall but is shorter than the child who received the picture.

3. Keith is shorter than the child who got the belt and one other child, but he is taller than both Gail and the child whose gift was wrapped in a plastic bag.

4. Ed is taller than the child whose present was packaged in a box.

5. The child who received a gift wrapped in newspaper is shorter than both Lisa and the child who received candy.

6. The child whose gift was enclosed in a pillowcase is not in the same immediate family as the child who received the belt.

3. **LEGO MY PYRAMID**

Of all the presents, the biggest hit was the Legos. None of the kids had played with Legos for years, and they all took turns building things. One of the really great structures built was a sort of pyramid made out of square Legos. It began with a single two-by-two square Lego. Then four squares were attached underneath it, which left a strip one bump wide showing all the way around the outside. This pattern was continued for a total of ten layers. Although the children could have built the pyramid by just putting Legos on the perimeter of each layer, they filled in each layer on the underside of the pyramid. How many bumps were showing, and how many Legos were used to make the pyramid?

4. THE HOLIDAY PARTY

After opening up all the presents, the Family family went to Grandma and Grandpa's house for the big Family family holiday party. Judy, who was interested in statistics, went around asking questions of everyone and recording their responses. She collected the following data: There were 80 people present. Every nephew was a cousin. Half of all uncles were cousins. One-third of all cousins were nephews. There were 30 uncles and 16 nephews. One-sixth of all uncles were nephews. She herself was a cousin but not a nephew or an uncle.

Judy wanted to know two things: She wanted to know how many cousins were neither nephews nor uncles. She also wanted to know how many uncles were also cousins but not nephews. And finally, how many people were neither uncles nor cousins? Can you help her?

5. HOLIDAY DINNER

The whole Family clan went out to dinner. The menu read as follows:

First course: chicken vegetable soup; shrimp cocktail; fruit salad

Second course: shrimp; vegetable casserole; beef; chicken

Third course: apple pie; carrot cake; ice cream

Papa wants everyone to order a different meal. (Note that soup, beef, and pie would be different from soup, beef, and cake.) Papa wants to know if there are enough different meals available that consist of one item from each course. However, he does not want any of the meals to include the same kind of food more than once during the meal. For example, if Ed ordered shrimp cocktail, then he could not order shrimp for the second course. You may assume that the items containing vegetables contain carrots and that the fruit salad contains apples. How many such meals are possible?

13

Convert to Algebra

Representing a problem algebraically allows you to use the tools of algebra to solve problems more efficiently. Combining the-guess-and-check strategy with algebra provides both problem-solving power and efficiency. Ice climbers use various tools, such as crampons and ice axes, to scale the sides of mountains. Using the tools together rather than separately makes a difficult climb much easier.

S o far you haven't needed much **algebra** to solve the problems in this book. This chapter focuses on algebra as a tool for problem solving. We saved this chapter for the later part of the book for these reasons:

- If you have taken any typical high school mathematics class, you are already familiar with algebra. We postponed having you solve problems algebraically because we wanted you to strengthen your confidence in your problem-solving abilities by using several strategies that didn't require algebra.

- In many texts, when algebra is used as a problem-solving strategy it is typically featured as part of a five-step problem-solving process: Read the problem, choose a variable, write the equation, solve the equation, and check the solution. This process gives the impression that writing an equation is an easy transitional step. It's not. Representing the problem with an equation is the hardest part of the process. Students often say, "I can solve the equation once I have it, but I can never set it up." The skills you developed in the earlier chapters of this book—drawing a diagram, solving subproblems, using guess-and-check, and so on—will help you build algebraic equations. Once you have written the equations, algebra is an efficient tool for solving the problem.

- As you advance in the study of mathematics, physical representations are impractical. Therefore, you need to become increasingly capable of representing problems with algebraic equations. This is a necessary skill that was probably overlooked when you were learning algebra.

- Algebra is not always the best problem-solving strategy. Algebra is only an effective tool when you can represent a relationship in a problem algebraically. In fact, many problems can be solved much more effectively when you use some other problem-solving strategy in conjunction with algebra.

Algebra is a language that can help you organize and communicate information. Like many other tools, it reaches its peak power when it's used in conjunction with complementary tools. This chapter focuses on using algebra in conjunction with the problem-solving strategies you've already learned in this book. We assume that you already have a working knowledge of how to solve equations.

We will revisit many problems from previous chapters, especially Chapter 6: Guess and Check. In that chapter we mentioned that an organized use of guess-and-check helps show relationships that can lead to equations that represent the problems. Once a problem is represented with equations, algebra provides a more efficient means to finding a solution than does guessing.

SATURDAY AT THE FIVE-AND-DIME GARAGE SALE

Cinci held a garage sale, during which she charged a dime for everything but accepted a nickel if the buyer bargained well. At the end of the day she realized she had sold all 12 items and had a total of 12 nickels and dimes. She had raked in a grand total of 95 cents. How many of each coin did she have?

See if you can re-create the guess-and-check chart for this problem without looking back at Chapter 6. Then make your guess a variable and try to write an equation.

Dimes	Nickels	Value of Dimes	Value of Nickels	Total Value	Rating
5	7	$0.50	$0.35	$0.85	low
8	4	$0.80	$0.20	$1.00	high

The chart ends before the right answer is reached. Now the objective is to set up an equation. Jana and Reina each solved this problem algebraically after they set up a guess-and-check chart.

Here's Jana's reasoning for using algebra instead of guess-and-check to solve the problem: "Sometimes guess-and-check takes too long. I'm usually pretty good at it, but I tend to get bored. So I made a couple of guesses, and then I tried to write an equation. Because I was guessing the number of dimes, I decided to call the number of dimes *d*. So I wrote down *d* in the Dimes column under the 8."

Dimes	Nickels	Value of Dimes	Value of Nickels	Total Value	Rating
5	7	$0.50	$0.35	$0.85	low
8	4	$0.80	$0.20	$1.00	high
d					

"Then I tried to figure out what to put in the Nickels column. My guesses in the Dimes column and the Nickels column have to add up to 12 because Cinci had a total of 12 coins, so the difference between 12 and *d* should be the number of nickels. I wrote down $d - 12$ in the Nickels column. My friend Denise pointed out that $d - 12$ was wrong. For instance, if $d = 8$, then $d - 12$ would be -4 and that wouldn't make sense. She said it should be $12 - d$. I checked that with the two guesses I had already made, and it made sense because $12 - 8$ is 4 and $12 - 5$ is 7. So I put $12 - d$ in the Nickels column."

Dimes	Nickels	Value of Dimes	Value of Nickels	Total Value	Rating
5	7	$0.50	$0.35	$0.85	low
8	4	$0.80	$0.20	$1.00	high
d	$12 - d$				

"Again I looked back at my guesses. Where did I get the amount of money in the Value of Dimes and Value of Nickels columns? I realized right away that they came from the number of dimes times 10 cents and the number of nickels times 5 cents. So all I had to do was multiply *d* times 10 cents and $12 - d$ times 5 cents. So I did that.

"It seemed like things were working really well so far. Finally, I figured out that my Total Value column came from adding the Value of Dimes and Value of Nickels columns. So I wrote that down."

Dimes	Nickels	Value of Dimes	Value of Nickels	Total Value	Rating
5	7	$0.50	$0.35	$0.85	low
8	4	$0.80	$0.20	$1.00	high
d	$12 - d$	$(0.10)d$	$(0.05)(12 - d)$	$0.10d + (0.05)(12 - d)$	

"Then I realized that this total value was supposed to be 95 cents. So I set the expression in the Total Value column equal to 95, and I had an equation.

$$0.10d + 0.05(12 - d) = 0.95$$
$$0.10d + 0.60 - 0.05d = 0.95$$
$$0.05d + 0.60 = 0.95$$
$$0.05d = 0.35$$
$$d = 7$$

Reread the question to make sure you answer it completely.

"It's funny, but when I used to solve problems in algebra class, I would always write down the value of the variable as the answer to the problem. So I would write down $d = 7$ as the answer. But guess-and-check really helps me answer the question asked in this problem: 'How many of each coin did she have?' I'd only figured out dimes so far, so I went back to my chart and figured out that the number of nickels she had was five because $12 - 7 = 5$. The answer is that she has seven dimes and five nickels."

Reina solved this problem in a similar way, but she decided to use two variables instead of one. Where Jana used $12 - d$ to represent nickels, Reina used n to represent nickels. She used cents for the units. Her chart looks like this:

Dimes	Nickels	Value of Dimes	Value of Nickels	Total Value	Rating
5	7	50	35	85	too low
8	4	80	20	100	too high
d	n	$10d$	$5n$	$10d + 5n = 95$	

Reina came up with the equation $10d + 5n = 95$. But she remembered that with two variables you need two equations. What is her other equation? Reina looked back at the problem and saw that the total number of coins was 12, but she had not used that in her equation. Then she wrote $d + n = 12$. Solving:

$$
\begin{array}{lllll}
d + n = 12 & \Rightarrow & -5d + -5n = -60 & & d + n = 12 \\
10d + 5n = 95 & \Rightarrow & \underline{10d + 5n = 95} & & 7 + n = 12 \\
& & 5d = 35 & & n = 5 \\
& & d = 7 &
\end{array}
$$

Again, the answer to the problem is seven dimes and five nickels.

FARMER JONES

Farmer Jones raises ducks and cows. She tries not to clutter her mind with too many details, but she does think it's important to remember how many animals she has and how many feet those animals have. She thinks she remembers having 54 animals with 122 feet. How many of each type of animal does Farmer Jones have? Again, re-create the guess-and-check chart for this problem. Then set up the equation or equations to solve the problem algebraically.

Imogen used the guess-and-check chart shown below. She guessed the number of ducks. To find the number of cows, she simply subtracted the number of ducks from 54 because there were 54 animals altogether. She easily figured out the number of feet for each kind of animal and checked to see whether there were 122 feet.

Ducks	Duck Feet	Cows	Cow Feet	Total Feet	Check
20	40	34	136	176	high
40	80	14	56	136	high
50	100	4	16	116	low
d	$2d$	$54 - d$	$4(54 - d)$	$2d + 4(54 - d)$	

Imogen was checking the Total Feet column, and she knew the total feet had to be 122 feet, so the equation was right there for the taking: $2d + 4(54 - d) = 122$. (Remember that when you set up an equation, you need to find two things that are supposed to equal each other.)

If two expressions have equal value, then they can be written as an equation.

$$2d + 4(54 - d) = 122$$
$$2d + 216 - 4d = 122$$
$$-2d + 216 = 122$$
$$-2d = -94$$
$$d = 47$$

Therefore, there are 47 ducks and 7 cows. She found the number of cows by subtracting the 47 ducks from 54 animals.

When you move from guess-and-check to algebra to solve a problem, you use exactly the same operations as you did in the guess-and-check version, but you use a variable instead of a specific number. In the Farmer Jones problem, you got from ducks to duck feet simply by multiplying by 2. The first guess was 20 ducks, so the number of duck feet was 2×20 or 40. This may seem like a very simple procedure, but some students have a difficult time translating the arithmetic of the guess-and-check language into the symbols of algebraic language. To move from one language to the other, ask the question "What did I do to get this number?" Whatever the answer is, the procedure for the arithmetic guess and the algebraic guess is the same. Because the arithmetic procedure was to multiply by 2, the algebraic procedure is also to multiply by 2. So, to move from ducks, *d*, to duck feet, you multiply *d* by 2.

Once you have the algebraic language you want, the next thing to do is write the equation. An **equation** is simply two **expressions** connected with an equals sign. If the two expressions are supposed to be equal, then you have a valid equation. In most problems, the equation jumps out of the last or next-to-last column of the guess-and-check chart. In the Farmer Jones problem, the equation came from setting the algebraic expression from the second to last column, Total Feet, equal to 122, the known total number of animal feet.

In the last line of the guess-and-check table we used *d* as a variable to represent all the possible guesses for the number of ducks. So the numbers *d* represented a variable because its value varied. We could have guessed any number. When we use a variable in an equation, as in $2d + 4(54 - d) = 122$, we often refer to it as an unknown, because in the equation, it represents a limited number of unknown values (often just one) that make the equation valid. Then we solve the equation to discover the unknown number(s).

ALL AROUND THE PLAYING FIELD

The perimeter of a rectangular playing field measures 504 yards. Its length is 6 yards shorter than twice its width. What is its area? Set up the guess-and-check chart for this problem, and then write the equation.

Sheri's chart is shown below.

Width	Length	Perimeter	
100	194	588	High
60	114	348	Low
80	154	468	Low

Sheri was guessing the width, so she decided to let that be her variable, and she called it w. Now, how did she get the length? "The problem says that the length is 6 yards shorter than twice the width. For instance, $100 \times 2 = 200$ would be twice the width. The length is 6 yards shorter than that, or $200 - 6 = 194$. If I'm letting w stand for width, then length must be $2w - 6$. The perimeter is twice the sum of the length and width, so it's 2 times the quantity $w + 2w - 6$, and it's supposed to be 504 yards."

Width	Length	Perimeter	
100	194	588	High
60	114	348	Low
80	154	468	Low
w	$2w - 6$	$2(w + 2w - 6) = 504$	

Now w is the unknown in the equation, and solving for w is a routine process.

$$2(w + 2w - 6) = 504$$
$$2(3w - 6) = 504$$
$$6w - 12 = 504$$
$$6w = 516$$
$$w = 86$$

So the length is 2(86) − 6, which equals 166. Check the perimeter: 86 + 166 = 252, and 252 × 2 = 504. The length is 166 yards, and the width is 86 yards. But the question asked for the area. The area of the field is 166 yards × 86 yards, or 14,276 square yards.

ORIGAMI

A group of exchange students from Japan went to a convalescent home to sing songs for the seniors and to demonstrate origami, the art of Japanese paper folding. Groups of students and seniors sat together at tables so the students could teach the seniors to fold origami models. As it turned out, at each table there was either 1 student at a table with 3 seniors, or 2 students at a table with 4 seniors. There were 23 students and 61 seniors in all. How many tables were being used? Work this problem before continuing.

Manmeet solved this problem. "I started out using guess-and-check because the problem really confused me. I guessed the number of each kind of table (T) and then figured out how many students (ST) and how many seniors (SN) were at each one. I checked by figuring out the total number of students and seniors and then comparing the totals to 23 and 61. I used the notation 1|3 to stand for tables with 1 student and 3 seniors, and 2|4 for the other kind."

1|3 tables			2|4 tables			Total ST 23	SN 61	Rating ST-SN
T	ST	SN	T	ST	SN			
10	10	30	10	20	40	30	70	high-high
5	5	15	5	10	20	15	35	low-low
8	8	24	8	16	32	24	56	high-low
5	5	15	9	18	36	23	51	right-low
5	5	15	10	20	40	25	55	high-low

"At this point, I was really frustrated. I was guessing two things, but I wasn't sure how to get closer to the right answer. I also thought that my first two guesses meant that the right answer was somewhere between 10 and 5 for each kind of table. But that didn't seem to work either. So I stopped using guess-and-check and switched to algebra. I called the number of 1|3 tables x and the number of 2|4 tables y. Then I just did the same thing to x and y that I'd done to the number guesses."

| 1\|3 tables | | | 2\|4 tables | | | Total | | Rating |
T	ST	SN	T	ST	SN	ST 23	SN 61	ST-SN
10	10	30	10	20	40	30	70	high-high
5	5	15	5	10	20	15	35	low-low
8	8	24	8	16	32	24	56	high-low
5	5	15	9	18	36	23	51	right-low
5	5	15	10	20	40	25	55	high-low
x	x	$3x$	y	$2y$	$4y$	$x + 2y$	$3x + 4y$	

"I knew that the total number of students had to be 23 and the total number of seniors had to be 61, so I just wrote two equations and solved them."

$$x + 2y = 23 \qquad \rightarrow \qquad -2x - 4y = -46$$
$$3x + 4y = 61 \qquad \rightarrow \qquad \underline{3x + 4y = 61}$$
$$x = 15$$

$$x + 2y = 23$$
$$15 + 2y = 23$$
$$2y = 8$$
$$y = 4$$

"There were 15 tables that held 1 student and 3 seniors, and 4 tables that held 2 students and 4 seniors. I checked this too. There are $15 \times 1 + 4 \times 2$ or 23 students and $15 \times 3 + 4 \times 4$ or 61 seniors, which works out."

Emi and Margit had stopped at the bottom of one of the highest waterfalls in Cascades State Park. As Emi looked up at the waterfall, she said, "Wow, I think the top of that fall is about 20 feet more than three times the height of that young redwood!" Margit, of course, had a different opinion. She said, "No, I think it's about 50 feet less than four times the height of the redwood." If both are approximately right, about how tall is the redwood and how high is the waterfall? Solve this problem by using a guess-and-check chart to set up the algebra before continuing.

Part of the guess-and-check chart from Chapter 6 is shown below. The first column—Height of the Redwood—was the number guessed.

Height of the Redwood	Emi's Estimate for the Waterfall	Margit's Estimate for the Waterfall	Difference Emi – Margit	Rating
50	170	150	20	low
30	110	70	40	lower
100	320	350	−30	high

To find an equation for this problem, add an algebraic guess. Since the number in the Height of the Redwood column is our list of guesses, the height of the redwood varies depending on what number we choose. We put the variable h in the first column to represent the height. Emi's estimate for the waterfall was 20 feet more than three times the redwood's height. Since the redwood's height is represented by h, then Emi's estimate of the waterfall is represented by $3h + 20$. Margit's estimate is represented by $4h - 50$.

Height of the Redwood	Emi's Estimate for the Waterfall	Margit's Estimate for the Waterfall	Difference Emi – Margit	Rating
50	170	150	20	low
30	110	70	40	lower
100	320	350	−30	high
h	$3h + 20$	$4h - 50$		

There were two ways to check this problem. One way is to find out where the two estimates were the same. In other words, when does column 2 equal column 3? That would lead to the equation $3h + 20 = 4h - 50$.

Solving this equation:

$$\begin{aligned} 3h + 20 &= 4h - 50 \\ 20 &= h - 50 \\ 70 &= h \end{aligned}$$

The second way to check the guess-and-check problem is to compute the difference column. In the table, subtract column 3 from column 2. In algebraic terms, this means $(3h + 20) - (4h - 50)$. Note that the parentheses are very important, because you must subtract entire expressions.

Height of the Redwood	Emi's Estimate for the Waterfall	Margit's Estimate for the Waterfall	Difference Emi − Margit	Rating
50	170	150	20	low
30	110	70	40	lower
100	320	350	-30	high
h	3h + 20	4h − 50	(3h + 20) − (4h − 50)	

In the guess-and-check chart, the problem was done when the difference column was zero. So the equation to solve this problem is $(3h + 20) - (4h - 50) = 0$.

Solving this equation is slightly different. Remember to distribute the minus sign.

$$\begin{aligned} (3h + 20) - (4h - 50) &= 0 \\ 3h + 20 - 4h - 50 &= 0 \\ -h + 70 &= 0 \\ -h &= -70 \\ h &= 70 \end{aligned}$$

In either equation, we get the answer $h = 70$. That means that the height of the redwood must be 70 feet. That means that both Emi's and Margit's estimates for the height of the waterfall are about 230 feet.

$$3(70) + 20 = 230 \quad \text{and} \quad 4(70) - 50 = 230$$

Whatever you do to check your guess becomes your equation. In this case there were two different checks you could have used, and thus two different equations. Both equations led to the same answer.

CHOCOLATE MILK

Augustus is trying to make chocolate milk. He has made a 10% chocolate milk solution (this means that the solution is 10% chocolate and 90% milk). He has also made a 25% chocolate milk solution. Unfortunately, the 10% solution is too weak, and the 25% solution is way too chocolaty. He has a whole lot of the 10% solution, but he has only 30 gallons of the 25% solution. How many gallons of 10% solution should he add to the 25% solution to make a mixture that is 15% chocolate? (Augustus is sure the 15% solution will be absolutely perfect.) Work this problem before continuing.

A portion of the guess-and-check and subproblems solution from Chapter 7 appears below.

Gallons of 10% Solution	Gallons of Choc in 10% Soln	Gallons of 25% Solution	Gallons of Choc in 25% Soln	Total Gallons of Choc	Total Gallons of Mix	% of Choc in Tot Mix	Rating
20 gal	2 gal	30 gal	7.5 gal	9.5 gal	50	0.19 = 19%	low
5 gal	0.5 gal	30 gal	7.5 gal	8 gal	35	0.229 = 22.9%	lower
30 gal	3 gal	30 gal	7.5 gal	10.5 gal	60	0.175 = 17.5%	low
50 gal	5 gal	30 gal	7.5 gal	12.5 gal	80	0.156 = 15.6%	low

Now we have to create the equation. In the guess-and-check chart, the guess appears in the first column: Gallons of 10% solution. Let's use g to represent the number of gallons of 10% solution. So put g in the first column. Now do to g what we did to the numbers. The second column was 10% of the first column. So the second column will be g times 10%. We know that 10% is equal to the decimal 0.10. Multiplied by g gives us 0.10g. And we can see in the guess-and-check chart that the third and fourth columns are always 30 and 7.5, respectively.

Gallons of 10% Solution	Gallons of Choc in 10% Soln	Gallons of 25% Solution	Gallons of Choc in 25% Soln	Total Gallons of Choc	Total Gallons of Mix	% of Choc in Tot Mix	Rating
20 gal	2 gal	30 gal	7.5 gal	9.5 gal	50	0.19 = 19%	low
5 gal	0.5 gal	30 gal	7.5 gal	8 gal	35	0.229 = 22.9%	lower
30 gal	3 gal	30 gal	7.5 gal	10.5 gal	60	0.175 = 17.5%	low
50 gal	5 gal	30 gal	7.5 gal	12.5 gal	80	0.156 = 15.6%	low
g	0.10g	30	7.5				

Now we need to figure out how to represent the next three columns. The fifth column—Total Gallons of Chocolate—was found earlier by adding the numbers in columns 2 and 4. For example, $2 + 7.5 = 9.5$ in the first guess. So add the second and fourth columns to get $0.10g + 7.5$. The sixth column is the sum of the first and third columns: $20 + 30 = 50$ in the first guess. So in the algebraic case, the sixth column will be $g + 30$.

Gallons of 10% Solution	Gallons of Choc in 10% Soln	Gallons of 25% Solution	Gallons of Choc in 25% Soln	Total Gallons of Choc	Total Gallons of Mix	% of Choc in Tot Mix	Rating
20 gal	2 gal	30 gal	7.5 gal	9.5 gal	50	0.19 = 19%	low
5 gal	0.5 gal	30 gal	7.5 gal	8 gal	35	0.229 = 22.9%	lower
30 gal	3 gal	30 gal	7.5 gal	10.5 gal	60	0.175 = 17.5%	low
50 gal	5 gal	30 gal	7.5 gal	12.5 gal	80	0.156 = 15.6%	low
g	0.10 g	30	7.5	$0.10g + 7.5$	$g + 30$		

Finally, what did we do to get the last column? In the first guess, we divided $9.5 \div 50$ to get 0.19. That's the fifth column divided by the sixth column. So in this case, it's $(0.10g + 7.5) \div (g + 30)$. It is a ratio of the two preceding columns. That is, the numerator is from the Total Gallons of Chocolate column, and the denominator is from the Total Gallons of Mixture column.

Gallons of 10% Solution	Gallons of Choc in 10% Soln	Gallons of 25% Solution	Gallons of Choc in 25% Soln	Total Gallons of Choc	Total Gallons of Mix	% of Choc in Tot Mix	Rating
20 gal	2 gal	30 gal	7.5 gal	9.5 gal	50	0.19 = 19%	low
5 gal	0.5 gal	30 gal	7.5 gal	8 gal	35	0.229 = 22.9%	lower
30 gal	3 gal	30 gal	7.5 gal	10.5 gal	60	0.175 = 17.5%	low
50 gal	5 gal	30 gal	7.5 gal	12.5 gal	80	0.156 = 15.6%	low
g	0.10 g	30	7.5	0.10g + 7.5	g + 30	$\dfrac{(0.10g + 7.5)}{(g + 30)}$	

We arrive at the algebraic expressions in the same way we arrived at the guess-and-check numbers. That is, we do the same operations with g that we did with all of the guesses. The final equation comes from the chart and from knowing that to get the percent of chocolate in the total mix we need to divide the total number of gallons of chocolate by the total number of gallons of mix to get 15%. So set the algebraic expression in the last column equal to 15%, which is actually 0.15. Then solve the equation for g to find the number of gallons of 10% solution added.

$$\frac{(0.1g + 7.5)}{(g + 30)} = 0.15 \qquad \text{then multiply both sides by } (g + 30).$$
$$0.1g + 7.5 = 0.15(g + 30) \quad \text{Distribute}$$
$$0.1g + 7.5 = 0.15g + 4.5$$
$$-0.05g = -3$$
$$g = \frac{-3}{-0.05}$$
$$g = 60 \text{ gallons}$$

So 60 gallons of the 10% mixture are required.

Note that if Augustus had wanted his mixture to be 14.65% chocolate, using guess-and-check would have taken a long time. Using algebra would have been as quick as it was when he was looking for a mixture of 15% chocolate. Guess-and-check is very useful for setting up an algebraic equation. You don't *have* to use guess-and-check to set up the algebra, but it isn't a bad idea and could save you a lot of grief. **If you are at all unsure of the equation you need, then set up a guess-and-check chart first.** Using guess-and-check can increase your understanding of a problem, making it easier to apply algebra later.

Here is another mixture problem on which to exercise your newfound skill.

SALT SOLUTION

A pet store sells salt water for fish tanks. Unfortunately, recently hired Flounder has mixed a salt solution that is too weak. He's made 150 pounds of 4% salt solution. The boss wants a 7% salt solution. Help Flounder out by giving him two options for reaching the 7% solution:

    a. Add some salt. How much?

    b. Evaporate some water. How much?

Work this problem before continuing.

Steve, Brenda, and Gil worked on this problem.

STEVE: Poor Flounder, this one sounds tough.

BRENDA: Come on, guys, we can do this. Let's set up a guess-and-check chart. What do we know, and what do we need?

GIL: We have 150 pounds of solution, and it is 4% salt. How many pounds of it is salt then?

STEVE: That sounds like a subproblem. Four percent of 150 is 6. So there are 6 pounds of salt and, therefore, 144 pounds of water in the original solution.

BRENDA: Which part should we do first?

GIL: It probably doesn't matter. Let's do part a. Here's my guess-and-check chart. I guessed the salt added. The first three columns never change.

Lbs Soln	Lbs Salt	Lbs Water	Salt Added	Total Salt	Total Soln	% Salt
150	6	144	1	7	151	$\frac{7}{151} = 4.6\%$
150	6	144	10	16	160	$\frac{16}{160} = 10\%$

BRENDA: I think I follow this. You guessed the salt added. Then you added that to the 6 pounds of salt you already had in the solution to get the Total Salt column. How did you get the Total Solution column?

STEVE: Oh, I get that. You're just pouring more into the solution, so you need to add how much you put in to how much was already there. He added the salt-added value to the 150 pounds of solution that was already there. That gave him how much total solution there was.

BRENDA: Okay, I get it. Then he divided the amount of salt by the total solution amount to get the percent salt. Of course, it comes out as a decimal, so he changed it to a percentage.

GIL: Right. And the correct answer has to be between 1 and 10 pounds of salt added.

BRENDA: I think we ought to use algebra. Guessing could take forever. It's probably a decimal answer anyway.

GIL: Okay, let's use algebra. How do we do that?

STEVE: Start by using a variable as a guess. Put x in the Salt Added column and do the same thing to x that you did to the numbers.

BRENDA: So the Total Salt column would be $x + 6$, the Total Solution column is $x + 150$, and the Percent Salt column is their ratio, just like with the numbers.

Lbs Soln	Lbs Salt	Lbs Water	Salt Added	Total Salt	Total Soln	% Salt
150	6	144	1	7	151	$\frac{7}{151} = 4.6\%$
150	6	144	10	16	160	$\frac{16}{160} = 10\%$
150	6	144	x	$x + 6$	$x + 150$	$\frac{(x + 6)}{(x + 150)}$

GIL: But where do we get the equation?

BRENDA: The Percent Salt column is supposed to be 7%, right? So set $\dfrac{(x+6)}{(x+150)}$ equal to 0.07. Then solve the equation.

GIL: Are you sure? This seems too easy for this kind of problem.

BRENDA: Try it. It works.

$$\frac{x+6}{x+150} = 0.07$$

$$x + 6 = 0.07(x + 150)$$
$$x + 6 = 0.07x + 10.5$$
$$0.93x = 4.5$$
$$x = \frac{4.5}{0.93}$$
$$x = 4.84 \text{ pounds (rounded)}$$

GIL: So he needs to add 4.84 pounds of salt. That's a lot of salt. Maybe he would be better off evaporating some water.

BRENDA: Well, let's try that one. Set it up by guess-and-check again. I'm just going to start with the third column, the 144 lbs of water, and not bother to write down the first two we used last time. I'll guess water evaporated.

lbs water	Water Subtracted	Water Total	Solution Total	% Salt
144 lbs	4 lbs	140 lbs	146 lbs	$\dfrac{6}{146} = 4.1\%$
144 lbs	20 lbs	124 lbs	130 lbs	$\dfrac{6}{130} = 4.6\%$

STEVE: Wow, this is taking forever. Let's try algebra. Put y in the Water Subtracted column.

lbs water	Water Subtracted	Water Total	Solution Total	% Salt
144 lbs	4 lbs	140 lbs	146 lbs	$\frac{6}{146} = 4.1\%$
144 lbs	20 lbs	124 lbs	130 lbs	$\frac{6}{130} = 4.6\%$
144	y	144 − y	150 − y	$\frac{6}{(150 - y)}$

BRENDA: We want $\dfrac{6}{(150 - y)}$ to be 7%, or 0.07.

$$\frac{6}{150 - y} = 0.07$$
$$6 = 0.07(150 - y)$$
$$6 = 10.5 - 0.07y$$
$$0.07y = 4.5$$
$$y = \frac{4.5}{0.07}$$
$$y = 64.29 \text{ pounds}$$

BRENDA: That's a lot of water to evaporate. Poor Flounder. It's going to be tough, whatever he does.

If the answer to the problem is not an integer, guess-and-check could literally take all day. Moving from guess-and-check to algebra in situations like these is the best way to solve the problem.

The next problem uses diagrams in conjunction with algebra.

A man 6 feet tall is walking away from a streetlight that is 15 feet tall. How long is the man's shadow when he is 10 feet away from the light?

Work this problem before continuing. Note that guess-and-check doesn't really help here, but a diagram helps a lot.

Suzanne drew the diagram shown below. "I drew the picture and saw where the shadow would be. I labeled the length of the shadow *x*."

Diagrams help you visualize a problem and set up an equation.

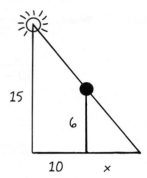

"It looked to me like the triangles were similar. The ratio of the height of the pole to the height of the man should be in the same proportion as what to what? Compare the big triangle to the small triangle: The vertical side of the big triangle and the vertical side of the small triangle should be in the same ratio as the base of the big triangle to the base of the small triangle. So I set up an equation. Solving it, I got the following for *x*."

$$\frac{15}{6} = \frac{10 + x}{x}$$

$$15x = 6(10 + x)$$
$$15x = 60 + 6x$$
$$9x = 60$$
$$x = \frac{60}{9} = \frac{20}{3}$$

"So the man's shadow is $\frac{20}{3}$ or $6\frac{2}{3}$ feet long."

Use Algebra

A guess-and-check chart often leads to an equation.

A diagram can also lead to an equation.

Algebraic equations can be solved using a set of routine procedures, which makes algebra an efficient problem-solving tool. The hardest part of using algebra as a problem-solving strategy is writing an equation to represent the problem.

The key to writing many algebraic equations is to start with a guess-and-check chart. A chart with a few checked guesses clarifies the relationships in the problem and can lead directly to an equation based on using a variable as the "guess."

Another problem-solving strategy that can lead to an equation is drawing a diagram. A diagram can reveal the relationship of the missing part to the rest of the problem, and then you can write an algebraic equation to represent this relationship.

Problem Set A

Directions: Write and solve an equation for each problem.

1. ALGEBRA THIS TIME

Go back to the problems from Problem Set A in Chapter 6 or to any Problem Set B problems you used guess-and-check to solve. Choose three of those problems, and write and solve equations for them.

2. MORE COINS

Juanita has $3.25 in nickels and dimes. She has eight more nickels than dimes. How many of each does she have?

3. SUPPLEMENTS

The measure of the larger of two **supplementary angles** is 6° more than twice the measure of the smaller of the two angles. What is the measure of each angle?

4. BIKE RIDE

Blaise rode his bike to his friend Elroy's house, which was 15 miles away. After he had been riding for half an hour, he got a flat tire. He walked his bike the rest of the way. The total trip took him 3 hours. If his walking rate was one-fourth as fast as his riding rate, how fast did he ride?

5. CHAMPIONSHIP GAME

A group of students was transported to the championship basketball game on buses and vans. When one bus and two vans unloaded, there were 55 students. A few minutes later, two more buses and one van unloaded 89 students. In all, three buses and eight vans drove students to the game. How many students went to the game?

6. FISHING POLES

Daniel and Gary are fishing. They each have several fishing poles, and each pole has several worms on its line. Daniel's poles each have 6 worms on their lines. Gary's poles each have 11 worms on their lines. Between the two of them, Daniel and Gary have 13 poles and 103 worms. How many poles does each boy have?

7. CAR WASH

Shea and Tucker are washing their father's car. Shea can wash it by herself in 20 minutes. Tucker can wash it by himself in 30 minutes. How long does it take them to wash the car if they work together?

8. INTEREST

Lakeitha earned $12,000 more in her accounting job this year because she received her CPA certification. She decided to invest the money and split it between two different savings accounts. One account was a certificate of deposit that paid 7.25% annual interest. The other was a money market account that paid 5.4% annual interest. At the end of 1 year, she had made $730. How much did she invest in the money market account?

9. CHEMISTRY

Dr. Loube, a chemist, mixes two solutions. One is 24% acid (the rest is water), and the other is 41% acid. About how much of each solution does she need to produce 50 gallons of a solution that is 31% acid? Answer to the nearest hundredth of a gallon.

10. TICKET PRICE INCREASE

With a student body card, tickets to the football game cost $1.00. Without a card, the price was $6. At the new computerized gates, each fan deposited his or her ticket to open the gate. As each gate reported, a running total was kept on a central computer. The computer was programmed to count the number of people who entered and keep a running total for the cost of the tickets. For the season opener, the total number of tickets was 12,438, and the total revenue was $36,798.

The university is pressuring the athletic director to raise the price of tickets. The athletic director needs to know how many tickets of each type were sold. Help her out.

11. LADDER

Stacy wants to put a ladder against the wall of her house and climb up to fix the gutter. The gutter is 14 feet above the ground. Unfortunately, there is an 8-foot-high retaining wall standing 3 feet away from her house. The ladder can't be placed between the retaining wall and the house, so it must be placed on the outside of the retaining wall. The ladder will go over the retaining wall (it can touch it) and then up to the roof. How long a ladder does she need?

12. MADAME XANADU'S CRYSTAL BALL

Madame Xanadu sets her crystal ball on a velvet pillow in the middle of a round table that sits in a corner of her apartment. One day, as she was watching TV, a small earthquake struck. Madame Xanadu, shocked that she had not predicted it, looked up to see her precious crystal ball rolling across the table toward the corner. The table has a diameter of 4 feet, and the crystal ball has a diameter of 10 inches. Should she fling herself across the room to stop the ball from reaching the edge of the table and falling? Predict the near future for her crystal ball.

13. STUPID NUMBER TRICK

Allyndreth showed Branwyn a number trick. She said, "Think of the number of times you bathed last week. Multiply by 2. Add 5. Multiply by 50. If you've already had your birthday this year, add 1,750. If you haven't already had your birthday this year, add 1,749. Add the last two digits of the current year—so if this year is 2012, you should add 12. Then subtract the four-digit year in which you were born." Branwyn did each step correctly and then told Allyndreth the answer: 722. Allyndreth said, "You bathed seven times last week and you are 22 years old." Branwyn was amazed, but Allyndreth told her it was just a stupid number trick. Try it for yourself. Then use algebra to figure out why it works.

14. TREADMILL

Mendy was working out on the treadmill. She worked out for 15 minutes. Part of the time she was walking at 3.3 miles per hour. Her heart rate during that time was 108 beats per minute. The rest of the time she was slowly jogging at 4.2 miles per hour. Her heart rate during that time was 129 beats per minute. During the 15 minutes, she covered 0.89 mile.

   a. How much time did she spend walking and how much time did she spend jogging?

   b. What was her average heart rate during the 15 minutes?

15. THOSE DARN BILLS KEEP PILING UP

Katinka's credit card bill in January was 1.5 times her food bill. Her apartment rent was 5 times her utilities bill. Her phone bill was $17 less than half of her food bill. Her phone bill was also $18 less than twice her utilities bill. The total of all five bills was $751. What was the amount of each bill?

16. WRITE YOUR OWN

Write your own algebra problem. Start in the same way as when you created your guess-and-check problem in Chapter 6, by first creating the situation and then making up the answer. Then write the rest of the problem. Consider creating a problem that has a decimal answer, such as a mixture problem.

CLASSIC PROBLEMS

17. BLENDING THE TEAS

A grocer buys two kinds of tea—one at 32¢ per pound and the other, a better quality, at 40¢ per pound. He mixes together some of each and proposes to sell the blend at 43¢ per pound, making a profit of 25% on the cost. How many pounds of each kind must he use to make a mixture that weighs 100 pounds?

Adapted from *536 Puzzles and Curious Problems* by Henry Dudeney, edited by Martin Gardner.

18. A DIESEL SHIP AND A SEAPLANE

A diesel ship leaves on a long voyage. When it is 180 miles from shore, a seaplane, whose speed is ten times that of the ship, is sent to deliver mail. How far from shore does the seaplane catch up with the ship?

Adapted from *The Moscow Puzzles* by Boris Kordemsky.

19. INVERNESS TO GLASGOW

In going from Inverness to Glasgow, a distance of 189 miles, I had a choice between looping the loops on a scenic railway or bumping the bumps on a lumbering old stagecoach. I selected the coach because it made the trip in 12 hours less time than did the train. From this fact I was able to jot down one of the most interesting puzzles of my globe-trotting tour.

My coach left Inverness at the same time that the train left Glasgow. When we met along the way, our distance from Inverness exceeded our distance from Glasgow by a number of miles that exactly equaled the number of hours we had been traveling. At the time we met the train, how far were we from Glasgow?

Adapted from *Mathematical Puzzles of Sam Loyd* Vol. 2, edited by Martin Gardner.

MORE PRACTICE

Directions: Write and solve an equation for each problem.

1. TRIANGLE

One side of a triangle is 47 cm, and another is 72 cm. The **perimeter** is five times the missing side. How long is the missing side of the triangle?

2. WAGONS AND TRIKES

Homer has a pile of broken down wagons and tricycles in the back corner of his yard. He's been collecting them for their wheels, which he uses to make scooters for kids. He knows there are a total of 22 trikes and wagons and that there are 74 wheels. How many trikes are there?

3. HOME FIELD ADVANTAGE

When the game started there were three times as many home team fans as away team fans in the stands, but then the buses pulled in with an additional 600 rival fans. Total attendance was 3,704. How many home team supporters were there?

4. ALL THE QUEENS' CHILDREN

Colonies of ants, bees, and wasps are similar in that each has a queen who lays all the eggs. On average, queen bees lay twice as many eggs per month as queen ants, and queen wasps lay 700 fewer than queen bees. The lab assistant, who was keeping track of the colonies, had lost his notes, and all he could remember was that the total for their two bee colonies, three ant colonies, and one wasp colony this month was 6,428 eggs. He needed to know the average for each queen. What is the average for each?

5. STATE INSECTS

Do you know your state insect? Most U.S. states have an official state insect. The three most popular are the monarch butterfly, the honeybee, and the ladybug. A total of 30 states have one of these as their official insect. Monarch butterflies and ladybugs are close to the same in popularity with only one more monarch butterfly state than ladybug state, but honeybees are just one less than three times as popular as ladybugs. How many states are honeybee states?

6. **EVERY LITTLE BIT COUNTS**

Angelita collects cans and bottles to take to the recycling center where she gets 7.5¢ for cans and 12¢ per bottle. Last week she got $11.91 for her total of 112 cans and bottles. How many cans and how many bottles were there?

7. **BUYING FRUIT**

Abigail went to Monterey Market and bought 7 mangos and 3 melons for $14.28. Her roommate didn't realize that Abigail had already shopped for fruit. She stopped at the same market on the way home from work and spent $16.05 on 4 melons and 5 mangos. What was the cost of a mango that day? What did a melon cost?

8. **A LOVELY BUNCH OF COCONUTS**

Coco Pine recently went to Hawaii on vacation and discovered how good fresh coconuts are. So she decided to buy some and take them home to the mainland to share with her friends. When she called her friend Shelly to tell her, Shelly asked Coco to buy some pineapples for her too. So Coco bought 19 pieces of fruit for $39. Pineapples cost $3 and coconuts cost half of that. When she got to the airport, she noticed a sign posted at the agriculture inspection station that stated that only 10 of each fruit could be taken out of the state. Did Coco have to leave any of her fruit behind?

This problem was written by Sierra College student Rachel Hamilton.

9. **HOURLY WORKERS**

Janeen works 20 hours a week for $24.92 less than Shari makes for working 12 hours a week. Shari makes one dollar less than twice the amount that Janeen makes per hour. How much does Janeen make per hour, and about how many hours per week would Janeen need to work to equal the amount that Shari makes in a week?

This problem was written by Sierra College student Shari Seigworth.

10. **CHARLOTTE TO CHARLESTON**

Noraly and Noemi drove from Charlotte to Charleston to visit their friend Nisha. On the way there they drove at an average speed of 57 miles per hour. On the way back, they went a slightly different way, which was 20 miles longer, but since they drove 67 miles per hour it took 15 minutes less time. What is the distance of the trip back and how long did it take?

At the local retirement home, a popular drink is Jim's juice, a mixture of cranberry juice and orange juice. Betty likes her mixture to be 30% cranberry juice and 70% orange juice. Her husband Don likes his mixture to be 60% cranberry juice and 40% orange juice. Ethel is very picky however, and she likes her mixture to be 53% cranberry and 47% orange. One day, Betty and Don visit Ethel's apartment. Betty brings 6 quarts of her mixture and Don brings 4 quarts of his mixture. Ethel decides to mix them together. She then wants to add either pure cranberry juice or pure orange juice to make the resulting mixture 53% cranberry and 47% orange. How much and which kind of juice does she need to add?

Problem Set B

1. VALLEY SPRINGS

Valley Springs Juicers bought two different kinds of grapes to make juice. The green grapes cost $500 per ton. The red grapes cost $200 per ton. The buyer ended up paying an average of $280 per ton for the grapes. What is the ratio of tons of green grapes to tons of red grapes?

2. ALL IN THE FAMILY

People of Hispanic heritage traditionally have two last names. The first one is from their father, and the second one is from their mother. For example, a woman named María Sánchez Jones would have had a father with the last name Sánchez and a mother with the last name Jones (other last names are irrelevant to this discussion).

Suppose a man named José García López marries a woman named María Sánchez Jones. She would then drop Jones and become María Sánchez de García, adding her husband's name after the word *de.* So she is now María Sánchez, "wife of García."

If that couple then has a child, the child will have the last names García Sánchez. Regardless of whether that child is a boy or a girl, the name that is passed on to his or her offspring is García. The sample family tree on the following page illustrates this.

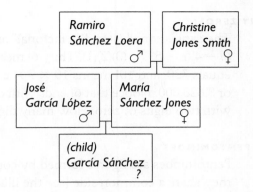

Write the names of the following people in the correct places on a family tree diagram. Notice that some people have been listed twice, once with their maiden names and once with their married names. (*F* indicates female, and *M* indicates male.)

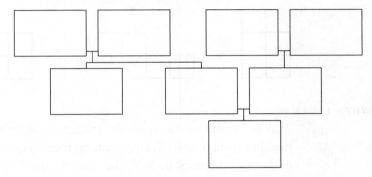

1. Dolores Lara Baez (F)

2. Alberto Rodijo Saenz (M)

3. Marisela Saenz de Rodijo (F)

4. Consuelo Baez García (F)

5. Rafael Lara Echeve (M)

6. Dolores Lara de Rodijo (F)

7. Ana Luísa Rodijo Saenz (F)

8. Marisela Saenz Vaquero (F)

9. Concepción Rodijo Lara (F)

10. Juan Carlos Rodijo Gómez (M)

11. Consuelo Baez de Lara (F)

3. HOW MANY ZEROS?

The expression $n!$ is read "n factorial" and means $n(n-1)(n-2)(n-3)$
$(n-4)(n-5) \ldots (3)(2)(1)$. Thus, 6! means $6 \times 5 \times 4 \times 3 \times 2 \times 1$, which
equals 720, and 10! means $10 \times 9 \times 8 \times 7 \times 6 \times 5 \times 4 \times 3 \times 2 \times 1$,
or 3,628,800. Notice that 6! ends with one digit of zero and that 10! ends
with two digits of zero. How many digits of zero does 5,000! end with?

4. FORMING PENTOMINOES

Pentominoes are figures formed by connecting five squares so that
they share a common side (see the illustrations of the valid and invalid
pentominoes below). How many different pentominoes are there?
Note that reflections and rotations are not considered to be different.

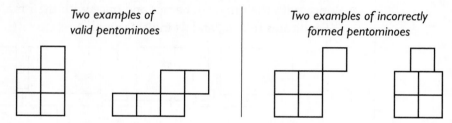

Two examples of valid pentominoes | *Two examples of incorrectly formed pentominoes*

5. WHEN I'M 64

This is a game for two people. Taking alternate turns, choose a whole
number from 1 to 8. Keep a running total. (For example, if you pick 2
and your opponent picks 6, the total is now 8. Then maybe you pick 5,
so now the total is 13, and so on.) The object of the game is to make
the score exactly 64 on your turn. (If it is your turn and the running total
at that point is 62, you would say 2, making the score 64, and you
would be the winner.) You are going to choose the first number. What
number should you pick to guarantee a win for yourself, and what
strategy should you follow?

14

Evaluate Finite Differences

Systematically calculating the differences in the outcomes of a pattern can lead to a general rule and a more efficient solution. Meterologists analyze small differences in quantifiable weather data at various places and times to establish a weather pattern that can aid them in making predictions.

he major problem-solving theme Organizing Information is evident in the strategy of **finite differences,** which ties together many of the strategies you've already learned: systematic lists, looking for patterns, subproblems, easier related problems, and algebra. The strategy of finite differences provides a method for organizing information that often leads to equations that you can use to solve problems. Several of the problems you looked at in earlier chapters can be solved with this new strategy.

You will use finite differences to find equations that represent a data set. To begin, consider the following two sets of data, called **functions,** which are displayed as **input-output** charts.

FAZAL'S FUNCTIONS

Fazal was creating input-output charts. He created these two functions, and he wanted to know the missing values. Can you help?

1.		2.	
x	y	x	y
0	3	0	−4
1	7	1	1
2	11	2	12
3	15	3	29
4	19	4	52
5	−?−	5	−?−
137	−?−	137	−?−

Your work with patterns has enabled you to find the pattern in the first chart and to use the pattern to determine the output for the input value 5. You can easily scan down the column of y-values and determine that they are increasing by 4 each time. So the y-value for x = 5 must be 23. But how would that help you figure out what y-value goes with x = 137? You don't really want to continue the pattern all the way to 137.

The second function has no obvious pattern. However, with a little pencil work, you can see that the y-value for $x = 5$ must be 81. Again, that does not help you find the y-value for $x = 137$.

x	y			
0	−4			
		+5		
			+6	
1	1			
		+11		
			+6	
2	12			
		+17		
			+6	
3	29			
		+23		
			+6	
4	52			
		+29		
5	81			

The method of finite differences will lead you to the equations for these two functions. You will then use your equations to find the y-value for any x-value you choose. The strategy of finite differences will work only for **polynomial functions.** It will not work for **exponential functions, trigonometric functions,** or other functions that are not **polynomial.**

Polynomial functions are functions that look like these:

$$y = 3x - 5 \qquad y = 4x^2 - 3x + 2 \qquad y = 5x^3 - 7x^2 - 9x + 1$$

The **degree of a polynomial** is the largest power of x that is in the polynomial. The first function above is a **linear function,** the second is a **quadratic function,** and the third is a **cubic function** because their degrees are 1, 2, and 3, respectively. In this chapter, we will consider only these three types of polynomial functions, but you can use the strategy of finite differences to find a rule for a polynomial function of any degree.

We will consider linear functions first. In algebra you learned the equation form $y = mx + b$. This equation is called the slope-intercept form of an equation of a line (hence *linear*). Note that most algebra classes use the letters m and b for linear equations in slope-intercept form. The form $y = ax^2 + bx + c$ is used for **quadratic equations.** The letter b is used differently in these two equations. To avoid confusion over what b represents, we've chosen not to use it. Instead, we use the letters e, f, g, and h for the equations in this book. We've also chosen to use these letters in reverse order for reasons that will become clear later in the chapter. You can actually use whatever letters you like—it doesn't make any difference. Our basic linear equation will be written $y = fx + e$. As you will see later, our basic quadratic equation will be written $y = gx^2 + fx + e$.

Anyone could create a function of the form $y = fx + e$ by simply making up numbers for f and e. Note that f is called the **coefficient** of the x-term. Suppose $f = 3$ and $e = 1$. Then the equation is $y = 3x + 1$. You can then set up an input-output chart for x and y and calculate the y-values that correspond to each x-value from 0 through 5. This chart and two others are shown below.

$f = 3 \quad e = 1$			$f = -2 \quad e = 9$			$f = 1 \quad e = -4$	
$y = 3x + 1$			$y = -2x + 9$			$y = 1x - 4$	
x	y		x	y		x	y
0	1		0	9		0	-4
1	4		1	7		1	-3
2	7		2	5		2	-2
3	10		3	3		3	-1
4	13		4	1		4	0
5	16		5	-1		5	1

Now suppose you were given the input-output chart and wanted to find the equation. Compute the differences between each set of y-values.

$f = 3 \quad e = 1$				$f = -2 \quad e = 9$				$f = 1 \quad e = -4$		
$y = 3x + 1$				$y = -2x + 9$				$y = 1x - 4$		
x	y			x	y			x	y	
0	1			0	9			0	-4	
		+3				-2				+1
1	4			1	7			1	-3	
		+3				-2				+1
2	7			2	5			2	-2	
		+3				-2				+1
3	10			3	3			3	-1	
		+3				-2				+1
4	13			4	1			4	0	
		+3				-2				+1
5	16			5	-1			5	1	

Note that each time the difference in the *y*-values is equal to the value of *f*. Also notice that the *y*-value when *x* equals 0 is equal to the value of *e*.

Why does this occur? Does this work every time?

LINEAR FUNCTION GENERAL CHART

Consider the general equation for a linear function, $y = fx + e$. Make a function chart for this equation, using the values 0 through 5 for *x*. Do this before reading on.

First, plug in the *x*-values 0 through 5 into the equation $y = fx + e$ to get the chart shown at right.

Next, figure out the differences between successive terms, as we did earlier. For this chart, you have to algebraically subtract each term from the term that follows it. For example, the first two *y*-values in the chart are *e* and $(f + e)$. Subtracting the first from the second gives $(f + e) - e = f$. The next difference is $(2f + e) - (f + e) = f$. The difference turns out to be *f* every time.

This general (or master) chart shows that the common difference will always be the value of *f* and that the value of *e* will always be given by the value of *y* when *x* equals 0.

$y = fx + e$

x	y
0	e
1	f + e
2	2f + e
3	3f + e
4	4f + e
5	5f + e

$y = fx + e$

x	y	
0	e	
		+f
1	f + e	
		+f
2	2f + e	
		+f
3	3f + e	
		+f
4	4f + e	
		+f
5	5f + e	

Find the equation for these two functions. (The first is Fazal's first function.) Then find value of y when x is 5 and when x is 137. Do this before reading on.

1.

x	y
0	3
1	7
2	11
3	15
4	19
5	—?—
137	—?—

2.

x	y
0	−2
1	5
2	12
3	19
4	26
5	—?—
137	—?—

Simone worked on this problem. "I found the differences between the numbers. I knew from the master chart that the common difference between the numbers was f and that the number opposite $x = 0$ was e. So the first equation is $y = 4x + 3$. For $x = 5$, the y-value is $4(5) + 3$ or 23. For $x = 137$, the y-value is $4(137) + 3$, or 551.

"For the second function, the common difference is 7, and the number opposite $x = 0$ is −2. So the equation is $y = 7x − 2$. For $x = 5$, $y = 7(5) − 2$, or 33. For $x = 137$, $y = 7(137) − 2$, or 957."

x	y	
0	3	>+4
1	7	>+4
2	11	>+4
3	15	>+4
4	19	>+4
5	23	
137	551	

x	y	
0	−2	>+7
1	5	>+7
2	12	>+7
3	19	>+7
4	26	>+7
5	33	
137	957	

Now go back and consider Fazal's other function from the beginning of the chapter.

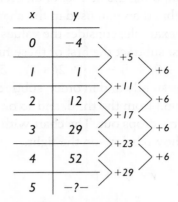

x	y
0	−4
1	1
2	12
3	29
4	52
5	−?−

This problem also has a common difference, but it does not show up until the second column of differences. This situation did not happen for linear equations, so the general form for this function must be different. Because it took two columns of differences to get a constant difference, the equation must be a second-degree polynomial. Why? Consider the well-known function in the following chart.

x	y
0	0
1	1
2	4
3	9
4	16
5	25

You probably recognize this function as $y = x^2$, the classic second-degree, or basic quadratic, equation.

QUADRATIC FUNCTION GENERAL CHART

The general form for a quadratic function is $y = gx^2 + fx + e$. As you did for the earlier Linear Function General Chart problem, make the general function chart for this equation by substituting x-values from 0 through 5. (Note that when $g = 0$, this chart is the same as the one for $y = fx + e$.) Do this problem before reading on.

The quadratic function general chart is shown at left below. Computing the first set of differences involves a little bit more algebra than you used in the chart for the linear function, $y = fx + e$. For example, consider the values of y for $x = 2$ and $x = 3$. You must subtract the first of these function values from the second: $(9g + 3f + e) - (4g + 2f + e) = 5g + f$. Be careful to compute the values consistently and in order, subtracting the first from the second, the second from the third, and so on. Note that with each difference the e term drops out. The chart with the first set of differences filled in is shown below on the right.

$$y = gx^2 + fx + e$$

x	y
0	e
1	$g + f + e$
2	$4g + 2f + e$
3	$9g + 3f + e$
4	$16g + 4f + e$
5	$25g + 5f + e$

$$y = gx^2 + fx + e$$

x	y	
0	e	
		$g + f$
1	$g + f + e$	
		$3g + f$
2	$4g + 2f + e$	
		$5g + f$
3	$9g + 3f + e$	
		$7g + f$
4	$16g + 4f + e$	
		$9g + f$
5	$25g + 5f + e$	

If differences are not constant, continue subtracting.

These differences are not constant, so you must subtract again. This time take the difference of the differences, as shown next.

$$y = gx^2 + fx + e$$

x	y		
0	e		
		$g + f$	
1	$g + f + e$		$+2g$
		$3g + f$	
2	$4g + 2f + e$		$+2g$
		$5g + f$	
3	$9g + 3f + e$		$+2g$
		$7g + f$	
4	$16g + 4f + e$		$+2g$
		$9g + f$	
5	$25g + 5f + e$		

The second difference turns out to be $2g$ every time. This is the constant difference we were seeking. In Fazal's second function, the second difference was constant. The second difference was also constant in the chart for the function $y = x^2$.

Now we want to find the equation for Fazal's second function. As we did for linear equations, we want to compare the particular chart to the general chart. It is very important to match up the corresponding parts of each chart.

FAZAL'S SECOND FUNCTION

Find the equation for Fazal's second function on page 380. Compare it with the general chart to find the values for g, f, and e. Then find the y-value for $x = 137$.

$y = -?-$

x	y
0	-4
1	1
2	12
3	29
4	52
5	81

+5 → +6
+11 → +6
+17 → +6
+23 → +6
+29

$y = gx^2 + fx + e$

x	y
0	e
1	$g + f + e$
2	$4g + 2f + e$
3	$9g + 3f + e$
4	$16g + 4f + e$
5	$25g + 5f + e$

$g + f$ → $+2g$
$3g + f$ → $+2g$
$5g + f$ → $+2g$
$7g + f$ → $+2g$
$9g + f$

Match up corresponding parts of the two charts.

Compare the parts of the charts that are the same. It's a good idea to find the value of g first, then f, and then e. The last column of the chart (that is, the second difference) is 6. This corresponds to $2g$ on the general chart, so $2g = 6$, which means that $g = 3$.

In the problem chart, the difference between output values for the x-values 0 and 1 is 5, which matches up with $g + f$ in the general chart.

$g + f = 5$
$3 + f = 5$ (Recall $g = 3$.)
$f = 2$

*Always check
your equation
with substitution.*

Finally, when x is 0, y is -4, which corresponds to e on the general chart. So $e = -4$.

Thus, the equation is $y = 3x^2 + 2x - 4$. Check it by substituting in $x = 4$. When $x = 4$, then $y = 3(4^2) + 2(4) - 4 = 52$, which is correct. You can use this equation to find the value of y for any x-value in this function. In Fazal's problem, he needed to find y when $x = 137$:
$y = 3(137^2) + 2(137) - 4 = 56{,}577$.

The **Big** Picture

To find the equation from a function chart, follow these steps:

1. Compute the differences between the y-values.

   a. If the difference is not constant, compute the difference of the differences.

   b. Continue computing differences until the difference is constant.

2. Determine the degree of the polynomial. The number of difference columns is equal to the degree of the polynomial. For example:

   a. One column of differences means first degree, a linear equation of the form $y = fx + e$.

   b. Two columns of differences means second degree, a quadratic equation of the form $y = gx^2 + fx + e$.

   c. Three columns of differences means third degree, a cubic equation of the form $y = hx^3 + gx^2 + fx + e$.

   d. Four columns of differences means fourth degree, and so on.

3. Choose the correct general chart.

4. To find the values of the coefficients, set up equations comparing corresponding positions of your function chart and the general chart.

5. Solve the equations. It's easiest to work from right to left. In a quadratic function problem, for instance, solve for g first, then f, and then e.

6. Check your equation by plugging in known values of x and y to see if it works.

7. Use your equation to find the y-value for any x-value.

- Make sure you have enough data to draw a conclusion, that is, make sure you use enough *x*-values to end up with at least three common differences.

- If you don't ever reach a common difference, then the equation is not a polynomial function. See the discussion of exponential functions at the end of the chapter.

MORE FUNCTIONS

Find the equation for each function displayed in the charts below. Then use your equations to find the missing *y*-values, indicated by –?–. Work these before going on.

x	y
0	−5
1	−2
2	5
3	16
4	31
5	50
48	−?−

x	y
2	14
3	11
4	8
5	5
6	2
7	−1
82	−?−

Sarah, Michele, and Iram worked on this problem.

MICHELE: Let's figure out what the differences are.

Compute first differences.

x	y	
0	−5	
		> −3
1	−2	
		> +3
2	5	
		> +11
3	16	
		> +15
4	31	
		> +19
5	50	

IRAM: Wait a second. You have the first two differences wrong.

MICHELE: I do? What is wrong with them?

SARAH: I see what Iram means. From -5 to -2 it goes up 3, not down 3. So the difference should be $+3$.

IRAM: Right. And from -2 to 5, it goes up 7. Think of owing someone two dollars and then earning seven dollars. You would pay off the person you owed and still have five dollars. So $-2 + 7$ equals 5.

Compute second differences.

Determine degree.

MICHELE: Okay, I see. (She changed her chart.) Now let's find the second differences. It comes out constant on the second differences. (See the chart at right.) That means it's a quadratic equation. So the general equation is $y = gx^2 + fx + e$.

x	y		
0	−5		
		+3	
1	−2		+4
		+7	
2	5		+4
		+11	
3	16		+4
		+15	
4	31		+4
		+19	
5	50		

SARAH: So we need the master chart for a quadratic.

IRAM: Wait, I think I have it in my notebook someplace.

MICHELE: Don't bother looking, Iram. We can just create it from scratch. I'm not into memorizing stuff. I just remember how to create it.

SARAH: I agree. I'll do it. (Here's the chart Sarah created.)

Use the correct general chart.

$$y = gx^2 + fx + e$$

x	y		
0	e		
		g + f	
1	g + f + e		2g
		3g + f	
2	4g + 2f + e		2g
		5g + f	
3	9g + 3f + e		2g
		7g + f	
4	16g + 4f + e		2g
		9g + f	
5	25g + 5f + e		

MICHELE: Now what do we do again?

IRAM: Compare the master chart with the chart for the problem we are doing.

Match corresponding parts.

MICHELE: Oh, yeah. (She put the problem chart next to the master chart.) Okay, 2g matches up with 4 in our problem.

$$y = gx^2 + fx + e$$

x	y			x	y		
0	-5			0	e		
		+3				g + f	
1	-2		+4	1	g + f + e		2g
		+7				3g + f	
2	5		+4	2	4g + 2f + e		2g
		+11				5g + f	
3	16		+4	3	9g + 3f + e		2g
		+15				7g + f	
4	31		+4	4	16g + 4f + e		2g
		+19				9g + f	
5	50			5	25g + 5f + e		

Solve equations.

IRAM: Right. So 2g must equal 4, and therefore g equals 2.

SARAH: No problem. The next one is a little trickier. You have to make sure you are comparing apples to apples.

IRAM: I know what you mean. The difference between 0 and 1 is g + f on the master chart. On our problem chart it is 3. So we have g + f = 3.

$$y = gx^2 + fx + e$$

x	y			x	y		
0	-5			0	e		
		+3				g + f	
1	-2		+4	1	g + f + e		2g
		+7				3g + f	
2	5		+4	2	4g + 2f + e		2g
		+11				5g + f	
3	16		+4	3	9g + 3f + e		2g
		+15				7g + f	
4	31		+4	4	16g + 4f + e		2g
		+19				9g + f	
5	50			5	25g + 5f + e		

MICHELE: I get it. We wouldn't want to match up g + f with 7, because 7 is the difference between x = 1 and x = 2.

SARAH: Right. That would be comparing apples and oranges.

MICHELE: Got it. Anyway, $g + f$ equals 3, and we already know that g is 2, so f has to be 1.

IRAM: Right. And e is obviously -5, because that is the y-value when x is 0.

MICHELE: So the equation is $y = 2x^2 + x - 5$. We'd better check it.

SARAH: That's a good idea. We need the check. Let's use $x = 4$. There is less risk of making a mistake there. We shouldn't check things with $x = 0$, 1, or 2 because it might turn out right and really be wrong. Using $x = 4$ is safe. Okay, so $y = 2(4^2) + 4 - 5$, which is 31, and 31 is what it is supposed to be.

MICHELE: Okay, now we need to know what y is when x is 48: $y = 2(48^2) + 48 - 5 = 4{,}651$.

IRAM: Now let's look at the next function. It's different.

MICHELE: It doesn't start at $x = 0$.

IRAM: Right. But I guess we can find the differences anyway. (The differences Iram calculated are shown next.)

x	y	
2	14	
		-3
3	11	
		-3
4	8	
		-3
5	5	
		-3
6	2	
		-3
7	-1	

MICHELE: How come those are -3 and not $+3$?

IRAM: It's just like in the last problem. The numbers are going down, so it's -3.

MICHELE: Okay, got it. It came out constant the first time, so it must be an easy one. Aren't these called linear equations?

SARAH: Yeah, they are. They represent lines.

IRAM: Let's compare this with the master chart for $y = fx + e$. I think I can generate it, even though I'm sure I have it in my notebook somewhere. (Here is what Iram wrote down next to the problem chart.)

$$y = fx + e$$

x	y			x	y		
				0	e		
				1	$f + e$	$> +f$	
2	14	> -3		2	$2f + e$	$> +f$	
3	11	> -3		3	$3f + e$	$> +f$	
4	8	> -3		4	$4f + e$	$> +f$	
5	5	> -3		5	$5f + e$	$> +f$	
6	2	> -3		6	$6f + e$	$> +f$	
7	-1	> -3		7	$7f + e$	$> +f$	

SARAH: I don't see why you added that stuff to the master chart like that.

IRAM: I was trying to follow your advice about apples and oranges. I am trying to show the comparison between two things that are the same.

MICHELE: Good idea. But how should we handle the holes in the problem chart?

IRAM: Well, either we don't worry about them and just compare the similar parts of both charts, or we could work backwards and figure out the y-values for $x = 0$ and $x = 1$.

SARAH: Or we could meet halfway and do both. Let's do it both ways and see if we get the same answers.

IRAM: Compare similar parts of the two charts first. That gives us $f = -3$. And opposite $x = 2$ we have $14 = 2f + e$.

$$14 = 2(-3) + e$$
$$14 = -6 + e$$
$$e = 20$$

So the equation is $y = -3x + 20$. This checks when x is 6 because $-3(6) + 20 = 2$.

MICHELE: If we work backwards on our chart, we have to *add* 3 to go *up* the chart, because we subtract 3 going down the chart. We get $e = 20$ anyway because that is the number opposite 0. And f would still be -3, because that is the common difference. So it must be right.

SARAH: Now we need y when $x = 82$: $y = -3(82) + 20$, which is -226. That seems right since it really starts going down.

x	y	
0	20	$\rangle$ -3
1	17	$\rangle$ -3
2	14	$\rangle$ -3
3	11	$\rangle$ -3
4	8	$\rangle$ -3
5	5	$\rangle$ -3
6	2	$\rangle$ -3
7	-1	
82	-226	

Notice that the trio was careful to compare parts of the master charts with corresponding parts of the problem charts, even when some of the intermediary values for x and y were missing. Also, they checked their equations after they'd found them.

HANDSHAKES

Suppose that at the first meeting of the House of Representatives, all 435 members shook hands with each of the other members. How many handshakes took place? Work this problem before continuing.

Draw a diagram for clarity.

Kowasky worked on this problem. "This problem seemed pretty hard, so I wanted to make it easier. So I figured, suppose only one representative showed up. Then there wouldn't be any handshakes. If two showed up, there would be one handshake. Let's say three showed up—call them U, S, and A. U shakes hands with S and then with A, and S shakes hands with A. That's three handshakes. When I got to four people, I drew a diagram. I counted lines in the diagram, which represented six handshakes.

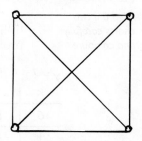

"Then I started making a chart and looked for a pattern."

Whenever you make a chart, look for a pattern.

People	Handshakes
0	0
1	0
2	1
3	3
4	6

Make sure you have enough data.

"I figured I'd better do five people. It was getting harder to make a diagram, but then I thought if there were four people there, they gave 6 handshakes. If a fifth person came into the room, he would shake hands with the four people who were there already. So that would make 10 handshakes. Then a sixth person would shake five hands, so there would be 15 handshakes for six people.

"I added this information to my chart and looked to use finite differences, since it would take forever to extend this pattern all the way to 435."

People	Handshakes
0	0
1	0
2	1
3	3
4	6
5	10
6	15

+0 > +1
+1 > +1
+2 > +1
+3 > +1
+4 > +1
+5 > +1

Align the two charts carefully and compare.

"I saw that the constant difference happened the second time. That meant that the equation was quadratic, of the form $y = gx^2 + fx + e$.

"I generated the master chart for that equation. I could have looked it up, but it's kind of fun to generate. I compared it to my chart for this problem."

$y = ?$

x	y
0	0
1	0
2	1
3	3
4	6
5	10
6	15

+0, +1, +2, +3, +4, +5

+1, +1, +1, +1, +1, +1

$y = gx^2 + fx + e$

x	y
0	e
1	$g + f + e$
2	$4g + 2f + e$
3	$9g + 3f + e$
4	$16g + 4f + e$
5	$25g + 5f + e$
6	$36g + 6f + e$

$g + f$, $3g + f$, $5g + f$, $7g + f$, $9g + f$, $11g + f$

$2g$, $2g$, $2g$, $2g$, $2g$

"Then I just had to compare similar parts of the two charts. The second constant difference is 1, which is equal to $2g$. If $2g$ equals 1, then g must equal $\frac{1}{2}$."

$y = ?$

x	y
0	0
1	0
2	1
3	3
4	6

+0, +1, +2, +3, +4

+1, +1, +1, +1

$y = gx^2 + fx + e$

x	y
0	e
1	$g + f + e$
2	$4g + 2f + e$
3	$9g + 3f + e$
4	$16g + 4f + e$

$g + f$, $3g + f$, $5g + f$, $7g + f$, $9g + f$

$2g$, $2g$, $2g$, $2g$

"Then I wanted to figure out what f was. The difference between $x = 2$ and $x = 3$ was 2, and that equals $5g + f$. I used these values because I wasn't sure this problem made sense for fewer than two people."

$$5g + f = 2$$
$$5(\tfrac{1}{2}) + f = 2 \quad \text{remember } g = \tfrac{1}{2}$$
$$2.5 + f = 2$$
$$f = -\tfrac{1}{2}$$

"Finally I wanted *e*. The *y*-value for *x* = 3 was 3. In the master chart, the value was $9g + 3f + e$."

$$9g + 3f + e = 3$$
$$9(1/2) + 3(-1/2) + e = 3 \quad \text{recall } g = 1/2 \text{ and } f = -1/2$$
$$4.5 - 1.5 + e = 3$$
$$3 + e = 3$$
$$e = 0$$

"I noticed that I would have gotten the same value of *e* if I had just used the value next to *x* = 0 in the chart. So my equation was $y = (1/2)x^2 - (1/2)x$, where *x* represents the number of representatives and *y* represents the number of handshakes. I checked this with six people."

$$y = (1/2)x^2 - (1/2)x$$
$$y = (1/2)6^2 - (1/2)6$$
$$y = (1/2)36 - 3$$
$$y = 18 - 3$$
$$y = 15, \quad \text{and that's what I got before.}$$

"Finally, I needed to find the number of handshakes for all 435 representatives."

$$y = (1/2)x^2 - (1/2)x$$
$$y = (1/2)(435^2) - (1/2)(435)$$
$$y = (1/2)(189,225) - 217.5$$
$$y = 94,612.5 - 217.5$$
$$y = 94,395$$

"There had to be 94,395 handshakes. That's a lot of handshakes. I think there must have been a lot of tired hands. And if the senators were there too, wow!"

"This problem used a lot of strategies. I started with an easier related problem, then a diagram, then a pattern, and finally finite differences. A lot of these strategies seem to work in concert really well."

Recall the checkerboard problem from Chapter 9: Solve an Easier Related Problem. To solve this problem, you will have to create the cubic general chart.

HOW MANY SQUARES?

Find a formula for the number of squares on any *n*-by-*n* checkerboard. Set up a pattern for this problem before continuing.

When we last left this problem, the chart below had been developed. Angie and Isaac are back to discuss this problem. Here, instead of 1-by-1, 2-by-2, 3-by-3, and so on, the first column shows just 1, 2, 3, . . . to represent the size of the checkerboard. See pages 239–243 in Chapter 9 for how this chart was developed.

SIDE SQRS	# OF SQRS
1	1
2	5
3	14
4	30
5	55
6	91
7	140
8	204
9	285
10	385

ANGIE: I think we can use finite differences to figure out the formula for this problem.

ISAAC: I agree. Let's go for it. First find all the differences.

SIDE SQRS	# OF SQRS
1	1
2	5
3	14
4	30
5	55
6	91
7	140
8	204
9	285
10	385

First differences: + 4, + 9, + 16, + 25, + 36, + 49, + 64, + 81, + 100

Second differences: + 5, + 7, + 9, + 11, + 13, + 15, + 17, + 19

Third differences: + 2, + 2, + 2, + 2, + 2, + 2, + 2

ANGIE: Wow, I've never seen a chart that took three differences to come out a constant.

ISAAC: Yeah. I wonder what we should do. It can't be linear or quadratic.

ANGIE: Huh?

ISAAC: You know, linear is $y = fx + e$ and quadratic is $y = gx^2 + fx + e$.

ANGIE: Yeah, because those take one difference and two differences. So what does that mean? Does it have to be a cubic equation?

ISAAC: I guess so. It has to have an x^3 in it. I wonder what that equation looks like?

ANGIE: Probably $y = x^3 + x^2 + x + e$.

ISAAC: Don't we need some coefficients in there?

ANGIE: Okay. How about $y = hx^3 + gx^2 + fx + e$?

ISAAC: Sounds good. Let's see if we can generate the master chart for the cubic equation.

Generate the chart for the general form of a cubic function, $y = hx^3 + gx^2 + fx + e$. Do this before continuing.

The general chart is the first one shown below.

ANGIE: Okay, I think I've got it.

ISAAC: Me too. Let's compare.

$y = hx^3 + gx^2 + fx + e$

x	y
0	e
1	$h + g + f + e$
2	$8h + 4g + 2f + e$
3	$27h + 9g + 3f + e$
4	$64h + 16g + 4f + e$
5	$125h + 25g + 5f + e$
6	$216h + 36g + 6f + e$

Differences:

$h + g + f$ → $6h + 2g$ → $6h$
$7h + 3g + f$ → $12h + 2g$ → $6h$
$19h + 5g + f$ → $18h + 2g$ → $6h$
$37h + 7g + f$ → $24h + 2g$ → $6h$
$61h + 9g + f$ → $30h + 2g$ → $6h$
$91h + 11g + f$

SIDE SQRS	# OF SQRS
1	1
2	5
3	14
4	30
5	55
6	91
7	140
8	204
9	285
10	385

Differences:

+4 → +5 → +2
+9 → +7 → +2
+16 → +9 → +2
+25 → +11 → +2
+36 → +13 → +2
+49 → +15 → +2
+64 → +17 → +2
+81 → +19 → +2
+100

ANGIE: We got the same thing. I think we really understand this stuff.

ISAAC: Now let's compare this master chart with our chart.

Compare the last differences column first.

ANGIE: Okay, let's start with the last column. In the master chart, the last difference is $6h$. In our chart, the last difference is 2.

ANGIE: So if $6h = 2$, then h is 3.

ISAAC: No it's not. It's one-third: $h = \frac{1}{3}$.

ANGIE: Oh right, sorry. I divided the wrong way.

Compare the second differences column next.

ISAAC: Now let's compare the second columns of differences. The top number in the second difference column of our chart is 5. The top number in the second difference column of the master chart is $6h + 2g$. So $6h + 2g$ equals 5.

ANGIE: No, $6h + 2g$ is the second difference related to the $x = 0$ and $x = 1$ values. But we never had $x = 0$ in our chart, so we need to compare 5 to $12h + 2g$.

Make the correct comparisons. It's tricky when you don't have a y-value for x = 0

$$y = hx^3 + gx^2 + fx + e$$

x	y
0	e
1	$h + g + f + e$
2	$8h + 4g + 2f + e$

$h + g + f$

$7h + 3g + f$ $6h + 2g$

$19h + 5g + f$ $12h + 2g$ $6h$

$18h + 2g$ $6h$

$6h$

SIDE SQRS	# OF SQRS
1	1
2	5
3	14
4	30

$+ 4$ $+ 5$ $+ 2$

$+ 9$ $+ 7$ $+ 2$

$+ 16$ $+ 9$ $+ 2$

$+ 25$ $+ 11$ $+ 2$

The positions must correspond exactly.

ISAAC: Oh, I see what you mean. We need to make sure the positions in the two charts correspond exactly. Okay, so $12h + 2g = 5$.

$$12h + 2g = 5$$
$$12\left(\tfrac{1}{3}\right) + 2g = 5 \qquad h = \tfrac{1}{3}$$
$$4 + 2g = 5$$
$$2g = 1$$
$$g = \tfrac{1}{2}$$

ANGIE: Great. Now let's go to the first column of differences and make sure we are comparing the same things.

$$y = hx^3 + gx^2 + fx + e$$

x	y
0	e
1	h + g + f + e
2	8h + 4g + 2f + e

$$h + g + f$$
$$7h + 3g + f$$
$$19h + 5g + f$$

$$6h + 2g$$
$$12h + 2g$$
$$18h + 2g$$

$$6h$$
$$6h$$
$$6h$$

SIDE SQRS	# OF SQRS
1	1
2	5
3	14
4	30

+ 4
+ 9
+ 16
+ 25

+ 5
+ 7
+ 9
+ 11

+ 2
+ 2
+ 2
+ 2

ISAAC: Okay, $7h + 3g + f = 4$. Those are the differences between $x = 1$ and $x = 2$.

ANGIE: Right. I can solve that equation for f.

$$7h + 3g + f = 4$$
$$7\left(\tfrac{1}{3}\right) + 3\left(\tfrac{1}{2}\right) + f = 4 \qquad h = \tfrac{1}{3},\ g = \tfrac{1}{2}$$
$$\tfrac{7}{3} + \tfrac{3}{2} + f = 4$$
$$\tfrac{23}{6} + f = 4$$
$$f = \tfrac{1}{6}$$

ANGIE: Okay, now what is e? Isn't it just the top number in the y column?

**When x = 0
y does not
have to be 0.**

ISAAC: It would be if we had a value for $x = 0$. It should be 0, because with no squares you wouldn't have any squares. But I'm not positive that y will be zero. So let's compare the y-values when $x = 1$.

$$y = hx^3 + gx^2 + fx + e$$

x	y
0	e
1	$h + g + f + e$
2	$8h + 4g + 2f + e$

$h + g + f$

$7h + 3g + f$

$19h + 5g + f$

$6h + 2g$

$12h + 2g$

$6h$

$6h$

SIDE SQRS	# OF SQRS
1	1
2	5
3	14

$+ 4$

$+ 9$

$+ 16$

$+ 5$

$+ 7$

$+ 9$

$+ 2$

$+ 2$

$+ 2$

$$h + g + f + e = 1$$
$$\tfrac{1}{3} + \tfrac{1}{2} + \tfrac{1}{6} + e = 1$$
$$1 + e = 1$$
$$e = 0$$

ANGIE: You were right—e is 0 but I bet we'll see problems later where it isn't zero. Okay, so what's our equation?

ISAAC: It's $y = (\tfrac{1}{3})x^3 + (\tfrac{1}{2})x^2 + (\tfrac{1}{6})x$, where x is the number of squares on a side and y is the total number of squares in the figure.

**Test your answer
at the end.**

ANGIE: Great! Let's test it. Suppose $x = 8$. We know that answer is supposed to be 204.

$$y = (\tfrac{1}{3})x^3 + (\tfrac{1}{2})x^2 + (\tfrac{1}{6})x$$
$$y = (\tfrac{1}{3})8^3 + (\tfrac{1}{2})8^2 + (\tfrac{1}{6})8$$
$$y = (\tfrac{1}{3})(512) + (\tfrac{1}{2})(64) + (\tfrac{1}{6})8$$
$$y = \tfrac{512}{3} + 32 + \tfrac{4}{3} = 204$$

ANGIE: It checks.

ISAAC: All right. So a checkerboard that is n-by-n will have $(\frac{1}{3})n^3 + (\frac{1}{2})n^2 + (\frac{1}{6})n$ total squares on it.

ANGIE: Finite differences are pretty cool. I think I could even figure out an equation that started with x^4 if I had to.

Calculate the Differences

The strategy of finite differences provides a way to find equations that are polynomial in nature. It is especially useful when used in conjunction with easier related problems, patterns, and sometimes diagrams. This strategy organizes the information in a problem in a new way and quickly leads to equations. But be careful: It doesn't work on all equations.

Finite Differences Do Not Apply to All Patterns

The strategy of finite differences works only for polynomial functions.

Consider the chart for the function $y = 2^x$ and what happens when we calculate the differences. Notice that the differences keep repeating themselves. This type of pattern is typical of exponential functions. For fun, you might want to investigate the pattern that arises from attacking the function $y = 3^x$ with finite differences. Be sure you consider x-values up to at least 9. The repeating pattern that shows up here with the function $y = 2^x$ doesn't occur, but something else interesting does happen.

x	y
0	1
1	2
2	4
3	8
4	16
5	32
6	64
7	128
8	256
9	512

Differences: +1, +2, +4, +8, +16, +32, +64, +128, +256

Second differences: +1, +2, +4, +8, +16, +32, +64, +128

Third differences: +1, +2, +4, +8, +16, +32, +64

Fourth differences: +1, +2, +4, +8, +16, +32

Be Sure There Is Enough Data

Be sure you can generate enough differences.

There are other traps in finite differences. Consider the following chart.

x	y				
0	4				
		+4			
1	8		+2		
		+6		+2	
2	14		+4		+2
		+10		+4	
3	24		+8		+2
		+18		+6	
4	42		+14		
		+32			
5	74				

It appears that there is a constant difference of 2 in the fourth column, which indicates that the equation is fourth degree, but you can see only two values. You should have more information to convince you of the validity of the pattern. If you're working on a problem for which you generated the data—as you did for the Handshakes and How Many Squares? Revisited problems—generate a few more pieces of data. Then you'll have a better idea of whether your pattern is correct. No pattern is guaranteed to go on forever, so be careful.

Calculate Carefully

Beware of calculation errors.

Another potential difficulty with the strategy of finite differences is making a mistake in a chart. Mistakes completely mask any patterns that are present, and the mistakes compound themselves as you move on to more columns. Any common difference that should be present will disappear because of one mistake. Scrutinize the following chart.

x	y				
0	−4				
		+6			
1	2		+4		
		+10		+2	
2	14		+8		−4
		+18		−2	+6
3	32		+6		+2
		+24		+0	
4	56		+6		
		+30			
5	86				

A simple subtraction mistake leads to all kinds of chaos. Be careful, especially when dealing with negative and positive numbers. By the way, did you find the subtraction mistake?

Used accurately, the strategy of evaluating finite differences ties together the strategies of systematic lists, looking for patterns, subproblems, easier related problems, and algebra. You can use finite differences to find the equation for any *polynomial* function. Enjoy the strategy. It can be a lot of fun.

Summary of General Charts

Here are the three general charts for linear, quadratic and cubic equations.

$y = fx + e$

x	y	
0	e	
1	$f + e$	$+ f$
2	$2f + e$	$+ f$
3	$3f + e$	$+ f$
4	$4f + e$	$+ f$
5	$5f + e$	$+ f$
6	$6f + e$	$+ f$
7	$7f + e$	$+ f$

$y = gx^2 + fx + e$

x	y		
0	e		
1	$g + f + e$	$g + f$	$2g$
2	$4g + 2f + e$	$3g + f$	$2g$
3	$9g + 3f + e$	$5g + f$	$2g$
4	$16g + 4f + e$	$7g + f$	$2g$
5	$25g + 5f + e$	$9g + f$	$2g$
6	$36g + 6f + e$	$11g + f$	

$y = hx^3 + gx^2 + fx + e$

x	y			
0	e			
1	$h + g + f + e$	$h + g + f$	$6h + 2g$	$6h$
2	$8h + 4g + 2f + e$	$7h + 3g + f$	$12h + 2g$	$6h$
3	$27h + 9g + 3f + e$	$19h + 5g + f$	$18h + 2g$	$6h$
4	$64h + 16g + 4f + e$	$37h + 7g + f$	$24h + 2g$	$6h$
5	$125h + 25g + 5f + e$	$61h + 9g + f$	$30h + 2g$	$6h$
6	$216h + 36g + 6f + e$	$91h + 11g + f$		

Problem Set A

1. EIGHT FUNCTIONS

Find the equation for each function. Then find the *y*-value for the *x*-value indicated.

a.

x	y
0	6
1	13
2	20
3	27
4	34
5	41
153	-?-

b.

x	y
0	6
1	12
2	20
3	30
4	42
5	56
94	-?-

c.

x	y
0	4
1	3
2	6
3	13
4	24
5	39
103	-?-

d.

x	y
0	5
1	1
2	−3
3	−7
4	−11
5	−15
141	-?-

e.

x	y
0	−2
1	3
2	20
3	55
4	114
5	203
82	-?-

f.

x	y
4	10
5	16
6	22
7	28
8	34
9	40
391	-?-

g.

x	y
2	3
3	10
4	19
5	30
6	43
7	58
175	-?-

h.

x	y
0	1
1	−1
2	9
3	43
4	113
5	231
67	-?-

2. TRIANGULAR NUMBERS

Find a formula for the triangular numbers. Then find the number of dots in the 87th figure.

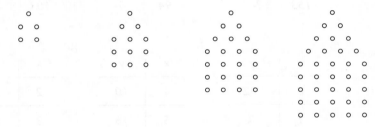

3. RECTANGULAR NUMBERS

Find a formula for the rectangular numbers. Then find the number of dots in the 73rd figure.

4. PENTAGONAL NUMBERS

Find a formula for the pentagonal numbers. Then find the number of dots in the 55th figure.

5. DIAGONALS

How many diagonals are there in a polygon with n sides?

6. GREAT PYRAMID OF ORANGES

A very bored grocer was stacking oranges one day. She decided to stack them in a triangular pyramid. She put one orange in the top layer, three oranges in the second layer, six oranges in the third layer, and so on. Each layer except the top formed an equilateral triangle. How many oranges would it take to build such a pyramid 50 layers high?

7. **WRITE YOUR OWN**

Create your own pattern that can be solved with finite differences. Start with the equation that you are going to use and then create the chart. If you feel ambitious, try creating a situation that will give you the numbers in your chart, like the situation in the Great Pyramid of Oranges problem in this problem set.

CLASSIC PROBLEM

8. **SUM OF CUBES**

What is the sum of the first 100 cubes?

From *How to Solve It* by George Pólya.

MORE PRACTICE

1. **A FEW MORE FUNCTIONS**

Find the equation for each function. Then find the *y*-value for the *x*-value indicated.

a.

x	y
0	2
1	9
2	16
3	23
4	30
5	37
521	-?-

b.

x	y
0	4
1	11
2	20
3	31
4	44
5	59
642	-?-

c.

x	y
0	50
1	44
2	38
3	32
4	26
5	20
3,190	-?-

d.

x	y
0	3
1	6
2	17
3	36
4	63
5	98
149	-?-

e.

x	y
0	8
1	7
2	2
3	−1
4	4
5	23
104	-?-

f.

x	y
0	5
1	8
2	15
3	26
4	41
5	60
276	-?-

g.

x	y
7	41
8	50
9	59
10	68
11	77
12	86
89	-?-

h.

x	y
0	−4
1	−7
2	4
3	41
4	116
5	241
123	-?-

i.

x	y
4	74
5	108
6	148
7	194
8	246
9	304
61	-?-

j.

x	y
2	9
3	71
4	203
5	429
6	773
7	1,259
75	-?-

2. DOTS IN ROWS

Find the formula for the dots below. Then find the number of dots in the 89th picture.

3. **TURNING SQUARES**

   Find the formula for the pictures of squares below. Then determine how many squares are in the 97th picture.

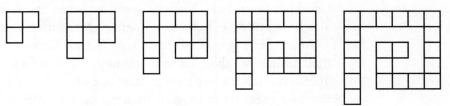

Problem Set B

1. **CELEBRATION TIME**

   Luke and Dicey decided to celebrate. They didn't know what they were celebrating, but it sounded like fun. They started with dinner. It cost them one-third of their money, plus $4.50 more for a tip. Then they spent $3.75 each on admission to the county fair, and they immediately purchased tickets for an open-air concert for half their money plus $1.00 more. They bought a bottle of antacid with $2.50 of their remaining money, and then spent one-third of what they had left on the trip home. At this point, feeling sick and exhausted, they had only $5.00 left. How much money did they spend on their celebration?

2. **WILSHIRE BOULEVARD**

   Ardith, Burris, Carmie, Dawn, and Eartha all live on Wilshire Boulevard, which is a very long street. Ardith lives at one end of the street. Driving down the street from Ardith's house, you would first get to Burris's house, then Carmie's, Dawn's, and finally Eartha's. The five of them had lunch one day and discussed the distances between their houses. The number of blocks between each pair of houses is different. These ten numbers, arranged in numerical order, are 15, 21, 27, 36, 42, 48, 63, 69, 84, and 111. If Ardith lives closer to Burris than Dawn lives to Eartha, determine the distance between each pair of adjacent houses.

3. FIVES AND ONES

Meghan Smith, Cameron Smith, Carol Jones, Beau Jones, and Tomás Ramirez each have some $1 bills. Meghan has one, Cameron has two, Carol has three, Beau has four, and Tomás has five. Each person may also have one $5 bill. None of them has any other kind of money.

At least one person has more money than Cameron and less money than Tomás. Meghan has more money than at least one of the Joneses. Tomás has less money than at least one of the Smiths. The person with the most money has $6 more than the person with the least amount of money. How much money does each person have?

4. REGIONS IN A CIRCLE

A circle can be separated into seven different regions by drawing three straight lines across the circle. The regions will not be the same size. Before you go on, be sure you can draw a circle and separate it into seven regions by drawing three straight lines.

What is the maximum number of regions that will be formed by drawing 100 straight lines across the circle? (Some of these regions will be very small!)

5. PRICEY PETS

The members of a boys' club went to a very large pet store. The club's advisor wanted to find out how much the pets cost, because some of the boys said they wanted one. He noticed some signs at various places in the store. The sign above the fish tank said "Buy 16 fish and 8 cats for the price of 7 dogs." The sign above the cat cage said "Buy 11 cats and 7 dogs for the price of 9 fish." The sign above the dog cage said "Buy 3 dogs and 9 fish for the price of 8 cats." The advisor ended up helping one of the boys buy a fish and a dog. The fish cost $42 less than the dog. At this time, the advisor discovered that exactly one of the three signs was incorrect. Which sign was wrong, and how much did a cat cost?

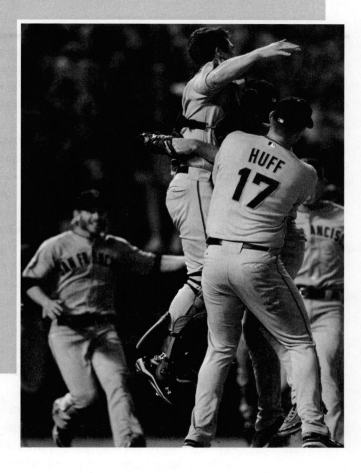

15

Organize Information in More Ways

Many of the strategies to this point have been about organizing information. This chapter looks at some innovative ways to make tables and use tree diagrams. Statstics for sports, including baseball, are organized in many different ways so fans, players, and coaches can evaluate different aspects of a player's and team's performance.

The problem-solving strategies you have explored in this book have been separated into three major problem-solving themes: Organizing Information, Changing Focus, and Spatial Organization. The last three chapters of the book feature additional strategies that fall under each of these three themes.

Throughout this book you've learned techniques for organizing information. The strategies of systematic lists, eliminating possibilities, matrix logic, looking for patterns, guess-and-check, unit analysis, algebra, and finite differences all require that you organize information in some way. (There are also elements of the Organizing Information theme in the strategy of working backwards, although working backwards fits primarily into the Changing Focus theme.) When you organize a problem's information in a meaningful way, the problem is usually much easier to solve. This chapter will explore two additional ways to organize information: two-dimensional arrays and tree diagrams.

Section 1: Two-Dimensional Arrays

At the California State Fair in Sacramento, there are two types of tickets. Discount adult admissions cost $6, and children's admissions cost $4. Before the fair opened, the assistant manager realized that with the large number of people who would crowd into the gates when they opened at 10 a.m., the time it would take the cashiers to add up the ticket price for each group would significantly slow down the line. So she devised an array to speed things up. (**Array** is another word for matrix.) All a cashier had to do was ask each customer, "How many adults, and how many children?" By finding the number of adults on the top of the array and the number of children down the side, the cashier could find the appropriate dollar amount to collect. Look at the array on the following page to determine the admission cost for five adults and seven children.

		0	1	2	3	4	5	6	7	8	9	10
	0	0	6	12	18	24	30	36	42	48	54	60
	1	4	10	16	22	28	34	40	46	52	58	64
	2	8	14	20	26	32	38	44	50	56	62	68
	3	12	18	24	30	36	42	48	54	60	66	72
	4	16	22	28	34	40	46	52	58	64	70	76
CHILDREN	5	20	26	32	38	44	50	56	62	68	74	80
	6	24	30	36	42	48	54	60	66	72	78	84
	7	28	34	40	46	52	58	64	70	76	82	88
	8	32	38	44	50	56	62	68	74	80	86	92
	9	36	42	48	54	60	66	72	78	84	90	96
	10	40	46	52	58	64	70	76	82	88	94	100

With the ticket information organized in this way, the cashiers can save a lot of time when they figure out what each family owes for its tickets. Did you get $58 for the family described above? How much would admission for your family cost?

～～～

You first encountered the next problem, The Three Squares, in Problem Set A of Chapter 3: Eliminate Possibilities. You probably solved it with a combination of strategies—making a systematic list and eliminating possibilities. In this chapter, we will look at a different way to organize the information in this problem.

Three cousins, Bob, Chris, and Phyllis, were sitting around watching football on TV. The game was so boring that they started talking about how old they were. Bob (the oldest) noticed that they were all between the ages of 11 and 30. Phyllis noticed that the sum of their ages was 70. Chris (the youngest) burst out, "If you write the square of each of our ages, all the digits from 1 to 9 will appear exactly once in the digits of the three squares." How old was each person? Do this problem before continuing. See if you can organize the information in the problem in a new way.

The solution for this problem was contributed by Ed Migliore, a mathematics instructor at Monterey Peninsula College in Monterey, California. It uses the strategy of eliminating possibilities as well as the strategy of organizing information in a two-dimensional array.

Make a list.

"First I made a list of numbers from 11 to 30 and their squares. I didn't bother to include numbers like $11^2 = 121$ because there are repeated digits, which obviously was not allowed."

AGE	SQUARE	AGE	SQUARE
13	169	23	529
14	196	24	576
16	256	25	625
17	289	27	729
18	324	28	784
19	361	29	841

Set up an array.

"I set up an array on graph paper. The numbers on the top row are the digits 1 through 9. The numbers down the side are the numbers between 11 and 30 that I squared. I just checked off the digits that appeared in each square. For example, 25^2 equals 625, so I put an $\times$ in each of the 2, 5, and 6 columns."

DIGIT

AGE	1	2	3	4	5	6	7	8	9
13	X					X			X
14	X					X			X
16		X			X	X			
17		X						X	X
18		X	X	X					
19	X		X			X			
23		X			X				X
24					X	X	X		
25		X			X	X			
27		X					X		X
28				X			X	X	
29	X			X				X	

To seek a
contradiction, make
an assumption.

Ed went on. "The only 3's appear in the squares for 18 and 19. So either 18 or 19 must be in the list. Assume it is 18 [an example of seeking contradictions]. This means that 3, 2, and 4 all appear in 18^2 and therefore can't appear in any other number. So eliminate all of the numbers that have 3's, 2's, or 4's in them. This eliminates 16, 17, 19, 23, 25, 27, 28, and 29 [as shown on the next array]."

DIGIT

AGE	1	2	3	4	5	6	7	8	9
13	X					X			X
14	X					X			X
~~16~~		*			*	*			
~~17~~		*						*	*
18		X	X	X					
~~19~~	*		*			*			
~~23~~		*			*				*
24					X	X	X		
~~25~~		*			*	*			
~~27~~		*					*		*
~~28~~				*			*	*	
~~29~~	*			*				*	

A contradiction means the assumption was false.

So 18 is eliminated, and 19 is correct.

"The only number left with a 7 is 24^2, which equals 576, so 24 must be one of the ages. But 576 also contributes a 5 and a 6, and the only remaining ages, 13 and 14, have squares that both contain 6's. So the assumption that 18 is one of the ages proves false, so 18 may be crossed off.

"This also means that 19 is one of the ages for sure, since it is the only other number whose square can contribute the 3. Now I had a bunch of crossed-out values from my false assumption. I had to clean up my chart before I could proceed."

DIGIT

AGE	1	2	3	4	5	6	7	8	9
13	X					X			X
14	X					X			X
16		X			X	X			
17		X						X	X
~~18~~		*	*	*					
19	X		X			X			
23		X			X				X
24					X	X	X		
25		X			X	X			
27		X					X		X
28				X			X	X	
29	X			X				X	

"Knowing that 19 is one of the ages means that the digits 1, 3, and 6 are contributed by 19^2 and I can cross off all of the other numbers whose squares contain 1, 3, or 6. This eliminates 13, 14, 16, 24, 25, and 29."

DIGIT

AGE	1	2	3	4	5	6	7	8	9
~~13~~	*					*			*
~~14~~	*					*			*
~~16~~		*			*	*			
17		X						X	X
~~18~~		*	*	*					
19	X		X			X			
23		X			X				X
~~24~~					*	*	*		
~~25~~		*			*	*			
27		X					X		X
28				X			X	X	
~~29~~	*			*				*	

"The only number left that contains a 5 is 23^2. So 23 must be one of the ages. I circled it and eliminated the other numbers that contained 2's, 5's, and 9's. This eliminated 17 and 27. The only number left in the list was 28, so it had to be in the final list also. So the ages are 19, 23, and 28."

	DIGIT								
AGE	1	2	3	4	5	6	7	8	9
~~13~~	✳					✳			✳
~~14~~	✳					✳			✳
~~16~~		✳			✳	✳			
~~17~~		✳						✳	✳
~~18~~		✳	✳	✳					
19	X		X			X			
23		X			X				X
~~24~~					✳	✳	✳		
~~25~~		✳			✳	✳			
~~27~~		✳						✳	✳
28				X			X	X	
~~29~~	✳			✳				✳	

"So Chris, the youngest, is 19; Phyllis is 23; and Bob, the oldest is 28."

Notice that in Ed's final chart each of the digits 1 through 9 that appeared in the squares appears in only one column. It is also interesting that Ed never used the clue that the ages had to add up to 70—this clue is superfluous to solving the problem. The way Ed organized the problem's information made the solution very easy to reach. His approach is somewhat similar to the matrix logic approach from Chapter 4: Use Matrix Logic.

Applied Problem Solving: Math Contest

The high schools in and around Sacramento, California, participate in a series of monthly mathematics competitions called Mathletes. Each Mathlete team consists of five members, each member must take three of the five tests given in the competition, and each test is taken by three team members. For many years the tests were given in arithmetic, Algebra 2, Algebra 1, trigonometry, and geometry (the test categories were recently revised and now include problem solving).

Ken was the coach of the Mathlete team at Luther Burbank High School for many years. When he first started coaching, for each event he made up a schedule like the one shown next.

Arith	Alg II	Alg I	Trig	Geom
John	Hoa	Khanh	Hoa	Hoa
Julie	Khanh	Julie	Khanh	John
Dahlia	John	Dahlia	Julie	Dahlia

The advantage of this schedule was that each team member could look at the schedule before each particular test started to see whether or not to take that test. However, it wasn't immediately apparent on the schedule which three tests each team member would take—they had to search for that information. In addition, Ken had a hard time figuring out whether each person was taking three tests.

Another Mathlete coach showed Ken an array that she used:

Design your array to allow for the easiest visualization.

	Arith	Alg II	Alg I	Trig	Geom
Hoa		X		X	X
Khanh		X	X	X	
John	X	X			X
Julie	X		X	X	
Dahlia	X		X		X

In her array it was easy to see which test was being taken by which person. It was also easy to see immediately which three tests each person would take. For the coach, it was evident that each person was taking three tests and that each test was being taken by three people. It was also really easy to change if necessary.

Section 2: Tree Diagrams

Another common way to organize information is to use a tree diagram. An example of a tree diagram is illustrated in Chapter 2: Make a Systematic List. You may also have used tree diagrams to solve some of the other problems in this book. Tree diagrams are often used in tournament scheduling. NCAA basketball tournaments and all tennis tournaments organize their schedules with tree diagrams.

These tournaments often feature seedings, or rankings. The best player in the tournament is seeded number 1, the next best player is seeded number 2, and so on. The tournament schedule is organized so that in the first round the best player plays the worst player, the second-best player plays the second-worst player, and so on. Additionally, the schedule is set up so that the number-1 and number-2 seeds cannot meet until the final round. Here is a typical tennis tournament diagram for eight players:

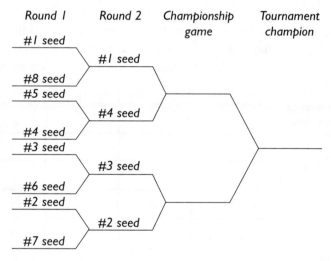

Notice that the sum of the seed numbers for each first-round game is 9 and, assuming that the higher seed wins each first-round game, the sum of the seed numbers for each second-round game is 5. Now assume that the higher-seeded player wins each second-round game, leaving the number-1 and number-2 seeds in the championship game. A similar pattern exists for a tournament of 16 players, 32 players, and so on.

The difference between tournament diagrams like these and typical tree diagrams is that in most tree diagrams the branches grow out to the right.

Steve Weatherly, who taught mathematics at Nevada Union High School in Grass Valley, California, for many years, wrote many of the tree-diagram problems in this chapter. The next problem illustrates how to use a tree diagram to solve problems that involve counting.

FOUR COINS

Liberty is going to flip four coins at once: a penny, a nickel, a dime, and a quarter. How many ways are there for the four coins to come up? Make a tree diagram of all possible results before continuing.

Liberty made the tree diagram shown below. The possible results for the coin flips are indicated in order: penny, nickel, dime, quarter.

Penny	Nickel	Dime	Quarter	Possible results P N D Q
		H	H	HHHH
			T	HHHT
	H	T	H	HHTH
			T	HHTT
H		H	H	HTHH
			T	HTHT
	T	T	H	HTTH
			T	HTTT
		H	H	THHH
			T	THHT
	H	T	H	THTH
			T	THTT
T		H	H	TTHH
			T	TTHT
	T	T	H	TTTH
			T	TTTT

Liberty then counted the ways, shown by the diagram, and found that there are 16 different results you can get when you flip four coins.

Tree diagrams are very useful for solving problems about probability, as demonstrated by the next couple of problems.

HEADS AND TAILS

After Liberty counted all the ways she could flip four coins, she wanted to know the probability of getting two heads and two tails when flipping four coins. What is that probability? Solve this problem before continuing.

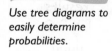

Use tree diagrams to easily determine probabilities.

Liberty looked back at the tree diagram she'd made to solve the Four Coins problem and counted all the places in the diagram where two heads and two tails came up. The diagram shows that two heads and two tails occurred 6 times out of the 16 total times. So the probability of two heads and two tails is $^6/_{16}$, which written as a decimal is 0.375, or 37.5%.

RED AND WHITE

A bag contains one white ball and two red balls. A ball is drawn at random. If the ball is white, then it is put back into the bag along with an extra white ball. If the ball is red, then it is put back into the bag with two extra red balls. Then another ball is drawn. What is the probability that the second ball drawn is red? Solve this problem before continuing.

Len decided to tackle the problem and started with the following tree diagram.

First draw   Second draw

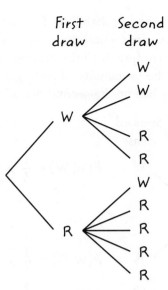

Len said, "First I thought the probability was ⁶⁄₉. But then I realized my diagram was wrong because there were two red balls to start with. I figured this diagram was getting way too confusing, so I tried just writing R and W once on each branch, and writing the probability of that branch along the line." He drew the new diagram shown below.

Len went on. "The probability of choosing a white ball on the first draw is 1 out of 3, since originally there were three balls in the bag—one white and two red. If a white ball is chosen the first time, it is put back into the bag along with an extra white ball. So now there are two white balls and two red balls in the bag. So the probability of drawing a white ball on the second draw is ²⁄₄, or ¹⁄₂. The probability of drawing a red ball on the second draw is also ²⁄₄, or ¹⁄₂.

"Then I looked at what happens if a red ball is chosen first. The probability of that happening is ²⁄₃, because originally there were two red balls and one white ball in the bag. If a red ball is chosen first, then it is put back into the bag along with two extra red balls. So now there are four red balls and one white ball in the bag. The probability of drawing a white ball on the second draw is ¹⁄₅, and the probability of drawing a red ball on the second draw is ⁴⁄₅.

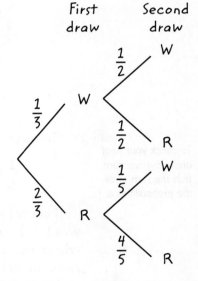

First draw   Second draw

"Now I wanted to find the total probabilities for each line. The probability that a white ball is drawn both times is ⅙. I found this by multiplying ⅓ times ½. I computed all the probabilities. The notation P(W,W) stands for the probability of getting a white ball on the first draw and a white ball on the second draw."

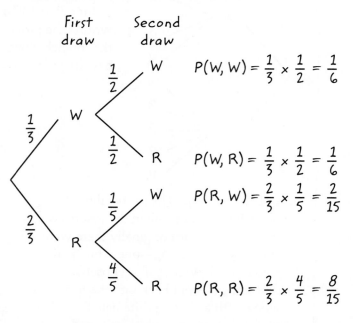

First draw   Second draw

$$P(W, W) = \frac{1}{3} \times \frac{1}{2} = \frac{1}{6}$$

$$P(W, R) = \frac{1}{3} \times \frac{1}{2} = \frac{1}{6}$$

$$P(R, W) = \frac{2}{3} \times \frac{1}{5} = \frac{2}{15}$$

$$P(R, R) = \frac{2}{3} \times \frac{4}{5} = \frac{8}{15}$$

To check your final answer, make sure that the sum of all the probabilities is 1.

"To check to see if I was doing it right, I added up the total probability."

$$\frac{1}{6} + \frac{1}{6} + \frac{2}{15} + \frac{8}{15} = \frac{5}{30} + \frac{5}{30} + \frac{4}{30} + \frac{16}{30} = \frac{30}{30} = 1$$

"I knew I was calculating correctly, because the total probability was 1. I also noticed that the sum of the top two branches was ⅓, which was the probability of getting a white ball the first time. Also, the sum of the bottom two branches was ⅔, and that was the probability of getting a red ball the first time.

"Finally, I calculated the probability of getting a red ball on the second draw. This could happen white, red (W, R) or red, red (R, R). The probabilities were ⅙ and 8/15, so I added ⅙ and 8/15 and got 21/30, which reduces to 7/10. So there is a 7 in 10 chance, or a 70% chance, of getting a red ball on the second draw."

Another excellent use of tree diagrams is in problems about conditional probability. Problems like the next two are common in statistics classes. This next problem may surprise you. It is an example of the mathematics of drug testing, greatly simplified for the purposes of this book.

DRUG TESTING

A company that conducts individual drug tests to identify the existence of certain substances in a person's system claims that its drug test is 90% accurate. That is to say, given a group of people who use drugs, the test will correctly identify 90% of the group as drug users. Futhermore, given a group of people who don't use drugs, the test will correctly identify 90% of the group as non-users. Those numbers sound pretty good on the surface, but suppose someone has just taken this drug test and the results are positive. What is the probability that he actually uses drugs?

To solve this problem, we need to know what percentage of the general population uses drugs. Let's use the company's numbers: To test the validity of its drug test, the company tested a sample group of people, knowing that 5% of this group did in fact use drugs. Solve this problem with a tree diagram before continuing.

Consider a hypothetical group of 10,000 people. If 5% of these people use drugs, then there are 500 drug users in the group (5% of 10,000 = 0.05 × 10,000 = 500). Suppose that all 10,000 people have taken this drug test. The test will correctly identify a drug user 90% of the time. So, presumably, of the 500 drug users, 450 of them will be identified by the drug test. But because the test is only 90% accurate, 50 of them will *not* be identified by the drug test—these test results are called false negatives. The following tree diagram shows this information.

Label your diagram.

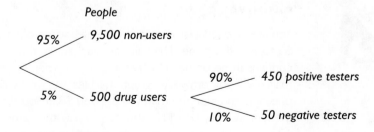

Now let's look at the 9,500 people in our group who do not use drugs. The test is 90% accurate, so it will correctly identify 8,550 people as non-users. However, 10% of the people will register as false positive, a test result indicating that they are drug users when they are not. Ten percent of 9,500 is 950 people, so 950 people will register as false positives. This information has been added to the tree diagram.

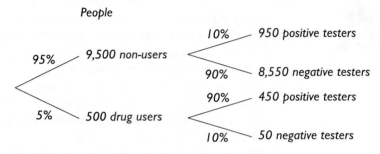

People

95% — 9,500 non-users
  - 10% — 950 positive testers
  - 90% — 8,550 negative testers

5% — 500 drug users
  - 90% — 450 positive testers
  - 10% — 50 negative testers

Now look at all the people who tested positive: 450 + 950 = 1,400 positive tests. But only 450 of those positive testers are actually drug users. The other 950 positive test results are false positives. So what percentage of the positive testers are actually drug users? If we divide 450 by 1,400, we get approximately 32%. That means roughly one in three people who tested positive for drugs with this test are actually drug users. What are the implications of this information for the people who use such drug tests to make decisions and for the people who are affected by those decisions? (Note that many drug tests now take into account these issues regarding false positives.)

Using tree diagrams to analyze a situation involving probability is as applicable to an entertainment situation, such as games of chance, as it is to the important situation presented in the Drug Testing problem. Consider the following problem.

CARNIVAL GAME

You are at a carnival. A man is offering a game of chance. He will charge you $3 to play the game. The rules are as follows: You are to reach into a bag that contains three orange marbles and seven green marbles and draw out two marbles. If you draw two green marbles, the man will pay you $2. If you draw one of each color, the man will pay you $3. If you draw two orange marbles, the man will pay you $12. Should you play this game? Remember, it costs $3 to play. Draw a tree diagram to analyze this situation before continuing.

Nichole solved this problem. "I drew a tree diagram of the situation. The probability of drawing orange the first time is $^3/_{10}$ and of drawing green is $^7/_{10}$. Then if an orange is drawn first, there are nine marbles left in the bag—two orange and seven green—so for the second draw the probability of drawing orange is $^2/_9$ and green is $^7/_9$. If a green ball is drawn first, there are nine marbles left in the bag—three orange and six green—so for the second draw, the probability of drawing orange is $^3/_9$ and green is $^6/_9$. I put all of this into my diagram. Then I multiplied to find the probability of each combination of oranges and greens."

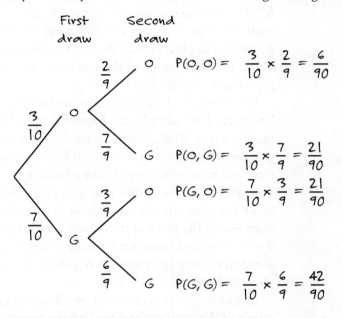

First draw   Second draw

$$P(O, O) = \frac{3}{10} \times \frac{2}{9} = \frac{6}{90}$$

$$P(O, G) = \frac{3}{10} \times \frac{7}{9} = \frac{21}{90}$$

$$P(G, O) = \frac{7}{10} \times \frac{3}{9} = \frac{21}{90}$$

$$P(G, G) = \frac{7}{10} \times \frac{6}{9} = \frac{42}{90}$$

"I then figured out the results of playing the game 90 times. In those 90 times, the following things would happen:

- Draw orange, orange (O, O) and win $12. This would happen 6 out of 90 times. So in those 6 times I would win $72 (6 × 12).

- Draw orange, green (O, G) or green, orange (G, O) and win $3. This would happen 42 out of 90 times ($^{21}/_{90}$ + $^{21}/_{90}$ = $^{42}/_{90}$). In those 42 times I would win $126 (42 × 3).

- Draw green, green (G, G) and win $2. This would also happen 42 out of 90 times. In those 42 times I would win $84 (42 × 2).

"I made a chart to summarize the winnings."

Outcome	Amount Won	Times Occurred	Total Winnings
O, O	$12	6	$72
O, G or G, O	$3	42	$126
G, G	$2	42	$84

"So in the 90 games I would win $72 + $126 + $84, which equals $282. The 90 games would cost me $3 each, which would be $270 total. In those 90 games I would come out $12 ahead, so it is to my advantage to play. I could also look at dividing 282 by 90. My **expected value** for each game would then be $3.13. Because it costs only $3 to play, this is a good deal."

Of course, commercial games of chance are, more often than not, bad deals. For example, the expected value for a state lottery ticket is usually about $0.50. Considering that lottery tickets cost $1.00 to play, you can expect to win only $0.50 for every dollar you pay, getting back only half of your money. That's a terrible investment. The contests that magazine publishers offer also give you a terrible expected value. For some contests, the expected value of an entry is only $0.15, which isn't even worth the price of the stamp you'd need to send it in. In general, if your expected value for winning is less than you have to pay, you shouldn't waste your money. A question to consider: Why do people play the lottery?

Examining all branches of a tree diagram gives you a very clear vision of the problem and the various possible outcomes. Therefore, tree diagrams are very effective in statistical analysis.

Try Different Ways to Organize Information

Ask yourself: Is there a better way to organize?

Often there is more than one way to organize information, as many of the examples in this chapter demonstrate. Experiment with different methods of organization as you do the problems in the problem set. Don't be afraid to throw out what you have and start over.

Tree diagrams are an alternative way to make an organized list. They are especially useful in solving problems that involve probability.

Problem Set A

I. COFFEE STAIN

At an amusement park, all the cashiers used a matrix that contained the prices for various combinations of children's and adults' tickets, similar to the matrix described at the beginning of this chapter. Toward the end of one day, entries to the park were getting rather slow. The manager, Patty, closed down all the entry gates except one. Unfortunately, the cashier spilled coffee on his matrix. All he could read were the numbers shown in the following chart, and he wasn't even sure that they were all correct, although he figured most of them were.

Re-create the matrix changing the fewest numbers possible, and determine what the prices are for adults' and children's tickets. Also determine which numbers in the matrix are incorrect.

ADULTS

		0	1	2	3	4	5	6	7	8
	0									
	1				59					
	2			58						110
	3		34				99			
CHILDREN	4			57		93				
	5				74			126		
	6		55							
	7					114				
	8									

2. THE OTHER THREE SQUARES

Three other cousins of the three cousins we met earlier noticed that their ages were all under 40 and that together the squares of their ages contained each of the digits from 0 to 9 exactly once. What were their ages?

3. TWO BILLS

I once met a father and a son, both named Bill. Bill Senior said, "I moved to Florida in 1965." Bill Junior said, "I was a sophomore in high school when my father moved to Florida." Bill Senior said, "Bill turned 16 in his sophomore year." The conversation took place in 1991. What were the possible ages for Bill Junior?

4. FAIR AIRFARE

On a recent airline flight, six people from the same company were seated in rows 1, 2, and 3. The plane had two seats in each row. Each of the six had booked the flight at different times, so they all may have been charged different amounts. The people in the window seats had fares that were $5 apart (such as $14, $19, and $24). One person in an aisle seat had a fare equal to her seat partner's. Another person in an aisle seat had a fare that was one and a half times her seat partner's. The third person in an aisle seat had a fare that was twice her seat partner's. The total fare for all six people was $1,025. All fares were in whole-dollar amounts. How much was the second most expensive fare?

5. AHSME

The Mathematical Association of America sponsors a mathematics contest every year called the American High School Math Exam. The contest consists of multiple-choice questions that include five possibilities for each question. For many years, the scoring system for the 30-question test was as follows: Each right answer was worth 4 points. Each wrong answer was worth −1 point. Each unanswered question was worth 0 points. Each participant started with a score of 30 points. Thus, for a person who got 6 right, 3 wrong, and left 21 unanswered, the score would be $30 + (6 \times 4) + (3 \times -1) + (21 \times 0)$, or 51 points. To qualify for the second round of competition, a person had to score 95 points.

Around 1988, the scoring system changed. The new scoring system became 5 points for a right answer, 0 points for a wrong answer, and 2 points for no answer, and each participant starts with 0 points. Thus, the example of 6 right, 3 wrong, and 21 unanswered would result in a score of $(6 \times 5) + (3 \times 0) + (21 \times 2)$, or 72 points. To qualify for the second round of competition, a person has to score 100 points.

Analyze these two different scoring systems and discuss which system you think gives a person a better chance of qualifying for the next round. For each system, decide on the best strategy for reaching the second round, given that you are sure of the answers to ten questions and have narrowed down ten other questions to two choices.

6. **TWO-INPUT FUNCTION**

The following tables all represent the same function. The function has two inputs. The first input is x, and the second input is y. What is the rule for calculating the output?

Input	Output	Input	Output	Input	Output
2, 3	17	1, 5	29	7, 4	44
4, 1	17	9, 2	40	1, 4	20
3, 7	61	0, 5	25	2, 5	33
4, 3	25	5, 2	24	4, 2	20
3, 9	93	2, 4	24	3, 8	76

7. **SIXTEEN-TEAM TOURNAMENT**

Draw a tournament diagram for a tournament with 16 teams, with each team seeded from 1 to 16.

8. **FOURTEEN-TEAM TOURNAMENT**

Draw a tournament diagram for a tournament with 14 teams, each one seeded from 1 to 14. The first round will need to include two byes. (A bye means that a team will not have to play a game but will automatically move on to the second round. Essentially that means that the team will not have an opponent in the first round—so they will receive a win without playing a game.)

9. **ORDERING PIZZA**

How many pizzas made from the following ingredients can be ordered? Choose one from each category.

Crust: thick, thin, or medium

Meat: sausage, pepperoni, hamburger, or Canadian bacon

Veggies: olives, onions, or mushrooms

10. INTERVIEW ORDER

In how many orders can four basketball players be interviewed after a big game?

11. BAG OF MARBLES

A bag contains three blue marbles, two red marbles, and five white marbles.

a. What is the probability that if you select one marble it will be white? Not red?

b. Now suppose you were going to draw two marbles out of the bag (without replacing either of them). What is the probability that both marbles are white? That one marble is red and the other is blue?

c. Now suppose that you are going to draw one marble out of the bag. If it is white, you will put it back into the bag along with two extra white marbles. If the marble you draw out of the bag isn't white, you will put it back into the bag but you won't add any additional marbles. Then you will draw a second marble. What is the probability that the second marble you draw is red?

12. BUSY SIGNAL

On any given evening, many high school students are on the phone, especially the students taking algebra. Of all the algebra students in one high school, 60% are ninth graders and 40% are tenth graders. On a given evening, 80% of these ninth graders are on the phone, and only 50% of these tenth graders are on the phone. If an algebra student is not on the phone, what is the probability that the student is a tenth grader?

13. LAWN MOWER

Jay has a business mowing lawns for the people in his neighborhood. He charges $13 for his basic lawn job. One neighbor offers to make him a deal: He will put two $10 bills and eight $5 bills in a bag and let Jay draw out two bills. Is it in Jay's favor to accept? Why?

14. FOUR ACES

The four aces from a deck of playing cards are lying facedown on a table. We turn over one card at a time until a red ace is found or until two black aces are found. What is the probability of finding a red ace?

15. X RAY

An X ray will reveal with probability 0.8 the presence of a certain lung condition in all affected patients. If the X ray is negative (fails to show the condition), then a second X ray is taken. If the results of the two X rays are independent, what fraction of affected patients will be diagnosed by this technique?

16. WORLD SERIES

Suppose the Cleveland Indians are playing the Chicago Cubs in the World Series (7 games). So far the Cubs have won two games, and the Indians have won one. If the odds for either team to win any game are even and the series ends when one team wins four games, what is the probability that the Cubs will win the Series?

17. SONS AND DAUGHTERS

A woman has three sons (Adam, Barry, and Cosmo) and six grandchildren. Adam has two sons, Barry has one son and one daughter, and Cosmo has two daughters. While talking to you over the telephone, the woman states that one of her sons has left his children in her care for the day. In a few minutes you hear a boy's voice in the background. What is the probability that the children at her house are Adam's?

18. DISEASE

Suppose the presence of a certain disease can be detected by a laboratory blood test with probability 0.94. If a person has the disease, the probability is 0.94 that the test will reveal it. But this test also gives a false positive in 1% of the healthy people tested. That is, if a healthy person takes the test, it will imply (falsely) that he or she has the disease, and this will happen with probability 0.01. Suppose 1,000 persons have been tested, of whom only 5 have the disease. If 1 of the 1,000 is selected at random, find the probability that he is healthy, given that the test is positive (that is, that the test says he has the disease).

CLASSIC PROBLEMS

19. NEW STATIONS

Every station on the N railroad sells tickets to every other station. When the railroad added some new stations, 46 additional sets of tickets had to be printed.

How many is "some"? How many stations were there before?

Adapted from *The Moscow Puzzles* by Boris Kordemsky.

20. THE MONTY HALL PROBLEM

Suppose you're on a game show and you're given the choice of three doors: Behind one door is a car, and behind the other two doors are goats. You pick a door, say door #1, and the host, who knows what's behind the other doors, opens another door, say door #3, which has a goat. He then asks you, "Do you want to switch to door #2?" Is it to your advantage to switch?

Published in the "Ask Marilyn" column in a 1990 issue of *Parade,* this puzzle generated great controversy. A form of this puzzle also appeared in Martin Gardner's "Mathematical Recreations" column in *Scientific American* in 1959.

MORE PRACTICE

1. WHICH DIGIT IS REPEATED?

Two cubes, each five-digit numbers with five different digits, contain all nine of the digits 1–9 with one digit repeated. What is the repeated digit?

2. ROOM RATES

In the "Van Winkle" wing of the 2nd floor of the Sleeptight Motel, unbeknownst to the guests, each of them may have paid a different amount for their identical rooms. The people on the south side of the building paid rates that were $10 different from one another (such as $43, $53, and $63). One of the people on the north side paid one-third more than the person across the hall. Another person paid twice as much as the person across the hall, and the last one paid the same amount as the person across the hall. All room rates were in whole-dollar amounts. The total of those six rooms for the night was $716. Two people paid the same amount for their room. What was that amount?

3. DOUBLE INPUT FUNCTION

The following tables all represent the same function. The function has two inputs. The first input is x, and the second input is y. What is the rule for calculating the output?

Input	Output	Input	Output	Input	Output
2, 3	26	7, 2	111	4, 1	40
2, 6	41	0, 4	23	9, 1	170
3, 3	36	8, 2	141	2, 7	46
3, 0	21	4, 4	55	6, 5	100
5, 2	63	1, 7	40	3, 5	46
5, 8	93	6, 1	80	7, 7	136

4. THE DILLY DELI SANDWICH SPOT

The Dilly Deli Sandwich Spot offers the following menu:

Meats	Bread	Extras	Condiments	Cheese
Pastrami	Sourdough roll	Tomato	Mustard	Provolone
Roast Beef	Wheat roll	Lettuce	Mayo	Swiss
Chicken Breast	Sliced rye	Sprouts	Both	
Turkey	Pita	Onion		
Tuna				

For $6.95 you can choose any combination of one item listed in each column. If you want to add more than one item from a column to your sandwich it will cost $0.50 for each extra item. If you decide not to order any item in a column, it will cost $0.50 less. Gerald orders a sandwich at the Dilly Deli every day, but he never messes around with extra items or fewer items, he always chooses exactly one item from every column. How many days could he order sandwiches before he had to repeat exactly the same sandwich as he had already ordered?

ABBY'S SPINNER

Abby, her friends, and her siblings were always asking each other, "What do you want to do today?" They got tired of wasting time and decided on a method to decide randomly. On Mondays, Wednesdays, and Fridays they would get together in the afternoon. Abby created a spinner that was divided into three equal sections and labeled them soccer, basketball, and lacrosse. At the beginning of each week they spun the spinner three times to make the decisions for them.

Make a tree diagram to show the probabilities, then answer these questions:

a. What is the probability that in a given week they will play lacrosse three times?

b. What is the probability they will play lacrosse at least once during the week?

c. What is the probability they won't play lacrosse at all (during a given week)?

6. **SHOVELING MANURE**

Neil gets paid for his labor by a couple of horse owners near his house. The training was quick, as the job simply involved shoveling manure. He usually gets paid $15 to clean out three stables in a barn. One time, however, Victoria makes him an interesting offer. She pledges to put four $20 bills, three $10 bills, and a $5 bill in a bag, and will pay him whichever bill he pulls out of the bag each time he comes to shovel the manure. Should Neil accept the offer?

7. **RED, WHITE, AND BLUE**

Keshila and Mikaela made up a game. They put 1 red, 1 white, and 2 blue marbles in a sock, and then pull one out at random. They both look at the color of the marble, then Keshila puts it back in. Mikaela puts in another marble of the same color, and they do a second, random draw.

Make a tree diagram to analyze the probabilities. What is the probability that the color of the marble on the second draw is blue?

8. PROTEIN SHAKE

I exercise every day—either jogging, yoga, or weight lifting. Seventy percent of the time I go jogging, 20% of the time I do yoga, and 10% of the time I weight lift. After jogging, I will drink a protein shake half of the time. After yoga I will drink a protein shake 30% of the time. And after lifting I will drink a protein shake 90% of the time. Last Saturday I drank a protein shake. What is the probability that I lifted weights that day?

9. CARNIVAL GAME

You are playing a carnival game. First you throw a dart at a balloon. You figure you have a 40% chance of popping the balloon. Next you shoot a basketball, and you figure you have a 1/3 chance of making the basket. If you pop the balloon and make the basket you get $8. If you pop the balloon and miss the basket you get $3. If you don't pop the balloon and make the basket you get $4. If you don't pop the balloon and miss the basket you win nothing. The game costs $2.75 to play. Should you play this game? (Note that you have to pay the $2.75 every time you play. A win of $3 means you come out $0.25 ahead.)

10. FLIP TO SPIN OR ROLL

Flip to Spin or Roll is a new game at the Grosspoint County Fair where you pay $3.00 for the possibility of winning a huge stuffed animal. First you spin a wheel, with the numbers 1–10. If you spin a prime number, you then get to flip a coin and you will win if it comes up heads, but lose if it comes up tails. If you don't spin a prime number you get to roll a die, and you will win if it comes up a 5 or 6 but lose if it comes up 1, 2, 3, or 4. The stuffed animal probably cost the people running the game $6.95. Is playing the game worth it?

11. ROCHAMBEAU

Armand, Barbara, and Carlos are playing a three-way variation of rock-paper-scissors where Armand wins when all three come up with the same result, Barbara wins when exactly two of them match, and Carlos wins when none of them match. Who is most likely to win? How could they award points to each player to make this a fair game?

12. CAT AND DOG LADY

Kayline loves cats and dogs. She loves them so much that she will often house sit for no pay if there is a cat or dog living there. She house sits every weekend and has arranged a regular rotation with Eddo, Teferi, and Geraldine so all have equal time. Eddo has two cats and a dog, Teferi has a cat and a dog, and Geraldine has a cat and two dogs. Her friend, Katrina, called Kayline one weekend and heard one of the animals meow. Since Katrina couldn't remember where Kayline was house sitting, what is the probability Kayline is at Eddo's house?

13. PIGEONS ON THE ROOF

Homer was working in his yard when he noticed a fantail pigeon on his roof. He knew three of his neighbors have fantail pigeons so he assumed that the pigeon belonged to one of his neighbors. Kip has three brown ones and a white one, Robin has a brown one and two white ones, and Paloma has two brown ones and two white ones. If the escaped bird has an equal chance of escaping from each house, what is the probability that the brown one on Homer's roof belongs to Paloma?

Problem Set B

1. MOVIE THEATER

Five adults and their five young children went to a movie. The adults all sat together and the children all sat together, with one empty seat between the two groups. As the lights dimmed and the movie began, the children realized they had sat behind a group of tall basketball players, and they asked the adults to trade places with them. However, to be considerate to the people behind them, they all decided to follow these rules:

1. They could move into the next seat if it was empty.

2. They could leapfrog over only one person into an empty seat.

3. Nobody would backtrack. All moves had to be toward their new seats.

How many moves were necessary for the two groups to switch places?

2. **CALCULUS AND FRENCH**

There are 530 students in the freshman class.

1. There are 200 students enrolled in both calculus and French.

2. There are 70 students who take a language but don't take math.

3. Two-thirds of all language students are taking French.

4. The number of calculus students not taking a language is one-third the number of math students not taking a language.

5. Three-fourths of all math students are taking calculus.

6. There are 420 language students and 330 calculus students.

7. There are 100 calculus students who take a language other than French.

8. The number of French students not taking math is 30 less than the number of math students not taking a language.

How many students are taking a math class other than calculus and a language class other than French? How many students are taking neither a math class nor a language class?

3. **LOTSA FACTORS**

Including 1 and itself, how many positive-integer factors does the number 1,746,360,000 have?

4. **DICEY DIFFERENCES**

You are playing a new dice game. You roll two regular dice and then subtract the smaller number from the larger number. (If the dice show the same number, then it doesn't make any difference which way you subtract.) What difference is most likely to occur?

5. **AREA AND PERIMETER**

Find the dimensions of all rectangles that have area equal to twice the perimeter, where both the length and the width are whole numbers. (Ignore units.)

16

Change Focus in More Ways

Changing focus generally means stepping outside the problem to change your point of view, solve the complementary problem, or change the representation of the problem. Like the correct path through a maze, the solution to a problem can be easy when you change your point of view.

T his chapter is about several strategies that fall under the major problem-solving theme Changing Focus. Some problems are best solved by looking at them in a completely different way. For example, subproblems direct your attention away from the question asked and onto parts of the problem that are more manageable. Easier related problems require that you temporarily suspend your work on the original problem and concentrate on easier versions of the same problem. When you work backwards, you go to the end of a problem to see how everything in the problem leads up to that point. This chapter explores three other ways to change focus: *change your point of view, solve the complementary problem,* and *change the representation.* Movie and television directors have often used a changing-focus strategy when filming a movie or a TV program. The old TV show *Mission: Impossible* used this technique. The camera would focus on a character in the foreground, then gradually move the focus to someone in the background—usually someone the person in the foreground was unaware of. If you've ever looked through a window on a rainy day, you have probably experienced a similar sensation. While looking out the window, you don't see the raindrops *on* the window until you focus your eyes on them.

Section 1: Change Your Point of View

You may have experienced some conflicts in your life that you and your family could not resolve. Possibly you sought the advice of a friend or a family counselor. This person was able to bring a new, fresh, and objective point of view to your situation and perhaps was able to help you resolve your conflicts. Sometimes you can solve your own individual problems by adopting a similar strategy. Looking at your situation the way someone else sees it may help you find a solution.

A story is told about a young boy who lost his contact lens while playing basketball in his own driveway. The game stopped while everyone looked for it, but the boys had no luck in finding it. Just then Mom came home. Her son said, "Mom, I've lost my contact lens." We've looked and can't find it anywhere." The mother got down on her hands and knees and in less than a minute announced "Here it is." The boys were amazed and asked, "How did you do that?" She replied, "Simple. You were looking for a piece of plastic, and I was looking for $150."

Similarly, being able to **change your point of view** may help you solve a mathematical problem that you think is impossible. You may have seen the next problem before. The solution to this problem is a good illustration of the strategy of changing your point of view.

NINE DOTS

Without lifting your pencil from start to finish, draw four line segments through all nine points.

The solution has nothing to do with how wide the dots are or with the idea that possibly the lines determined by them are not parallel. The points are mathematically defined—they have no width, and they determine sets of parallel lines.

Work this problem before continuing.

The solution to this problem is very simple, but finding the solution is not easy. Generally people feel constrained to make the solution fit within the confines of the square determined by the nine dots. The key here is to be flexible enough to allow your thinking to diverge from the zone indicated by the dots. You must change your point of view, moving your focus away from the structure of the square and allowing yourself to draw outside it. This type of thinking is commonly referred to as **thinking outside the box.** (You can find the answer to this problem before the applied problem solving feature at the end of this chapter.)

You could use a computer to solve the next problem, but that would be taking a sledgehammer approach. This chapter is about finding creative ways to look at problems in a different light. Think creatively as you solve this problem.

Mayra is a human computer. She has appeared on talk shows to show off her amazing ability with numbers. One type of problem Mayra is very adept at solving is this: A person from the audience will give Mayra a number, and Mayra will immediately be able to tell how many one-digit **factors** that number has. For example, if you were in the audience and you said 50, Mayra would say 3, because 50 has 3 one-digit factors (namely, 1, 2, and 5).

One day, Mayra was on a well-known talk show, and some wise guy in the audience asked Mayra to tell him how many one-digit factors the numbers from 1 to 100 had. The answer was not 9, because Mayra had to count each factor for a particular number, then add that count to the number of factors for each of the other numbers from 1 to 100. For instance, even though the factor 5 appears in the factorization of 50, it also appears in the factorization of 45, so it must be counted once for each of those numbers. Mayra quickly "programmed" her brain to give her the answer, and she had it in a few moments. What was her answer?

Work this problem before continuing.

This problem encompasses many strategies. Arlene took one possible approach to this problem. "I decided to list all the numbers from 1 to 100 and write down all their factors. Then I counted up the factors that were only one digit."

A portion of Arlene's list appears at right.

At this point, Jessica joined her. As Jessica would say later, "It seemed like this was going to take an awfully long time. So I suggested to Arlene that maybe there was an easier way. We looked at her list together to see if we might notice something."

Number	Factors
1	1
2	1, 2
3	1, 3
4	1, 2, 4
5	1, 5
6	1, 2, 3, 6
7	1, 7
8	1, 2, 4, 8
9	1, 3, 9
10	1, 2, 5, 10
11	1, 11
12	1, 2, 3, 4, 6, 12
13	1, 13
14	1, 2, 7, 14

ARLENE: When Jessica came over, I resented it at first. I mean, I had a perfectly reasonable way to do this problem, and I was doing fine. I knew it was going to take a long time, but I tend to be a diligent, hardworking student and I figured I could do it. Sometimes, when I see a way that will work, I just continue with that method, even though it may take a while. It's better than not doing anything.

JESSICA: I tend to be lazy, so I wanted to find an easier way. I had a feeling that Arlene didn't like my interfering, but lots of times two heads are better than one. I asked her if she noticed anything about her list.

ARLENE: At first I was a little irritated, but right away I noticed that I had written down 1 every time, 2 every other time, 3 every third time, and so on.

JESSICA: After Arlene pointed this out, all we had to do was count up the number of 1's, 2's, 3's, 4's, 5's, 6's, 7's, 8's, and 9's that appeared in the list. But we didn't have to make the whole list to do that. There were obviously going to be 100 ones and 50 twos.

ARLENE: We had a little trouble with the number of threes. Then I realized that the last 3 would occur in the number 99, so there would be 33 threes.

JESSICA: All we had to do was divide each number into 100. If it didn't come out evenly, we had to round down. For example, how many 6's are there? Well, 100 divided by 6 is $16\frac{2}{3}$. The last 6 therefore occurs as a factor of 96, and we can ignore the remainder because the next 6 shows up as a factor of 102 and we weren't supposed to go that far.

Factor	Times It Appears
1	100/1 = 100 times
2	100/2 = 50 times
3	100/3 = 33 times
4	100/4 = 25 times
5	100/5 = 20 times
6	100/6 = 16 times
7	100/7 = 14 times
8	100/8 = 12 times
9	100/9 = 11 times
Total	281 times

ARLENE: So there are 281 one-digit factors of the first 100 numbers. Jessica was right, there was an easier way.

JESSICA: We verified that two heads are better than one.

Changing the focus can help you organize the solution.

Jessica and Arlene changed their point of view to solve the Human Factor problem. They completely changed their approach to the problem when it became apparent that the original approach was going to take a long time. Instead of attacking the problem by checking each number to see how many factors it had, they changed their focus to the number of times a given factor appeared in the list of the first 100 numbers. This method counted exactly the same thing but organized the count in a much more manageable fashion, which made the problem a lot easier.

You must be "a little lazy" to use this strategy effectively. If you are willing to proceed with "the long way," you probably won't look for an easier way. However, a little laziness is a good motivator for stepping back and surveying the scene to see if there is a shorter way.

Jacques left his home in Austin and drove to San Antonio. On the way there, he drove 40 miles per hour (there was a lot of traffic). On the way back, he drove 60 miles per hour. What was his average speed? Work this problem before continuing.

Use a variety of strategies: change your point of view, use an easier related problem.

Because 50 is the average of 40 and 60, it appears that 50 miles per hour is the answer, but this answer is incorrect. Try changing your point of view. Instead of looking at the speed, concentrate on the time and the distance. Notice that there is no stated distance in this problem. This causes some difficulty, so use an easier related problem. One way to make this problem easier is to make up a distance. What you learn from the easier problem could help you solve the original problem. Because you might not be sure this approach works, make up at least two distances and figure out the problem for both.

A good distance to use would be 120 miles. One hundred twenty is the **least common multiple** of 40 and 60. This should keep some of the other numbers in the problem simple.

If you use 120 miles as the distance from Austin to San Antonio, then it took Jacques 3 hours to make the trip to San Antonio traveling at 40 miles per hour. On the way back, he traveled 120 miles at 60 miles per hour. This trip took 2 hours. The total distance traveled was 240 miles, and it took 5 hours to travel that far.

$$\frac{240 \text{ miles}}{5 \text{ hours}} = 48 \text{ miles per hour}$$

Verify this answer by using a distance different from 120 miles. The answer comes out the same. If you shift your point of view away from speed and instead focus on time, this problem becomes easy to solve.

In the movie *Patch Adams,* the character of Patch Adams (played by Robin Williams) has a discussion with Mendelson (played by Harold Gould). Mendelson holds up four fingers and asks Patch how many fingers Patch sees. He responds, "Four." But Mendelson claims that the answer is incorrect. "You are looking at the problem," he says. "Look past the problem. See the solution." When Patch looks past the problem (the fingers), he no longer sees four fingers. He sees eight.

Try this great example of changing your point of view. Hold up four fingers at arm's length and look past them to an object across the room. Now how many fingers do you see? By focusing on an object past your fingers, your fingers will blur and you will actually see *eight* fingers. The point here is that when you are confronted with a problem, you sometimes have to change your point of view to be able to solve it. In the movie, Patch has a breakthrough in his thinking when he realizes that doctors need to change their focus from the disease to the patient. By making this change, Patch becomes a better doctor.

Section 2: Solve the Complementary Problem

Some problems are more easily approached by a "comin' around the back" approach. That is, instead of solving the problem that is asked, try finding the opposite. Consider a problem from probability: How many ways are there to roll two dice and have them sum to a one-digit number? Instead of answering this question, answer the opposite question: How many ways are there to get a sum of 10, 11, or 12? This problem is much easier. A 10 can be rolled as (6, 4), (5, 5), or (4, 6). An 11 can be rolled as (5, 6) or (6, 5). A 12 can be rolled as (6, 6). So there are six ways to roll 10, 11, or 12. Therefore, because there are 36 outcomes when rolling two dice, there must be 30 ways to roll a one-digit number.

This approach, to solve the **complementary problem,** is another way to change focus. You are probably familiar with the word *compliment,* as in "Your hair looks really nice today." The word *complement,* on the other hand, refers to two things that are different yet fit together. Here are some of Webster's definitions of **complement:**

- That which completes or brings to perfection.

- Something added to complete a whole; either of two parts that complete each other.

- In music, the difference between a given interval and the complete octave.

- In geometry, **complementary angles** are two angles whose measures sum to 90°.

The key in all these definitions is the idea of two parts making a whole. In the dice problem, the sums of 2 through 9 complement the sums of 10 through 12, and vice versa, because the complete set of possible sums includes those from 2 through 12.

Keep in mind that the strategy of solving a complementary problem provides another way to change focus. Rather than solving the stated problem, solve the complementary problem if it appears to be easier to solve. Then use the result to solve the original problem. Use this strategy to solve the next problem.

TERM PAPER

Seiko had to read five books for her 20th-Century American Lit. class. She then had to write a paper about the role of the family as presented in the books. She had narrowed down her choices to six books.

Of Mice and Men by John Steinbeck

The Bean Trees by Barbara Kingsolver

The Joy Luck Club by Amy Tan

The Color Purple by Alice Walker

A Lesson Before Dying by Ernest J. Gaines

Native American Testimony, edited by Peter Nabokov

In how many ways could Seiko choose the five books? Work this problem before continuing.

Seiko explained, "I thought about making a systematic list of all the ways I could choose five books. But then I realized that each entry in my list would always contain five of the books and leave one out. Figuring out the number of ways to leave out a book is a lot easier to do. Since there are six books, each could be left out once. So there are six ways to choose five books. In this case, the complementary problem is much easier to solve."

Complementary events come up in many different areas of life. A baseball manager attempting to determine which of his pitchers will be relievers must first determine who will be starting. Mechanics can figure out what is broken by noting what is working. Teachers count heads on field trips to determine whether anybody is missing.

Some other strategies and problems in this book incorporated the idea of complementary situations. In matrix logic problems, your objective was to determine how different things matched up. You often did this by eliminating the complementary possibilities. In Chapter 13: Convert to Algebra, the pet-store employee in the Salt Solution problem was asked to raise the concentration of salt in a solution. He could accomplish this task by either of two complementary actions: adding salt or evaporating water.

AREA

Find the area of the shaded region. Work this problem before continuing.

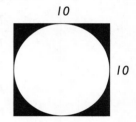

The shaded area is a complement of the circle when both are viewed as parts of the square. This section focuses on complementary problems, and this problem clearly illustrates the concept of two parts that fit together as complements. Nobody would try to find the area of the shaded section directly (though it can be done). Instead, it is far easier to take the subproblem-complementary problem approach to it. That is, rather than finding the area of the regions that *are* shaded, it is much easier to find the area of the region that is *not* shaded and subtract it from the area of the square.

In this case, the area of the square is 100 square units. The area of the circle is $\pi(5^2)$, which is approximately 78.54 square units. The complementary problem here is finding the area of the circle, which complements the area of the shaded regions. By subtracting, we find that the answer to the original problem is approximately 21.46 square units.

THE TENNIS TOURNAMENT

A big regional tennis tournament in New Orleans drew 378 entries. It was a single elimination tournament, in which a player was eliminated from the tournament when she lost a match. How many matches were played to determine the champion? Work this problem before continuing.

Henrick solved this problem with a diagram and easier related problems. "I looked at this problem and wanted to give up. It was way too hard. So I changed it to only two people in the tournament. That was a much easier problem, and it had an easy solution: one match. Then I used three people. This was a little harder. If the players were A, B, and C, I figured A could play B while C had a bye. Then the winner would play C for the championship. That would be two matches. Four players seemed harder. I decided to make one of those elimination charts like they use in tennis tournaments and the NCAA basketball tournaments."

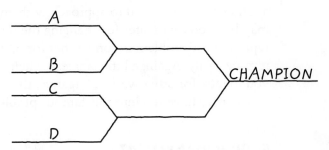

"From this I could see that there would be three matches. I suspected that the number of matches needed was one less than the number of players, but I really wasn't sure. That's when LaVonn helped me."

LaVonn explained the solution. "I saw that Henrick had a good pattern going, but it seemed as if he would never really be sure if he was right. It suddenly occurred to me that maybe we could look at the complementary problem. Instead of determining the number of matches necessary to determine a winner, I figured out how many matches were necessary to determine all the losers. There is only one winner, so everyone else must lose. Each match produces a winner and a loser. The winner goes on, but the loser is eliminated. When the tournament is over, all players but one have lost. So there must be exactly that many matches: one less than the number of players in the tournament. In a tournament that has 378 players, there must be 377 losers and therefore 377 matches to determine a champion."

(For more about tree diagrams and tournament charts, see Chapter 15: Organize Information in More Ways.)

Section 3: Change the Representation

Change the representation is a strategy that combines changing focus with some other strategy. Generally, you use the strategy of drawing a diagram when you change the representation, but other strategies are possible. For instance, you might use a systematic list to organize your problem information, but you could convert your list to a matrix. In the new representation, all sorts of hidden relationships become clear.

One problem may seem to lend itself to a particular strategy, and another may seem hard to approach with any strategy. Such problems may be good candidates for changing the representation. If you can represent the problem in some other form, you may be able to solve it more easily. Again, a little laziness is helpful because it motivates you to look for other ways to organize the information in the problem.

You may have run into this famous problem about cards before.

CARD ARRANGEMENT

You have ten cards numbered 1, 2, 3, 4, 5, 6, 7, 8, 9, and 10. Your task is to arrange them in a particular order and put them in a stack, hold the stack in your hand, and then do the following: Put the top card on the table faceup, put the next card on the bottom of the stack in your hand, put the next card on the table, put the next card on the bottom of the stack, and so on, continuing to alternate cards that go on the table and under the stack, until all ten cards are on the table.

That, of course, is easy to do. The trick is to lay the cards on the table in numerical order. In other words, the first card you put on the table will be number 1, the next card you put on the table will be number 2, the next card you put on the table will be number 3, and so on, until the last card placed on the table will be number 10. In what order should the cards be arranged in the original stack so that this will happen?

See if you can represent this problem in a completely different way (we used a diagram) and solve it more easily. Work this problem before continuing.

Changing the way a problem is represented on paper can make it easier to solve.

If you worked on this problem before, you probably solved it with a physical representation in conjunction with working backwards. This problem can also be solved by changing the representation. Draw ten lines in a circle, as shown at right. Label one of them the first (top) card in the stack. Write the number 1 on that line. Then go to the next line. (This solution will proceed in a clockwise direction, but you could just as easily go counterclockwise.) The next line represents the second card in the stack. This card will be put under the stack, so skip this line. The next line will be the third card in the stack. This will be card number 2. Label this line 2. Skip the next line, because that card will go under the stack. Label the next line 3. Skip the next. Then label 4. Skip the next. Then label 5. Skip the next.

The next line you come to is line 1, but this line represents a card that has already been put on the table, so it would no longer be in the stack. The next empty line will be card 6. Skip the next empty line, because this card will be put under the stack. Continue in this way until you reach card 10.

The final order proceeds from the top (card 1) clockwise to the last card (card 8). So the order is 1, 6, 2, 10, 3, 7, 4, 9, 5, 8. Try it. Note that completely changing the way the problem was represented made the problem much easier to solve.

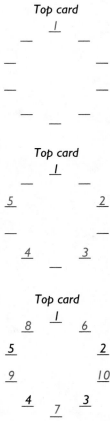

Ed also used the strategy of changing the representation on this problem, but he used a different representation. "I wrote down ten blanks in a row to represent the ten cards. Then I numbered the first blank 1, the third blank 2, the fifth blank 3, and so on. The second, fourth, sixth, eighth, and tenth blanks I labeled A, B, C, D, E."

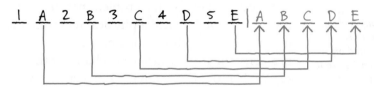

"The cards numbered 1 through 5 would be the first five cards laid down on the table. The cards labeled A through E would be the five cards put under the stack. So I drew arrows to show them being put under the stack."

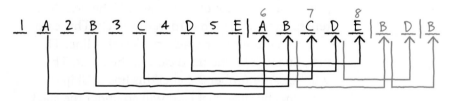

"After card E, the next card should be 6. That means that card A should be 6. Then card B goes under the stack again, card C becomes 7, card D goes under the stack, and card E becomes 8."

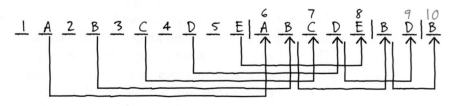

"Then card B goes under the stack again, and card D becomes 9. So card B is 10."

"Finally I went back to the original list and labeled A–E with the numbers they turned out to be. Now I knew the order of the cards."

1 6 2 10 3 7 4 9 5 8

Changing the representation is another way of changing the focus. It can be a very good strategy, though it may be difficult to apply. By the way, here's the solution to the problem that opens the chapter.

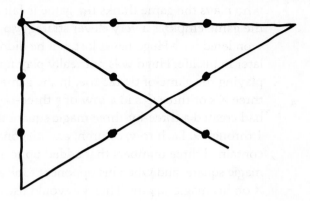

Applied Problem Solving: Con Game in Central Park

At the Annual Meeting of the National Council of Teachers of Mathematics in San Diego, California, in April 1996, mathematics instructor Alan Barson from Beaver College in Glenside, Pennsylvania, told the story of a man in New York City's Central Park who played the following card game: The game was played with nine cards, each showing one of the digits 1 through 9. Two players took turns choosing one of the cards. The object of the game was to be the first player who had three cards that added up to 15. Consider the sample game below.

PLAYER 1 TAKES	PLAYER 2 TAKES
4	6
2	9
8	5
3	

At this point the game is over and player 1 wins the game because he has 3, 4, and 8, and these sum to 15. Note that player 2 did not win when he had 6 and 9, because you need *three* cards that add up to 15. Try playing this game with someone.

The man playing the game—let's call him Hugo—never lost this game. Either he would always tie the game or he would win. As you'll see as you read on, Hugo had an unseen edge. The unsuspecting person who plays the game thinks the game is fair, but the person running the game employs a very clever strategy so that he will not lose. Alan wondered why Hugo never lost, but he didn't figure it out until much later in his life. Hugo was not really playing a card game; he was playing the game of tic-tac-toe. In the game of tic-tac-toe, you need three X's or three O's in a row of a three-by-three grid to win. Hugo had created a three-by-three magic square that contained the digits 1 through 9. Each row, column, and diagonal of his magic square contained three numbers that added up to 15. He'd memorized this magic square, and when his opponent took a number he made a mental X on his magic square. Then he would use a mental O to indicate his own move. In this way, he could always force either a tie (cat's game) or a win.

If you've played a lot of tic-tac-toe in your life, you probably have strategies you use that prevent you from losing. But usually when you play tic-tac-toe, your opponent knows that she is playing tic-tac-toe too. Imagine playing tic-tac-toe against someone who doesn't even know she's playing tic-tac-toe! She'd make some very strange moves, but you could always force a tie or a win.

If you wonder how this works, make the three-by-three magic square for this game, using the digits 1 through 9. Each row, column, and diagonal of your square must add to 15.

Now try playing this game with someone. Create the nine cards that contain the digits 1 through 9. Then sit at a table, opposite your opponent. Alternate taking cards. Hold your magic square in your lap and use it to play tic-tac-toe while you're also playing the card game. It will take some practice, but you can do it.

This game is a perfect example of changing the representation. Hugo had changed a relatively difficult game into a game of tic-tac-toe, which is much easier to play. In mathematics, this type of situation is called an **isomorphism** (a one-to-one correspondence of mathematical sets). The tic-tac-toe game is said to be isomorphic to the card game in the park.

Change Your Focus

Rise above the problem or stand outside it to see different sides or even what is not there.

The three strategies in this chapter all require insight, imagination, and creativity:

- Change your point of view.

- Consider the complementary problem.

- Change the representation.

To use these strategies, you must be able to see a problem from a completely different perspective. The result of changing the focus is that the problem becomes easier to solve in its new form than it was in the original. If a problem seems too difficult and an easier related problem doesn't seem to help, then try changing your focus.

Problem Set A

1. FEARLESS FLY

This is a famous problem. Two bicyclists, Frances and Fred, rode toward each other. Each traveled 20 miles per hour, and they started 10 miles apart. A frivolous fly flew furiously fast at 50 miles per hour from Frances to Fred. The fly started on Frances's handlebars as Frances and Fred started riding. It flew to Fred's handlebars and then back to Frances's and back to Fred's, always following the bike path. (No self-respecting fly flies a beeline.) It continued in this way until Frances and Fred reached each other. How far did the fly travel?

2. TOOTHPICKS

Using six toothpicks, make four triangles of the same size.

3. PERFECT SQUARES

How many perfect squares are there between 2,000 and 20,000?

4. MORE DOTS

Without lifting your pencil from start to finish, draw six line segments through all 16 points. The solution has nothing to do with how wide the dots are or with the idea that possibly the lines determined by them are not parallel. The dots are mathematically defined as points—they have no width, and they determine sets of parallel lines.

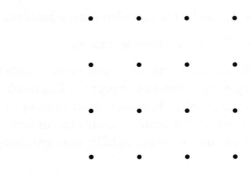

5. COMPLEMENTARY EVENTS

What is the complementary event for each given situation?

a. You roll a die one time and get a 6.

b. You roll a die one time and get an odd number.

c. You roll a die two times and don't get any 6's.

d. You roll a die five times and get a 6 every time.

e. There are 40 people in the same room, and they all have different birth dates.

6. PAYDAY

Marissa recently got a smaller-than-usual weekly paycheck from her job. She normally works from 8:00 a.m. to 5:00 p.m., with an hour for lunch, five days a week. Her pay is $7.50 per hour, but she gets paid only for the hours she works. On Monday she left at 3:30 to go see her son play softball. On Tuesday she was 45 minutes late getting back from lunch because she had to change a flat tire. On Wednesdays she leaves at 2:15 to go to her class. On Thursday she had a doctor's appointment in the morning and didn't get to work until 10:30. On Friday she took a 3-hour lunch break to pick up her brother at the airport. Then at 4:10, she left early to beat weekend traffic. How much did she get paid for the week?

7. THE LIKELIHOOD OF BEING LATE

The airline you're flying on has an on-time rate of 98% for the route you are taking. The bus from the airport to the convention center has an on-time rate of 95%, but if you are late, of course, you will miss the bus and be late arriving at the convention. What are your chances of being late to the convention?

8. ANOTHER CARD ARRANGEMENT

This problem is similar to the Card Arrangement problem in this chapter. Start with all the cards in one suit in the deck: ace, 2, 3, 4, 5, 6, 7, 8, 9, 10, jack, queen, and king. Arrange the cards in a particular order in your hand. Then spell the name of each card, putting one card on the bottom of the stack for each letter in the name of the card. For an ace, you would say "A-C-E ace," putting three cards on the bottom of the stack and laying the fourth card faceup on the table (this card will be the ace). Then continue by saying "T-W-O two," putting three cards on the bottom of the stack and laying the fourth card (a 2) on the table just as you say "two." Continue in this way all the way through jack, queen, and king, each time ending the spelling by saying the card you just spelled as you lay it faceup on the table. What order do the cards have to be in originally for this to work out?

KNIGHT MOVES

This problem comes from our friend and mentor Tom Sallee. Suppose you arranged chess knights on a three-by-three chess board as shown.

White Knight 1		White Knight 2
Black Knight 1		Black Knight 2

A knight's move is one space in any direction (horizontally or vertically), then two spaces in either perpendicular direction. (That's equivalent to two spaces and then one space in the perpendicular direction—knights can jump over pieces in their path.) Two knights may not occupy the same space at the same time. Only one knight may move at a time.

a. How many moves does it take for the white knights to change places with the black knights?

b. How many moves does it take for White Knight 1 to change places with Black Knight 1?

10. **NEW GAME IN CENTRAL PARK**

When Alan Barson figured out what Hugo was doing (see the applied problem solving feature in this chapter), he went back to Central Park to see if Hugo was still there. Hugo wasn't, but his son was, and he was playing a related game. This time, each of the nine cards contained a word. The words were

BRIM   FORM   HEAR   HOT   SHIP   TANK   TIED   WASP   WOES

Players alternated choosing words. The object of the game was to be the first person to possess three words that contained the same letter. For example, if you chose *form, hot,* and *woes,* you would have three words that contained the letter *o.* Hugo's son was also using the strategy of change the representation and was actually playing tic-tac-toe. He'd made a three-by-three magic square, but he had put a word instead of a number into each square. Each row, column, and diagonal of the magic square contained three words that shared a common letter. Make this magic square.

CLASSIC PROBLEMS

11. WOULD HE HAVE SAVED TIME?

Our man Ostap was going home from Kiev. He rode halfway by train—15 times as fast as he goes on foot. The second half he went by ox team. He can walk twice as fast as that. Would he have saved time if he had gone all the way on foot? If so, how much?

Adapted from *The Moscow Puzzles* by Boris Kordemsky.

12. THE CAMEL PROBLEM

A camel driver has 3,000 bananas. The camel can carry up to 1,000 bananas at a time. The driver wants to deliver bananas from one town to another town that is 1,000 kilometers across the desert. However, his camel refuses to move unless fed a banana for every kilometer that it travels. What is the maximum number of bananas the driver can get across the desert to the other town? (These are special bananas, the bananas will not spoil in the desert.)

Source unknown. Originally posed to the authors by math teacher Don Gernes.

13. MARTINI GLASS

Move two toothpicks and the olive will be outside the glass, but the glass will still be the same shape.

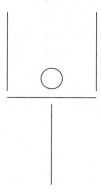

Source unknown. Martin Gardner says this is one of his favorite problems.

MORE PRACTICE

1. MORE DOTS

Without lifting your pencil from start to finish, draw 5 line segments through all 12 dots. The solution has nothing to do with how wide the dots are or the possibility that the lines determined by the dots are not parallel. The dots are mathematically defined—they have no width, and they determine sets of parallel lines.

2. SUM OF DIGITS

There are 90 two-digit numbers from 10 through 99. What is the probability that a two-digit number will be divisible by the sum of its digits? For example, the number 42 is divisible by 6 (the sum of $4 + 2$), but the number 71 is not divisible by 8 (the sum of $7 + 1$).

3. INTERESTING SETS OF THREE

There are two sets of three consecutive two-digit numbers that have the following property. Each number in the set is the product of a one-digit prime number and a two-digit prime number. For example, 38 and 39 are consecutive numbers and each has this property, since 38 is 2×19 and 39 is 3×13. However, 38 and 39 are only a pair, and we are looking for a set of three. There are two such sets of three two-digit numbers. What are they?

4. MOVIE REVIEWS

Christy is a movie reviewer. There are 6 new movies opening this weekend. She only has time to see 4 of them, so she asks her colleague Ignatiy to review two of them. How many sets of 4 out of the 6 movies can Christy choose to review?

5. RAIL PASS

Bizu and Ghenet have rail passes for traveling around Europe. They are on a tight budget, so instead of sleeping in hostels, they find a late-night train and head off to a new city, sleeping as best they can on the train. Their passes are good for the next five days, so they can only visit five of these cities: Brussels, Paris, Copenhagen, Vienna, Prague, Berlin, and Amsterdam. In how many ways can they select a set of five cities to visit?

6. SPRINKLERS ON A RECTANGULAR LAWN

There are two sprinklers that spread water in a circular sector (90 degrees). They only spray in an 18-foot radius. (See the diagram: One sprinkler is in the top left corner and the other is in the bottom right corner. Part of the lawn is not covered by either sprinkler.)

The lawn measures 18´ by 36´, and each sprinkler has a spray radius of 18´.

What is the area of the section of lawn that is not watered by either sprinkler?

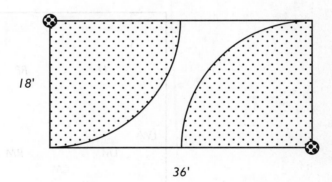

18'

36'

7. WASHING DISHES

Chelsea has made a deal with her brother. He will wash the dishes if she can roll at least one 3 on five rolls of a die. What are the chances that Chelsea gets out of washing the dishes?

Pia is the coach of a soccer team. She believes that the ten field players (not the goalkeeper) will learn a lot more if they constantly change positions. So during a practice scrimmage against another team she sets up a rotation system. She will blow a whistle every 4 minutes. When the players hear the whistle they will switch positions as follows. **Backs (3)**: The left back will move to center back, the center back will move to right back, and the right back will move to left back. **Wings and Midfielders (5)**: The left wing will move to left midfielder, the left midfielder will move to center midfielder, the center midfielder will move to right midfielder, the right midfielder will move to right wing, and the right wing will move to left wing. **Forwards (2)**: The left forward and the right forward will switch positions. After another 4 minutes the switch will happen again. These switches will continue until **everyone** is back in their original positions. At that point, practice is over. How long is practice?

See the field below for positions. The goalkeeper always stays the same.

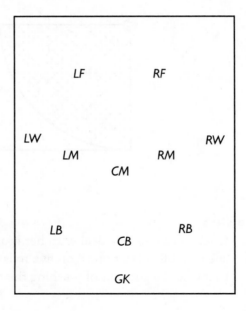

Problem Set B

1. COMPUTER ERROR

A computer was printing a sequential list of positive integers, but because of a glitch in the software it neglected to print any numbers that were **integral** powers of integers (the powers were larger than 1). Thus, the list began 2, 3, 5, 6, 7, 10, 11, 12, 13, 14, 15, 17, ... How many numbers in this list had fewer than four digits?

2. PALINDROME CREATOR

How many three-digit numbers have the following special property? Take a three-digit number. Reverse the digits to create a new three-digit number (this number may be the same as the original number). Add the two numbers together. The result is a *palindrome*, a number that reads the same backwards and forwards. Neither the original number nor the new number can start with 0.

3. THE AMAZING RESTIN

The Amazing Restin is a psychic. She can figure out anything anybody is thinking, as long as she is lying down. She gives performances all over the country. One of her favorite gimmicks features words: Her assistant thinks of a five-letter word. Then the Amazing Restin proceeds to guess five-letter words, and her assistant tells her how many letters the word he thought of shares with the word she guessed. One day, she guessed five words and he told her that each of her words shared exactly two letters with the word he had in mind. Here are the five words she guessed:

BLUNT   VOTER   SPICE   BUOYS   MADLY

What word did the assistant have in mind?

Place the digits 0 through 9 into the circles below, subject to the following rule: Each pair of digits that are joined by a line must form a two-digit number (in either order) that is divisible by either 8 or 13. So, for example, the digits 4 and 8 could be connected with a line because 48 is divisible by 8. (Note that 84 is divisible by neither 8 nor 13, but that doesn't matter. As long as it works with the digits in one of the orders it's okay.) There are two solutions, but they are basically the same—just two digits are switched.

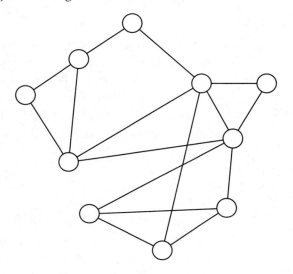

Emile, who lives in the South Pacific, is planning a rowboat trip. His rowboat will hold him plus enough food to last 3 weeks. By a strange coincidence, several small islands about 1 week's rowing apart are near the island he lives on. He plans on rowing to the fifth island, which is a large island with grocery stores similar to those on the main island he lives on now. The four small islands between his island and the other big island have no stores, and he will not be able to buy any food on those islands. So Emile realizes that he will have to row out to an island, store some food, and return to the main island for more supplies. Show how Emile should organize his food supplies so he can get to the fifth island. How many weeks will his trip take altogether?

17

Visualize Spatial Relationships

Graphs and scale drawings provide insights into relationships that might be obscured by numbers, words, or formulas. A graph can be a dramatic and persuasive way to show that a product is earning a profit or losing money for its manufacturer.

e have already discussed strategies that fall under the major problem-solving theme Spatial Organization: diagrams, physical representations, and Venn diagrams. This chapter will explore two more strategies that involve spatial organization: graphs and scale drawings.

Section 1: Use a Graph

Some problems are best represented with a **graph.** For example, newspapers are full of graphs. A graphs can clearly show a relationship between **variables** that may otherwise be difficult to understand.

CHICKEN NUGGETS

A local fast-food vendor sells chicken nuggets for the following prices:

SERVING SIZE	PRICE
6 nuggets	$2.40
10 nuggets	$3.60
15 nuggets	$5.10
24 nuggets	$7.80

Draw a graph of this information. Then answer the following questions:

   a. What is the equation for this graph?

   b. What is the slope for the graph? What is the real-world significance of the slope in relation to the price of chicken nuggets?

   c. What is the *y*-intercept for this graph? What is the real-world significance of the *y*-intercept?

   d. How much would it cost to buy a serving size of 50 chicken nuggets?

   e. Another restaurant sells 13 nuggets for $4.25 and 20 nuggets for $6.75. Use the graph to find out whether these are good deals compared to those of the restaurant above. Then use the equation for the graph to check your answer.

A group of students worked on this problem.

BETTY: Okay, where do we start? I guess we should draw the graph. But what should we put on the *x*-axis?

SHAUNA: I think we should put the price on the *x*-axis.

SHELLY: No. The price depends on the number of nuggets purchased. Always put the **dependent variable** on the *y*-axis and the **independent variable** on the *x*-axis.

BETTY: So *x* is the number of nuggets, and *y* is the total price. I can draw a line through all the points.

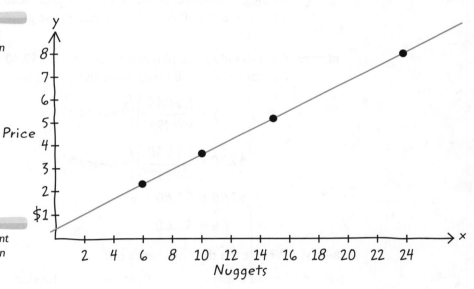

The dependent variable goes on the y-axis.

The independent variable goes on the x-axis.

BARBARA: What does the line mean?

BETTY: The points on the line give all possible combinations of the number of nuggets and the corresponding price. So, for instance, 8 nuggets would probably cost $3.00.

BARBARA: I see. Let's try to figure out the equation.

SHELLY: We need the **slope,** and we need the **y-intercept.** Then we can write the equation, using the form $y = mx + b$.

NANCY: I can figure out the slope. It's just rise over run. If we choose the run from 6 nuggets to 10 nuggets, the price goes up, or rises, $1.20. A price change of $1.20 for 4 nuggets is $0.30 per nugget. That's the ratio of rise over run.

SHAUNA: That must be what the question means about the real-world significance of the slope. It's just price per nugget. That's interesting. I never thought of that before.

SHELLY: Since *m* stands for the slope, the equation must be $y = 0.30x + b$. But how do we get *b*?

BARBARA: Well, *b* is just the *y*-intercept. Let's look at the graph. It looks as though the line goes through the *y*-axis somewhere around $0.50.

NANCY: Let's work backwards. If it costs $0.30 per nugget, then 6 nuggets should cost $1.80. But 6 nuggets really cost $2.40. That's an extra $0.60, which is where the graph crosses the *y*-axis.

BETTY: Or we could have plugged 6 nuggets and $2.40 into the equation that Shelly suggested and solved for *b*.

$$y = \left(\frac{\$0.30}{\text{nugget}}\right)\left(x \text{ nuggets}\right) + b$$

$$\$2.40 = \left(\frac{\$0.30}{\text{nugget}}\right)\left(6 \text{ nuggets}\right) + b$$

$$\$2.40 = \$1.80 + b$$

$$b = \$0.60$$

BETTY: Wow, even the units work out.

SHELLY: So our equation is $y = 0.30x + 0.60$. The slope is the price per nugget, but what is the significance of the *y*-intercept?

NANCY: Well, if you bought zero nuggets, it would cost $0.60. That's what I was trying to say before. That's weird. How come it costs $0.60 to buy nothing? You must be paying for something.

SHAUNA: Maybe you're paying for the privilege of buying your chicken nuggets there.

BETTY: Or maybe you're paying for the forks and spoons and napkins. Or maybe it's just profit. Or the packaging.

NANCY: There could be a lot of things you're paying for. I wonder if all restaurant food is priced like that.

SHELLY: Question d asks for the price of 50 nuggets. We can figure that out with our equation or our graph.

BETTY: I didn't use paper big enough for the graph, but I think 50 nuggets would cost between $15 and $16.

NANCY: According to the equation, the price is $y = 0.3(50) + 0.60$, which is 15.60. That matches your estimate, Betty.

BARBARA: How are we going to answer the last question?

BETTY: Let's plot those points on the graph. The point (13, 4.25) is below the line, and the point (20, 6.75) is above the line.

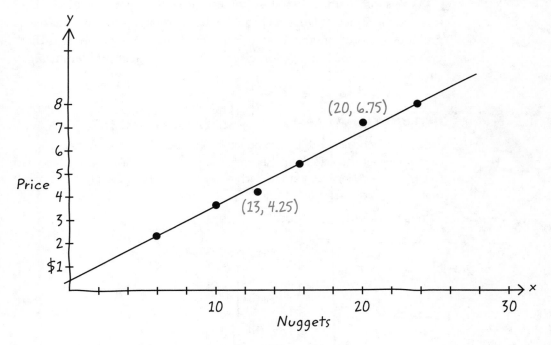

BARBARA: What does that mean?

SHELLY: Well, if the point is below the line, the cost is less than it should be, so it's a good deal. And if the point is above the line, the cost is more than it should be, so it's a bad deal.

BARBARA: That makes sense. Can we also use the equation to figure that out?

SHAUNA: Yes, we can. Use the equation to figure out how much 13 nuggets would cost at the first restaurant: $y = 0.3(13) + 0.60 = 4.50$. That means $4.25 is a good deal, better than $4.50.

BETTY: And 20 would cost $0.3(20) + 0.60$, or $6.60. That's cheaper than the other restaurant's price of $6.75, which is a bad deal.

SHELLY: Wow, this was a good problem. The graph and the algebra together are pretty convincing.

As this discussion illustrates, you can use either graphing or algebra to solve the Chicken Nuggets problem, but combining the two strategies provides a more convincing solution. An advantage of graphing is that you can represent information visually. Work the next problem using similar methods.

INTERNATIONAL CALLS

Aletta has friends in both Egypt and Libya. She uses a pay-as-you-go calling card whenever she calls them. Recently she made three calls to Egypt and two to Libya. The calling card rate to each country is different. Calls to Libya cost more than calls to Egypt.

TIME (IN MINUTES)	COST
11	$2.26
17	$4.57
6	$1.26
22	$4.46
7	$2.07

Determine the connect fee and the cost per minute for the two rate schedules Aletta called under: calls to Egypt and calls to Libya. (A connect fee is the charge levied the instant a phone conversation begins.) Note that both the Egyptian call rate and the Libyan call rate are linear relationships. Work this problem before continuing.

The following graph is going to play a crucial, though not complete, role in solving this problem. Here are the graph points plotted with time in minutes (t) on the horizontal axis and cost (c) on the vertical axis.

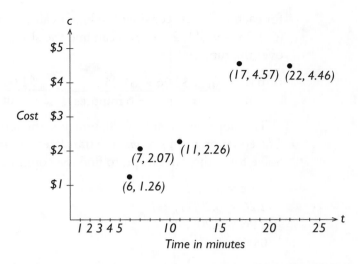

Because we know that Aletta called under two different rate schedules, we can reasonably assume that we need to determine the graphs of two linear functions. Three points in particular (the 6-, 11-, and 22-minute calls) line up easily. The remaining two points also form a line, slightly steeper than and above the first. The slope of each line represents the cost per minute. The *c*-intercept of the upper line reflects the higher connect fee. Because calls to Libya cost more than calls to Egypt, we can assume that the upper and steeper line represents the calls to Libya.

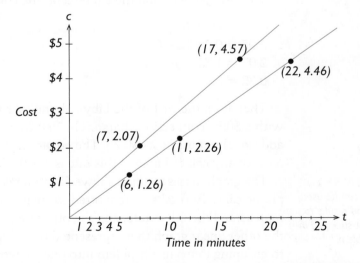

Now that we've identified the calls associated with the Libyan rates and the Egyptian rates, we can determine the slope and the *c*-intercept for each line. First consider the Egypt calls, using the two points (6, 1.26) and (11, 2.26). You can find the slope by calculating the rise over the run:

$$\frac{\text{rise}}{\text{run}} = \frac{\$2.26 - \$1.26}{11 \text{ minutes} - 6 \text{ minutes}} = \frac{\$1.00}{5 \text{ minutes}} = \frac{\$0.20}{\text{minute}}$$

The slope, m, must be 0.20, which is the rate per-minute to Egypt. The equation form $y = mx + b$ translates to $c = mt + b$ in this problem. Solve for b, the c-intercept, to find the connection charge.

$c = mt + b$
$2.26 = 0.20(11) + b$
$2.26 = 2.20 + b$
$0.06 = b$

The connection charge is $0.06, so the first-minute rate for a call to Egypt is $0.26, and each additional minute costs $0.20. The equation you can use to find the cost for any call to Egypt at this rate is $c = 0.20t + 0.06$, where c is the cost of the call and t is the number of minutes talked.

To find the rate for calling Libya, we find the slope and the c-intercept for the upper line.

$$\frac{\text{rise}}{\text{run}} = \frac{\$4.57 - \$2.07}{17 \text{ minutes} - 7 \text{ minutes}} = \frac{\$2.50}{10 \text{ minutes}} = \frac{\$0.25}{\text{minute}}$$

We solve for b to find the c-intercept, the connection charge.

$c = mt + b$
$2.07 = 0.25(7) + b$
$2.07 = 1.75 + b$
$0.32 = b$

Therefore, the cost of the Libya phone calls is $0.25 for each minute, with a $0.32 connection charge. The first minute costs $0.57, and each additional minute costs $0.25. The equation you can use to find the cost for any call to Libya at this rate is $c = 0.25t + 0.32$.

The graph in this solution showed us two rates that applied to Aletta's phone calls: Two calls were plotted to create the linear graph that represented the Libya rate, and the other three calls were plotted to get the other linear graph that represented the Egypt rate. In addition to grouping items in a problem into one category or another, you can use a graph to extend the results to data not specified in the problem.

Graphs can be used to extend results to data not specified in the problem.

Note that the linear model for this particular function is not entirely accurate. Phone costs do not rise at a constant per-minute rate, so the costs do not actually correspond to linear graphs. Rather, they are examples of what are called **step functions.** In the problem, the charge for Aletta's 11-minute call is given as $2.26. This would also be the charge for a 10-minute-1-second call, a 10-minute-2-second call, and so on, all the way up to an 11-minute call. As soon as the twelfth minute begins, the charge goes up to $2.46. The next graph shows this step function for the calls to Egypt. On the graph, note that an open circle means that the left-hand endpoint does not belong to the line segment. A closed circle means that the right-hand endpoint does belong to the line segment. The endpoints indicate rate changes, so if a call is exactly 3 minutes, for example, it will be charged at the lower rate. The line from the original graph, shown in orange, connects the right-hand endpoints of the steps.

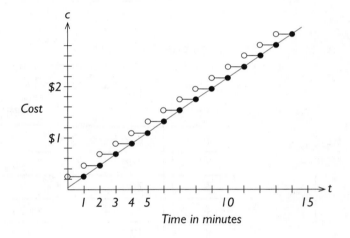

Time in minutes

VACATION

The Family family wants to take another vacation. They have decided to drive their van to a destination 600 miles away. If they drive at a reasonable speed, how much time should they set aside for the trip? Plot a graph that shows various driving speeds (in miles per hour) on the x-axis and driving time (in hours) on the y-axis. Work this problem before continuing.

Mohamed solved this problem with a systematic list and a graph. "I didn't really understand this problem at first, so I started making a list of possible speeds and times. I realized right away that if they traveled 60 miles per hour, it would take 10 hours of driving. And if they traveled 30 miles per hour, it would take 20 hours of driving. I decided to make a systematic list showing a lot of the possibilities—but not all of them because then the list would be endless."

SPEED	TIME
1 mph	600 hrs
2	300
3	200
4	150
5	120
6	100

"At this point, I realized that I should try to put this information in a graph. I knew I needed more varied data or my graph was going to be pretty dull. So I decided to skip some possible speeds.

Make a systematic list.

SPEED	TIME	SPEED	TIME
1 mph	600 hrs	50 mph	12 hrs
2	300	60	10
3	200	70	8.57
4	150	75	8
5	120	80	7.5
6	100	100	6
10	60	120	5
20	30	300	2
30	20	600	1
40	15		

"Although some of these pairs of numbers were unrealistic, I didn't think my graph would be complete without them. As the problem directed, I put speed on the *x*-axis and time on the *y*-axis, because the time it takes the Familys to reach their destination depends on how fast they drive."

Draw a graph.

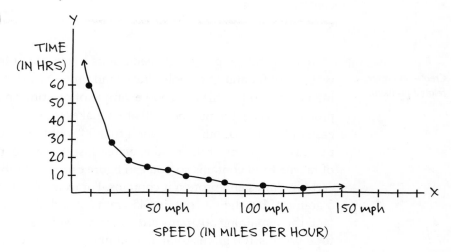

"The curved graph clearly shows that the faster you drive, the sooner you'll get there. A more interesting graph would show the time you'd spend waiting for the highway patrol to write out your tickets if you'd been driving 90 miles per hour."

FAT CONTENT

Many ads for food indicate either the percentage of fat the food contains or to what percentage the food is fat-free. For example, an ad might say "Our burgers are 85% fat-free" or perhaps "Our lean hamburger is only 10% fat." A person reading these ads might assume that the percentage of calories from fat is also only 15% in the first case and 10% in the second case. But this assumption wouldn't be accurate, because the ads indicate the percentage of fat by *weight*, not in *calories*.

Fat has 9 calories per gram. Carbohydrates and protein each have only 4 calories per gram. In addition, virtually all foods have moisture content (as well as minerals, and so on), which contains 0 calories per gram. For example, 2% milk (2% of the weight of the milk comes from fat) is about 89% water. When this water weight is taken into account, about 34% of the calories come from fat.

Draw a graph that shows a food's percentage of fat by weight on the x-axis and its percentage of fat in calories on the y-axis. Assume that this food is 50% water and that the remaining percentage of the food represents carbohydrates and protein. Then use your graph to find what percentage of fat by weight gives 50% of the calories from fat. Work this problem before continuing.

Create an easier related problem.

To draw the graph, you'll need points to plot. Start by inventing weights of fat and of carbohydrates and protein for 100 grams of imaginary food. (This is a good example of creating an easier related problem—using the number 100 often makes working with percentages easier.) Because of our assumption that 50% of the food is water, in this case 50 grams of the food is water. List the number of grams of fat the food contains so you can figure out the number of grams of carbohydrates and protein it contains. For example, if the food contains 5 grams of fat, then it contains 45 grams of carbohydrates and protein (and the remaining 50 grams are water, as we assumed). Because our food weighs 100 grams total, the number of grams of fat will also be the percentage of fat by weight.

Now calculate the number of fat calories. Look again at 5 grams of fat and calculate how many calories of our food that represents.

$$5 \text{ grams} \times \frac{9 \text{ calories}}{\text{gram}} = 45 \text{ calories}$$

Because our food is made up of fat *and* of carbohydrates and protein (and water), we also need to calculate the number of calories of carbohydrates and protein.

$$45 \text{ grams} \times \frac{4 \text{ calories}}{\text{gram}} = 180 \text{ calories}$$

So our food contains a total of 225 calories: 45 + 180 = 225. Now we can figure out the percentage of calories from fat compared to the total calories.

$$\frac{45 \text{ calories from fat}}{225 \text{ calories total}} = 20\% \text{ fat}$$

To generate a sufficient list of points to plot, continue in this way, choosing numbers for grams of fat—which are also the percentages of fat by weight—and figuring out the appropriate percentages for calories. Then draw a graph like that shown to the right of the chart.

% fat by weight	% fat in calories
0	0.0
5	20.0
10	36.0
15	49.1
20	60.0
25	69.2
30	77.1
35	84.0
40	90.0
45	95.3
50	100.0

The graph shows that with about 15% fat by weight, the food contains 50% fat in calories. Remember that we assumed our sample food had 50% moisture weight. Many foods are more than 50% moisture. The moral of the story is to be cautious when you hear that a food is 90% or 85% fat-free, because a lot more of the calories come from fat than you might expect.

DUCKS AND COWS

Farmer Brown has ducks and cows. The animals have a total of 12 heads and 32 feet. How many ducks and how many cows does Farmer Brown have? Solve this problem with a graph before continuing.

We've already solved this problem by drawing a diagram, guessing and checking, and converting to algebra.

Eric solved this problem with a graph. "I set up two equations, one representing heads and the other representing feet. I let x represent ducks and y represent cows. My equations were

$x + y = 12$ (the total number of heads is 12)

$2x + 4y = 32$ ($2x$ is the number of duck feet and $4y$ the number of cow feet, and the total number of feet is 32)

"Then I graphed each equation, which gave me two straight lines. They intersected at the point (8, 4), which is 8 ducks and 4 cows."

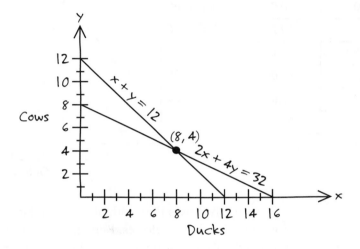

MAXIMUM AREA

A farmer with 100 feet of fencing wants to build a rectangular garden. What should the dimensions of the garden be in order to enclose the maximum area? Work this problem before continuing.

You could solve this problem in any number of ways. You could use guess-and-check or you could generate a systematic list, but you might not be sure you had the best answer. You could also use calculus, a very good technique for solving problems such as this. But you can also solve this problem by graphing.

First set up the equation to be graphed. Start by drawing a picture of the rectangular garden. The total fencing available measures 100 feet. If the top and bottom sides of the garden each measure x feet, then the left- and right-hand sides each measure $(100 - 2x)/2$ feet, which can be written as $50 - x$ feet.

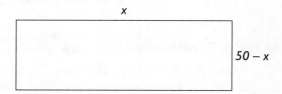

The area of the garden is $x(50 - x)$ square feet. By hand, or with a graphing calculator or a computer, graph the equation $y = x(50 - x)$, or $y = 50x - x^2$.

Notice how effectively this graph conveys the maximum value.

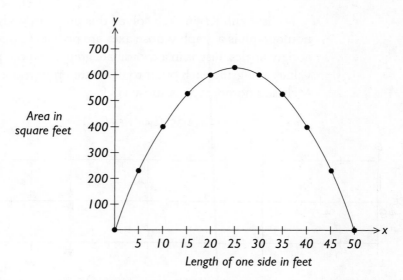

The y-value represents the area of the garden. The x-value represents the length of one side of the garden. Look on the graph for where the y-value is at its highest point (called a **maximum value**). It appears that when x measures 25 feet, the area is at a maximum at 625 square feet.

Graphs are powerful visual tools for solving problems. A graph can provide a visual image of the situation given in a problem, making both the problem and the solution easier to understand.

Section 2: Make a Scale Drawing

Consider the Ducks and Cows problem one more time.

DUCKS AND COWS

Farmer Brown has ducks and cows. The animals have a total of 12 heads and 32 feet. How many ducks and how many cows does Farmer Brown have? Read the following explanation of how to solve this problem with a special type of scale drawing.

Melissa and Kevin each solved this problem with a **nomograph.** A nomograph is a graph whose axes are parallel to one another instead of at right angles. Just as in a coordinate graph, you can place corresponding values alongside each other to illustrate proportions and relations. Melissa's nomograph is shown below.

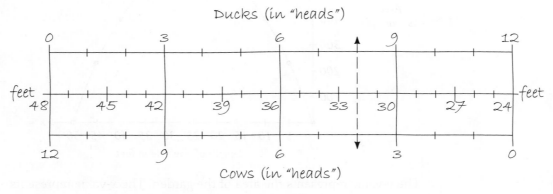

Melissa explained her diagram: "Because the total number of heads is 12, the nomograph is set up so that a vertical line drawn anywhere in the graph will yield a solution of x ducks + y cows = 12 heads. The maximum number of feet possible for 12 heads is 48, and the minimum is 24. Therefore, the range of the number of feet is between 24 and 48.

"To find a solution, move along the feet-axis until the desired number is reached—in this case, 32. Draw a vertical line through that point, and the corresponding numbers on the ducks-axis and on the cows-axis are the solution.

"The answer to '32 feet' is 4 cows and 8 ducks."

Kevin took a similar approach to this problem. He used a constant of 32 feet and looked for the number of heads that added up to 12.

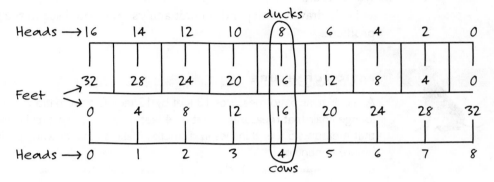

Kevin explained: "With this nomograph, I can locate the correct number of ducks and cows by finding which vertical line intersects numbers that add up to 12 heads, because the number of feet for each line equals 32. Therefore, the answer is 8 ducks and 4 cows."

The two nomograph solutions to the Ducks and Cows problem combine the strategies of using a graph and making a **scale drawing.** In Section 1 of this chapter you learned about drawing graphs to solve problems. In this section you'll learn about making scale drawings to solve problems.

Who uses scale drawings? Matt Tsugawa is a landscape architect. He frequently uses scale drawings to plan landscapes for his clients. He makes a scale drawing of the basic area he'll be working with, such as a backyard or a park. Then, to check the plans for aesthetics, that walkways are the appropriate width, for desired foliage density, and so on, he draws various plants and trees to scale on transparent overlays and places them over the main drawing. If he doesn't like the results, he creates a new transparency so that he doesn't have to eliminate the main drawing.

Police investigators also make scale drawings of traffic-accident scenes. They use their scale drawings—along with some algebra, geometry, trigonometry, unit analysis, and subproblems—to determine the original speeds of the vehicles involved. This process helps them determine whether either driver was speeding, which in turn may indicate who was at fault.

Scale drawings depend on unit analysis, as you'll see in the next problem.

PATIO FURNITURE

A rectangular patio measures 18 feet by 12 feet. On the patio is a rectangular lounge chair that measures 2 feet by 4 feet 3 inches and a circular picnic table that measures 5 feet 8 inches in diameter. Make a scale drawing of the patio before continuing.

The first step in creating a scale drawing is to choose a scale. It is often easier to use centimeters rather than inches to represent feet, so let's pick the scale 1 cm equals 2 ft. You must then do some unit analysis to convert real-world measurements to scale measurements. For example, to find the scale measurement of the picnic table's diameter, you must convert 5 ft 8 in. to feet and then multiply by the scaling fraction to get centimeters.

$$8 \text{ in.} \times \frac{1 \text{ ft}}{12 \text{ in.}} \approx 0.67 \text{ ft}$$

$$5.67 \text{ ft} \times \frac{1 \text{ cm}}{2 \text{ ft}} = 2.83 \text{ cm}$$

So the table with a diameter of 5.67 ft has a diameter of 2.83 cm in the scale drawing. Using an approximation of 2.8 cm will work well for this problem. A drawing of the patio is shown next.

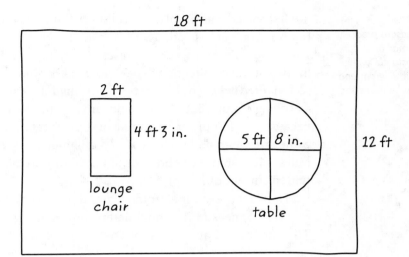

18 ft

2 ft

4 ft 3 in.

5 ft | 8 in.

12 ft

lounge
chair

table

1 cm = 2 ft

Note the difference between unit analysis and scaling. In unit analysis, the fraction you multiply by has the value 1 (for example, $^{1\,ft}/_{12\,in.}$). The scaling fraction does not have the value 1 (such as $^{1\,cm}/_{2\,ft}$), but you choose to make 1 cm and 2 ft correspond for the purposes of scaling.

MAYDAY

"Mayday, Mayday!" The call startled Ned, who was stationed in the Coast Guard office. He immediately radioed: "Coast Guard here. What is your position? Over."

"I'm not sure. We left the port at Miami at 7:30. We sailed due southeast for 2 hours at 35 knots. Then we turned about 30° to starboard [right] and sailed for 4 hours at 25 knots. Then we lost our engines, and we have been adrift for about an hour and a half. We would have called earlier, but our radio was out. Can you send us some help?"

Ned replied, "I'll work out your position and send out a chopper right away. Over." Ned knew that the ocean current at that time of day was approximately 5 knots due south. A knot (kn) is 1 nautical mile per hour. One nautical mile (NM) is about 1.15 land miles. The helicopter speedometer measures land miles per hour. The likely speed of the helicopter is 80 miles per hour (that is, land miles). In what direction should Ned send the helicopter, and how many minutes will it take it to get to the stranded boat?

Make a scale drawing and solve this problem before continuing.

Ned solved this problem as follows: "I knew I could probably use trig for this, but instead I made a scale drawing. I let 1 millimeter equal 1 nautical mile. From a dot representing Miami, I measured a 45° angle in the southeast direction. The 2 hours at 35 knots is 70 nautical miles, so I marked off 70 millimeters with a big dot. From there I measured a 30° angle representing the right turn and marked off 100 millimeters because the 4 hours at 25 knots is 100 nautical miles. From there I drew a line due south and marked off 7.5 millimeters to represent 7.5 nautical miles, the distance the boat would have drifted in an hour and a half with a current of 5 knots.

"Finally, I drew a line from Miami to the boat's last position and found it to measure 171 millimeters, representing 171 nautical miles. The line formed an angle about 64° south of east."

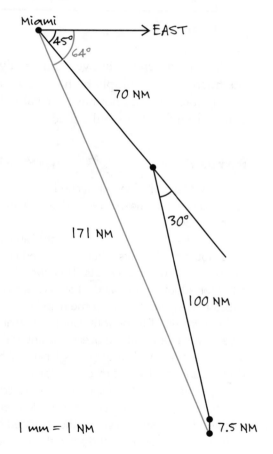

$$171 \text{ NM} \times \frac{1.15 \text{ mi}}{1 \text{ NM}} = 196.65 \text{ mi}$$

"Then I figured out the time, given that the helicopter travels 80 miles per hour."

$$196.65 \text{ mi} \times \frac{1 \text{ hr}}{80 \text{ mi}} \approx 2.46 \text{ hr}$$

$$0.46 \text{ hr} \times \frac{60 \text{ min}}{1 \text{ hr}} = 27.6 \text{ min}$$

"So it would take the helicopter about 2 hours 28 minutes to reach the boat. That is, it would take that long to reach the boat's last location. Of course, in the time it took the helicopter to get there the boat would have drifted a little farther south, which means the helicopter wouldn't find the boat at that spot. It would have to turn due south at that point and look for the boat."

Many people use scale drawings to solve real-world problems that deal with space or to visualize problems that involve trigonometry. You can get a much clearer overview of a situation when you make a scale drawing of it.

Use a Graph or a Scale Drawing

Graphs and scale drawings are powerful organizational tools.

You could have solved all the problems in this chapter with methods other than graphs or scale drawings. The purpose of this chapter was to show you that visual representation sometimes has advantages over symbolic manipulation. Drawings bring your visual experience to problems, and vision is the most highly developed sense for many people. Graphs and scale drawings enable you to integrate the mathematical with the visual to give a clear picture of the whole process.

Problem Set A-1

Draw a graph for each problem.

1. REFRIGERATOR

Cher visits a store to buy a new refrigerator and finds two recommended models. The Major brand is $1,000 and is expected to cost $6 per month for energy use. The Minor brand is $800 and is expected to cost $16 per month for energy use. Based on cost, which refrigerator would you advise Cher to buy?

2. SODA

Two vendors next door to each other sell soda. One is a hamburger restaurant, the other is a gasoline station and mini-mart. The prices each charges for take-out sodas are shown in the chart below.

Size (ounces)	8	12	16	20	32
Restaurant	$1.29	$1.59		$2.19	
Mini-mart			$1.29	$1.59	$2.49

Each vendor is considering offering a 24-ounce soda. What would be the probable price at each establishment?

3. LETTUCE

Fatima is the produce manager for a local grocery store. Information provided by the grower and shipper states that iceberg lettuce will stay fresh for 21 days at 35°, for 15 days at 40°, for 10 days at 45°, and for 1 day at 70°. During a heat wave Fatima knows that the store will not be able to keep the produce as cool as she would like. How long can she expect the lettuce to last if the store keeps it at about 60°?

4. APPLE ORCHARD

Ms. Pomme has an apple orchard. If she hires 20 workers to pick the apples, it will take them 6 hours. If she hires 30 workers, it will take them 4 hours. Draw a graph that shows workers on the x-axis and time on the y-axis. Use the graph to estimate how long it would take if she hired 50 workers and how long it would take if she hired 6 workers.

If you were Ms. Pomme, how many workers would you hire? Why?

5. JEANNE'S ORIGAMI BOOK

Jeanne wrote a booklet showing people how to do origami folds. She then photocopied it (each booklet cost her $1.45 to copy) and sold the copies for $2.00 each. She also bought a small ad, costing $24, to advertise the book. Not considering the cost of her labor, how many books would she need to sell before her endeavor turns profitable?

6. CHRISTMAS TREE LOT

Jocelyn owns a store and wants to sell Christmas trees in her parking lot. She wants to fence in three sides of a rectangular area with 200 feet of fencing and use the wall of her store as the fourth side. What is the maximum area she can fence in?

7. BOX

Cut same-sized squares out of the corners of an 8.5-by-11-inch piece of paper. Fold the paper up to form an open-topped box. What is the maximum volume for such a box?

8. JAWS

The movie *Jaws* was a big hit and grossed $130 million. The sequel didn't do quite as well, and neither did *Jaws III* or *Jaws IV*. Using the information in the chart below, draw a graph and predict how much money *Jaws V* might have grossed if the producers had filmed it.

Movie	Jaws	Jaws II	Jaws III	Jaws IV	Jaws V
Gross (in millions)	$130	$50	$26	$12	–?–

9. MORE PHONE CALLS

Below is a list of several phone calls made to different places using a prepaid phone card. The calls were either placed in-state (the least expensive per minute), to Canada, or to France (the most expensive per minute). Each type of call has different first-minute charges and subsequent-minute charges. Determine the first- and subsequent-minute charges for each type of call.

Time (in minutes)	5	7	8	9	10	11	13	18
Cost	$0.70	$0.64	$0.71	$0.45	$1.15	$0.92	$1.42	$0.90

10. BIG PROBLEMS

Lewis and Jerry are trying out for the football squad and want to put on weight. Lewis is eating and working out and has found that he gains about 1 pound each week. At this point he weighs 180 pounds. Jerry, on the other hand, weighs 167 pounds and is eating, working out, and then eating more. He is gaining about 5 pounds every 3 weeks. How long will it take for Jerry and Lewis to weigh the same?

11. SAILING

Diane was sailing. She had been timing herself and charting the average reported wind speed to determine what kind of times to expect if she were sailing a race. On five previous trips, she had recorded the following times and wind speeds:

Wind speed	20 kn	8 kn	12 kn	23 kn	11 kn
Time	45 min	112.5 min	75 min	39 min 8 sec	81 min 6 sec

Today the weather report predicts an average wind speed of 17 kn. What time should she expect to have on the course?

Problem Set A-2

Make a scale drawing for each problem.

12. YOUR BEDROOM

Make a scale drawing of one room in your place of residence. Choose a scale that allows you to fit your drawing nicely onto a piece of paper. Include all your furniture. Label all the interesting parts of your diagram with their actual measurements.

13. CELL PHONE TOWER

You want to measure the height of a cell phone tower. Standing 40 feet away from the pole, you measure the angle between the ground and the top of the tower to be 57°. How tall is the tower?

14. **STADIUM POLE**

You want to measure the height of a stadium light pole. However, it is on the other side of a fence, and you don't want to test the local trespassing laws. You try the following method instead: You stand at point *A* and measure the angle to the top of the pole and find it measures 40°. You then walk 50 feet from point *A* in a direction directly away from the pole and measure the angle to the top again. This time it measures 28°. How tall is the pole?

15. **HOW WIDE IS THE RIVER?**

Mai Khanh wants to measure the width of a river. She stretches a 100-yard-long string parallel to the river along the ground. (The river is completely straight for these 100 yards.) Directly across the river from one end of the string is a tree on the riverbank. From the other end of the string, she sights to the tree and finds that the angle between the string and the line of sight to the tree measures 35°. What is the approximate width of the river?

16. **FRISBEE ON THE ROOF**

Julia is standing on the ground, looking up at Thu in the second-story window. Jason is on the roof retrieving a Frisbee. From where Julia is standing, the angle of elevation to Thu measures 22° and the angle of elevation to Jason measures 40°. Thu is 12 feet above the ground. How far from the building is Julia standing and how high off the ground is Jason?

17. **KITE STRING**

It's Hai-Ting's turn to go on the roof after the kite—this time it got stuck on the chimney. The string is caught on the chimney about 25 feet off the ground. Wesley has the end of the string below and can sight Hai-Ting at about a 15° angle from the ground. How long is the kite string from Wesley to Hai-Ting?

CLASSIC PROBLEM

18. SAID THE SPIDER TO THE FLY

A room is 30 feet long, 12 feet wide, and 12 feet high. A spider on the centerline of the west wall of the room and 1 foot above the floor sees a fly asleep on the centerline of the east wall and 1 foot below the ceiling. The spider wants to get to the fly as soon as possible. What is the shortest path for the spider to take to get to the fly, and what is the length of this path?

Adapted from *Mathematical Teasers* by Julio Mira.

MORE PRACTICE

1. SPORTS CAR COMPETITION

Ignacio just bought a BMW for $40,000 and was showing it off to his friend, Natalio, who had just purchased a Miata for $25,000. Natalio was a little annoyed at being one-upped by his friend, so he pointed out that his Miata would be worth more than Ignacio's car in a few years because the Miata was only depreciating at 20% per year while the BMW was losing value at the rate of 30%. In how many years will Natalio's car be worth more than Ignacio's, assuming they are both in good running condition and no one has an accident?

2. SCIENCE FICTION BOOK SERIES

Greer is the editor of a popular series of science fiction books, featuring the boy astronaut Perry Haughter. So far there have been six books in the series. Each book is a little longer than the previous book, but Greer doesn't like to increase the number of pages by too much as she is afraid that if the books get too long people will stop reading them. So each book gets a gradual increase. The chart below shows the volume and the pages. What would be a reasonable number of pages for book 7?

Volume	1	2	3	4	5	6	7
Pages	250	350	425	475	500	520	???

3. TV RATINGS

Grant is the president of a television company and is the person who determines whether or not a TV series gets renewed for the next season or gets cancelled. The chart below shows the ratings history of one of the network's shows. So far the show has been on for five years. Even though the ratings have dropped every season, the show is still profitable and Grant will keep it on the air. However, if the ratings drop below 12 million people per week in a season, then the show will be cancelled at the end of that season before the next season starts. How many more years can the show sustain enough viewers so that it doesn't get cancelled?

Years	1	2	3	4	5
Avg Viewers (in millions)	30	22	18	16	14.5

4. BROKEN TELEPHONE POLE

A telephone pole snapped at a point about 10 feet above the ground and fell so that it remained attached at the point where it snapped. The top of the pole hit the ground at a distance of 15 feet from the base. The pole is planted 3 feet deep in the ground. The repairmen need to order a replacement pole. How long should it be?

5. ROOFTOP ANTENNA

The Flat family live in a square house with a flat roof that is 50 ft by 50 ft. Mr. Flat is a ham radio operator and he has a 40-ft antenna dead center on the top of his house. The antenna is secured at a point 20 ft above the roof by guy wires from each corner of the roof and from the center of each side. He needs to replace the guy wires and wants to know about how much wire to get. The wire comes in spools of 100 feet. How many spools of wire should he buy?

6. LAKE TRIP

The skipper left the lake dock and sailed due north for 7 hours. He then sailed due northeast for $4\frac{1}{4}$ hours. Then he sailed due east for 3 hours. He sailed the same speed the entire time, and there is no current in the lake. In order to get back to the dock, he plans on sailing due southwest for a certain amount of time, and then due south to get back to the dock. How many hours total will it take him to get back?

7. **THE NEIGHBOR'S TREE**

I have a redwood tree in my front yard and my neighbor across the street has a redwood tree in his back yard. I was curious who had the taller tree, so I measured the following. I have a clinometer that measures the angles from the ground to the tops of things. Standing under my tree, I measured the angle from the ground to the top of my neighbor's tree at 40 degrees. Then I walked 60 feet toward my neighbor's tree and again measured the angle to the top of my neighbor's tree. This time it was 50 degrees. From this point (60 feet from my tree) I also measured the angle to the top of my tree and it was 72 degrees. Whose tree is taller and by about how much?

Problem Set B

1. **SODA JERK**

As a clerk at a soda counter, Jared had to take orders, get the right-flavor soda in the right-sized cup, and get the drink to the right customer. On his first day on the job, he didn't fill a single order correctly. He did, however, get the flavor right on 13 orders and the drink size right on 12 orders. He also managed to get 9 drinks to customers who'd actually ordered drinks, but the size, the flavor, or both were wrong. In defending himself to his manager, he said, "At least there were 14 drink orders where only one thing was wrong, and only once did I give a wrong-flavor soda in a wrong-sized cup to a customer who had not ordered a drink." What is the minimum number of orders that Jared got wrong? What is the highest possible percentage of drinks that were salvageable (the right flavor was in the right cup and it just needed to be given to the right customer)?

2. **LICENSE PLATES**

License plates are issued systematically. In a state where the license plates contain three letters followed by three numbers, the following sequence of license plates could be issued: CMP998, CMP999, CMQ000, CMQ001, and so on. The most recent license plate issued in this state is AZY987, which has no repeated letters or digits. How many license plates must be issued after this one before a license plate will again have no repeated letters or digits?

3. **KAYAKING**

Irka and Boris both live on the lake. To get to Boris's house by land, Irka has to take Mountain View Road 2 miles south and then take Lakeshore Drive 3 miles east. Boris called Irka and suggested that they go watch a soccer game at the end of School Road. To get to the end of School Road, Boris simply needs to kayak due north across the lake. If Irka kayaks directly from her house to the end of School Road, it turns out to be exactly the same distance that Boris has to go. How far is it from Boris's house to the end of School Road?

4. **THE DIGITAL CLOCK AND THE MIRROR**

As I look in the bathroom mirror while I'm shaving, I can see the reflection of the digital clock in my bedroom. Sometimes the time I see in the mirror (a reflection of the actual time) is the same as the actual time. How many times does this happen per day? (Ignore the colon, because I can't see that anyway—the clock is too far away.)

5. **THE LATTICE**

Veronica and Archie are building a lattice out of wood. The lattice will be in the shape of a right triangle. The frame for the lattice measures 18 feet along the base and 12 feet in height. They are going to put in vertical strips of 1-inch-wide wood every 9 inches, connecting the base to the hypotenuse. How many feet of 1-inch-wide wood will they put into the lattice?

Appendix

Converting Metric Measurements

Consider the basic unit of length in the metric system—meters. Now consider this question: How many centimeters are in a dekameter?

To answer the question, you must convert dekameters into centimeters. To convert from one metric measurement to another, you must first know the metric prefixes and what they represent. Metric prefixes are combined with the basic unit names of metric measurement (meter, liter, and gram) to indicate a multiple or submultiple of the unit measurement. Six common metric prefixes are kilo-, hecto-, deka-, deci-, centi-, and milli-. One way for you to remember these prefixes is to memorize a mnemonic, a device that helps you remember something. A mnemonic can be something like a rhyme, an acronym, or a sentence. To help you remember metric prefixes, use the sentence "Kind hearts don't use dirty crummy manners." Here's how this mnemonic sentence corresponds to the prefixes:

kind	**h**earts	**d**on't	**u**se	**d**irty	**c**rummy	**m**anners
kilo-	**h**ecto-	**d**eka-	**u**nit	**d**eci-	**c**enti-	**m**illi-

The prefixes for the multiples of the unit are Greek, and the prefixes for the fractions of the unit are Latin. What do the prefixes mean?

kilo-	hecto-	deka-	unit	deci-	centi-	milli-
1,000	100	10	1	0.1	0.01	0.001

Note that each prefix represents a multiple of 10 that is 10 times as large as the one to its right. The word "unit" in the middle refers to the basic unit of measure in the metric system (meter for length, liter for volume, or gram for mass). Each prefix in the chart is combined with the basic unit name to indicate a new measure. For example, 1 kilometer equals 1,000 meters, 1 centigram equals 0.01 grams, and 1 milliliter equals 0.001 liters.

As you move from one space to the next in the chart, you either multiply or divide by 10, depending on whether you're moving left or right. Let's return to the question asked earlier: How many centimeters are in a dekameter? In the chart, the deka- position is three places to the left of the centi- position, so multiply 10 times 10 times 10 (that's 1,000). So there are 1,000 centimeters in 1 dekameter. Sometimes it is easy to get this statement backwards and say that there are 1,000 dekameters in 1 centimeter. To figure out the correct conversion, think about which unit is greater. Centimeters are very small, so it takes a whole lot of them to make up a dekameter.

You can also find the answer to the question by canceling units:

$$\frac{1\ \cancel{\text{dekameter}}}{1} \times \frac{10\ \cancel{\text{meters}}}{1\ \cancel{\text{dekameter}}} \times \frac{10\ \cancel{\text{decimeters}}}{1\ \cancel{\text{meter}}} \times \frac{10\ \text{centimeters}}{1\ \cancel{\text{decimeter}}}$$

Or you can try this:

$$\frac{1\ \cancel{\text{dekameter}}}{1} \times \frac{10\ \cancel{\text{meters}}}{1\ \cancel{\text{dekameter}}} \times \frac{100\ \text{centimeters}}{1\ \cancel{\text{meter}}}$$

Or perhaps this:

$$\frac{1\ \cancel{\text{dekameter}}}{1} \times \frac{10\ \cancel{\text{meters}}}{1\ \cancel{\text{dekameter}}} \times \frac{1\ \text{centimeter}}{0.01\ \cancel{\text{meter}}}$$

In all three cases, all the units cancel except centimeters and the arithmetic comes out to 1,000. So the answer is 1,000 centimeters.

Abbreviations of Units of Measure

Abbreviations of metric measures—other than the basic units of meter, liter, and gram—are formed by adding the abbreviations of metric prefixes to the abbreviations of the metric units. For example, to abbreviate kilogram, you combine the abbreviation for kilo (k) with the abbreviation for gram (g). The result is the abbreviation kg.

The following chart of standard metric abbreviations comes from Wikipedia.org in their article on the metric system and the international system of units (SI).

METRIC PREFIXES IN EVERYDAY USE

TEXT	SYMBOL	FACTOR	POWER OF 10	NAME
peta	P	1,000,000,000,000,000	10^{15}	quadrillion
tera	T	1,000,000,000,000	10^{12}	trillion
giga	G	1,000,000,000	10^{9}	billion
mega	M	1,000,000	10^{6}	million
kilo	k	1,000	10^{3}	thousand
hecto	h	100	10^{2}	hundred
deca	da	10	10^{1}	ten
(unit)	(unit)	1	10^{0}	one
deci	d	0.1	10^{-1}	tenth
centi	c	0.01	10^{-2}	hundredth
milli	m	0.001	10^{-3}	thousandth
micro	μ	0.000,001	10^{-6}	millionth
nano	n	0.000,000,001	10^{-9}	billionth
pico	p	0.000,000,000,001	10^{-12}	trillionth
femto	f	0.000,000,000,000,001	10^{-15}	quadrillionth

Here are lists of commonly used abbreviations of measure, many of which appear in this book. Note that whether the units of measure are in singular or plural form, the abbreviations for the units are the same. For example, the abbreviation for both mile (singular) and miles (plural) is mi.

DISTANCE (ENGLISH)		TIME		VOLUME	
foot	ft	hour	hr	gallon	gal
mile	mi	minute	min	quart	qt
yard	yd	second	sec	fluid ounce	fl oz
inch	in.	day	dy	cubic foot	ft^3
		year	yr		

Common Conversions

Here are lists of common conversions from one measurement to another. Many of these conversions will come in handy as you work through the problems in Chapter 8: Analyze the Units.

DISTANCE (ENGLISH)

12 inches = 1 foot

3 feet = 1 yard

5,280 feet = 1 mile

VOLUME (ENGLISH)

1 gallon = 4 quarts

1 quart = 4 cups

1 cup = 8 fluid ounces

DISTANCE (METRIC)

100 centimeters = 1 meter

1,000 millimeters = 1 meter

1 kilometer = 1,000 meters

TIME

1 day = 24 hours

1 hour = 60 minutes

1 minute = 60 seconds

1 week = 7 days

1 year = 365 days*

*Note that if you're converting four or more years to days, you must multiply the number of years by 365.25 days to account for leap years.

METRIC TO ENGLISH

Distance: 1 meter ≈ 3.281 feet

Volume: 3.79 liters ≈ 1 gallon

ADDING, SUBTRACTING, MULTIPLYING, AND DIVIDING FRACTIONS

To add or subtract fractions, find the common denominators of the fractions, then add or subtract the numerators.

$$\frac{1}{6} + \frac{2}{3} = \frac{1}{6} + \frac{4}{6} = \frac{5}{6} \qquad \frac{2}{3} - \frac{7}{12} = \frac{8}{12} - \frac{7}{12} = \frac{1}{12}$$

To multiply fractions, multiply the numerators together and multiply the denominators together.

$$\frac{3}{4} \times \frac{1}{2} = \frac{3}{8} \qquad \frac{1}{2} \times \frac{3}{5} \times \frac{3}{7} = \frac{9}{70}$$

To divide a fraction by a fraction, multiply the first fraction by the reciprocal of the second fraction.

$$\frac{5}{8} \div \frac{2}{3} = \frac{5}{8} \times \frac{3}{2} = \frac{15}{16}$$

AREA AND VOLUME FORMULAS

Area

Triangle: $A = \frac{1}{2}bh$
where b is the base and h is the altitude to that base.

Square: $A = s^2$
where s is the side of the square.

Rectangle: $A = bh$
where b is the base and h is the height of the rectangle.

Parallelogram: $A = bh$
where b is the base of the parallelogram and h is the altitude to that base.

Trapezoid: $A = \frac{1}{2}h(a + b)$
where a and b are the parallel sides of the trapezoid and h is the altitude between them.

Circle: $A = \pi r^2$
where r is the radius of the circle.

Volume

Cube: $V = b^3$
where b is the edge of the cube.

Rectangular prism: $V = abc$
where a, b, and c are nonequal edges of the rectangular prism.

Cylinder: $V = Ah$
where A is the area of the base and h is the height of the cylinder.
When the base is a circle, $A = \pi r^2$ and $V = \pi r^2 h$.

Pyramid or Cone: $V = \dfrac{1}{3} Ah$
where A is the area of the base and h is the altitude to the base.

Sphere: $V = \dfrac{4}{3} \pi r^3$
where r is the radius of the sphere.

PROPERTIES OF TRIANGLES

Similar Triangles

The corresponding sides of similar triangles share a common ratio.
So $\dfrac{a}{d} = \dfrac{b}{e} = \dfrac{c}{f}$.

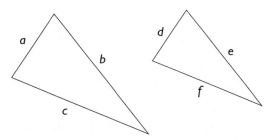

Pythagorean Theorem

If a and b are the legs of a right triangle and c is the hypotenuse, then
$a^2 + b^2 = c^2$.

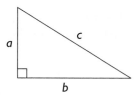

Divisibility Rules

There are divisibility rules for all the numbers less than 13, but the most useful ones are for 2, 3, 4, 5, 6, 8, 9, and 10. You might think, "Why do we need these? We can just use a calculator." But there are many times when you need to identify whether a number is divisible by 3 or 5, and you need a quick way to do it.

Divisibility by 2: A number is divisible by 2 if the number is even. That is, if the last digit is 0, 2, 4, 6, or 8. Examples: 12 and 624 are divisible by 2. The numbers 43 and 579 are not divisible by 2.

Divisibility by 3: A number is divisible by 3 if the sum of the digits is divisible by 3. Examples: 57,624 is divisible by 3 because the sum of the digits $5 + 7 + 6 + 2 + 4 = 24$, which is divisible by 3. The number 6,057,529 is not divisible by 3 because the sum of the digits $6 + 0 + 5 + 7 + 5 + 2 + 9 = 34$, which is not divisible by 3.

Divisibility by 4: A number is divisible by 4 if the last two digits of the number form a number that is divisible by 4. So all you need to do is look at the last two digits. Examples: 65,480 and 724 and 15,548,768 are all divisible by 4 because the number formed by the last two digits of each number (80, 24, and 68) is divisible by 4. The numbers 325 and 847,654 are not divisible by 4 because the numbers formed by the last two digits (25 and 54) are not divisible by 4.

Divisibility by 5: A number is divisible by 5 if it ends in either 0 or 5. Examples: 255 and 84,736,985 and 8,420 are divisible by 5, whereas 67 and 7,834 are not.

Divisibility by 6: A number is divisible by 6 if it is divisible by both 2 and 3. So it must satisfy the divisibility rules for 2 and for 3. It must be an even number *and* the sum of the digits must be a multiple of 3. Examples: 684 is divisible by 3 because the sum of the digits $(6 + 8 + 4)$ is 18, which is divisible by 3, and the number 684 is an even number (and therefore divisible by 2). The number 786 is also divisible by 6, because the sum of the digits is 21 (divisible by 3) and 786 is an even number.

Divisibility by 8: A number is divisible by 8 if the last three digits form a number that is divisible by 8. Example: 49,878,098,544 is divisible by 8 because the number composed by the last three digits, 544, is divisible by 8.

Divisibility by 9: A number is divisible by 9 if the sum of the digits is divisible by 9. Example: 1,264,873,920,921 is divisible by 9 because the sum of the digits is 1 + 2 + 6 + 4 + 8 + 7 + 3 + 9 + 2 + 0 + 9 + 2 + 1 = 54, which is divisible by 9.

Divisibility by 10: A number is divisible by 10 if the last digit is 0. Examples: 540 and 8,765,820 are both divisible by 10.

Determining Whether a Number Is Prime

A prime number is divisible only by itself and 1. In other words, it only has two factors. To determine whether a number is prime, you should try to factor it. To illustrate this, consider a number that is not prime, and factor it; for example, 24.

The factors of 24 all come in pairs: 1 and 24, 2 and 12, 3 and 8, and 4 and 6. Write an ordered list of these numbers: 1, 2, 3, 4, 6, 8, 12, 24. Notice the numbers pair up from the outside in.

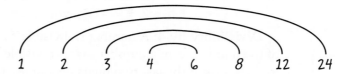

Note that the middle of this diagram is not half of 24, but rather roughly the square root of 24.

So consider the number 113. We want to know whether 113 is prime. Try to factor it. One pair of factors is obviously 1 and 113. Try 2. 113 is not divisible by 2. It is not divisible by 3 (the sum of the digits is 5, which is not divisible by 3). You don't need to try 4, because 2 did not work. It is not divisible by 5 (it doesn't end in 0 or 5). It is not divisible by 7 (try on your calculator). You don't need to try 6, 8, 9, or 10, because they aren't prime. The square root of 113 is between 10 and 11. So you do not need to try any larger numbers, because if any number larger than 11 was a factor of 113, then there would have to be a number smaller than 11 for it to be paired with. So the number 113 is prime.

In summary, to determine whether a number is prime, divide it by each prime less than or equal to the square root of your number. If none of these primes will divide your number, then your number is prime.

Glossary

Numbers in brackets refer to chapters in which the term is first defined or significantly covered.

act it out A problem-solving strategy in which people play roles in the problem. [10]

adjunct list A list of connected information made in conjunction with a logic matrix. [4]

algebra A strategy for solving a problem using variables and equations.

area The measure in square units of a bounded region.

array See **matrix.**

assumption An unfounded idea about a problem, which can be tested to be true or false. [4]

average Generally used to represent the mean average. The sum of a group of numbers, divided by the number of numbers. [9]

bouncing A substrategy through which information on one part of a matrix is aligned with different, connected information to produce a connection with previously unconnected information. See also **cross-correlating.** [4]

canceling See **unit conversion.**

change the representation A strategy that combines changing focus with some other strategy, often a diagram. [16]

change your point of view A strategy for solving a problem by looking at it in a different or unexpected way. [16]

Changing Focus An overall theme to several problem-solving strategies, in which a shift in perspective is intrinsic to the strategy. [7, 16]

circumference The distance around a circle. Given by the formula $C = 2\pi r$, where r is the radius of the circle.

coefficient A number by which a variable is multiplied in an algebraic expression. [14]

complement Something added to complete a whole; either of two parts that complete each other. [16]

complementary angles Two angles whose sum is 90°.

complementary problem A problem that, taken with an original problem, completes a whole picture. For example, computing the number of days someone works in a week may be computed by subtracting from 7 the number of days not worked. [16]

composite A whole number that has more than two factors; examples: 9, 24, and 38.

compound unit Two or more units combined by multiplication or division. [8]

cross-correlating A substrategy through which information on one part of a matrix is aligned with different, connected information to produce a connection with previously unconnected information. See also **bouncing.** [4]

cryptarithm An arithmetic problem based on letters, usually with a clever combination of words, such as "SEND + MORE = MONEY." Each letter stands for a different digit, 0 through 9. [3]

cubic function See **polynomial.**

degree of a polynomial For a polynomial expression in one variable, it is the highest exponent. [14]

denominator The bottom number in a fraction; it divides the top number, or numerator. [9]

dependent variable A variable that represents the set of output numbers to a function. [17]

diagonal A line connecting two non-adjacent vertices in a polygon. [9]

diagram A pictorial representation used in solving a problem. [1]

diameter A chord drawn through the center of a circle; the diameter has length two times the radius.

digit Any of the figures 0, 1, 2, 3, 4, 5, 6, 7, 8, 9. [Intro, 3]

dimensional analysis Another name for **unit analysis.** [8]

disjoint A description applied to sets that have no elements in common. [12]

dogs Mythical creatures alternately credited with human-like intelligence or a wide range of erratic behavior. Reported sightings are frequent, though there is no proof that they exist.

easier related problems A problem-solving strategy in which the solver considers smaller, simpler versions of the problem to be solved in order to develop a process for solving the original problem. [9]

element See **set.**

eliminate possibilities A problem-solving strategy in which all elements or outcomes are listed and then as many elements or outcomes as possible are removed from consideration through logical deduction. [12]

empty See **set.**

equation A mathematical statement asserting the equality of two expressions. [13]

even number A number that is divisible by 2.

exercise One of a number of repetitive tasks. [Intro]

expected value In probability, an estimation based on the average output of a function. [15]

exponent A number placed above and after another number to indicate how many times to use that number as a multiplier. For example: 2^5 indicates that the number 2 should be used as a multiplier 5 times. $2 \times 2 \times 2 \times 2 \times 2 = 32$. So 2 to the fifth power is 32.

exponential function A function in which a number is raised to the power of the input variable. If consecutive integers are used as input values in an exponential function, the output values form a geometric sequence. [14]

expression An algebraic quantity. [13]

factor One of two or more numbers that multiply together to create a given number. For example: 3 and 6 are both factors of 18. The factors of 24 are 1, 2, 3, 4, 6, 8, 12, and 24. See also **prime factor.**

factorial Shorthand notation for the product of a list of consecutive positive integers from the given number down to 1; that is, $n! = n \times (n - 1) \times (n - 2) \times (n - 3) \times \ldots \times 3 \times 2 \times 1$. For example, $5! = 5 \times 4 \times 3 \times 2 \times 1 = 120$.

Family family An imaginary family consisting of a father, mother, two daughters, a son, and five **dogs.**

Fibonacci number Any number of the Fibonacci sequence: 1, 1, 2, 3, 5, 8, 13, 21, 34, 55, 89, . . . , where each term in the sequence is the sum of the previous two terms. [5]

finite differences A strategy for finding an equation for a table of input/output values by subtracting pairs of outputs. [14]

freezing Leaving a value in a systematic list unchanged until all options for that value are exhausted.[1] [2]

function A relationship between sets in which, for each and every element of the input set, there is exactly one element of the output set. A function with a rule or equation is a predictable, consistent relationship between the elements of the input set and the elements of the output set. [5]

graph A picture clearly showing the relationship between variables. [17]

[1]Definition by Sierra College student Patrick Pergamit.

greatest common factor Given two or more numbers, the greatest common factor is the largest number that is a factor of all of the given numbers. For example: 2 is the greatest common factor of 12 and 14. 6 is the greatest common factor of 12, 24, and 30.

guess-and-check A strategy for solving a problem by making, refining, and keeping track of estimations in an organized chart. [6]

heptagon See **polygon.**

heuristic Based on discovery, rather than taught. [Intro]

hexagon See **polygon.**

independent variable A variable that represents the set of input numbers to a function. [17]

indirect proof See **seeking contradictions.**

input A number that can be substituted for the independent variable in a certain function. [14]

integer A whole number or its opposite: A member of the set {. . . , -4, -3, -2, -1, 0, 1, 2, 3, 4, . . .}.

integral Pertaining to or being an integer; not fractional.

intersecting In a Venn diagram, a set that has characteristics of two or more distinct sets. Also known as **overlapping.** [12]

isomorphism A one-to-one correspondence between two sets that preserves relations between elements of the sets. [16]

knot One nautical mile per hour.

least common multiple Given two or more numbers, the least common multiple is the smallest number that appears in the lists of multiples for each of the numbers. For example: 15 is the least common multiple of 3 and 5. 24 is the least common multiple of 4, 6, and 8.

linear function See **polynomial.**

logic matrix A matrix used to organize information in which related information is entered in consecutive cells of a row or column. [4]

manipulative An object that can be moved or positioned and is used to represent an element of a problem. [8, 10]

matrix An organization of information in rows and columns, in which each and every element in any given row or column is related to the rest of the information in that same row or column. Also called an **array.** [4]

matrix logic A strategy for using a process of elimination by organizing the information in a matrix or in a chart composed of several matrices. [4]

maximum value The largest value for a function.

member See **set.**

mixture problem A common application problem, solvable by subproblems or algebra, in which two or more substances are mixed together to form a blend, the two substances differing in some characteristic such as tint, concentration, or economic value. [7, 13]

model A physical representation of a problem that can be helpful in finding a solution. [10]

mpg Miles per gallon; should be written as a fraction: miles/gallon.

mph Miles per hour; should be written as a fraction: miles/hour.

multiple A number that contains another number an integral number of times with no remainder. For example: 35 is a multiple of 5, and 36 is a multiple of 3.

nautical mile One minute of arc of the earth's circumference equal to 6076.103333 feet. [8]

negative number A number that is less than zero.

nomograph A graph with parallel scales on which you can join values on two scales with a straightedge in order to read a value on a third scale. [17]

*n***th term** See **term**.

numerator The top number in a fraction; it is divided by the bottom number, or denominator. [9]

octagon See **polygon**.

odd number A number that has a remainder of one when divided by 2.

one-n-o A fraction that equals 1, used in converting units. [8]

one-to-one correspondence A linking of elements such that for each and every element of one set, there is exactly one element of another set, with no duplication in either set. [4]

Organizing Information An overall theme to several problem-solving strategies, in which the organization of information is intrinsic to the strategy. [7, 15]

output The number that is a result of substituting a number into a certain function. [14]

overlapping sets See **intersecting**.

pattern A coherent rule or characteristic shared by elements of a set. [5]

pentagon See **polygon**.

per Divided by. [8]

perimeter The total length of the boundary of a region.

physical representation A strategy for solving problems by using physical objects. [10]

pipeline A diagram used to represent unit conversions in which lines connect related units for which a conversion relationship is present. [8]

polygon A closed figure in a plane consisting of vertices and edges. **Triangles, quadrilaterals, pentagons, hexagons, heptagons,** and **octagons** are all polygons, with three, four, five, six, seven, and eight sides, respectively. [9]

polynomial A sum of terms in which each term is the product of a coefficient and a power of the variable. For instance, $x^3 - 2x^2 + 3x - 1$ is a cubic polynomial. A **polynomial function** pairs **inputs** with **outputs** given by a polynomial. For a **linear function,** the greatest power of a variable is 1, for a **quadratic function** the greatest power is 2, and for a **cubic function** the greatest power is 3. [14]

position The location of a term within a sequence. [5]

positive number A number that is greater than zero.

power A power of a number or expression is an exponent. See **exponent.**

prime factor Any composite number can be written as a product of prime factors: For example, 24 can be written as $2 \times 2 \times 2 \times 3$.

prime number A whole number that has only two factors, the number itself and 1. For example, 3, 17, and 31 are all prime.

prism A three-dimensional shape with parallel bases that are congruent polygons and sides that are parallelograms. [9]

problem A task that may not have a clear path to the solution. [Intro]

product An indicated multiplication or the result of multiplying two or more numbers together. For example, 2×3 and $5x$ are products, and 14 is the product of 2 and 7.

proof by contradiction See **seeking contradictions.**

quadratic equation An equation in which 2 is the highest power of any variable.

quadratic function See **polynomial.**

quadrilateral See **polygon.**

ratio A comparison of two numbers by division. [8]

reciprocal The multiplicative inverse of a fraction. A fraction and its reciprocal multiplied together equal 1. [8]

region In Venn diagrams, a bounded area of the diagram containing elements that share a characteristic. [12]

scale drawing A drawing in which the shape and relative proportions of lengths and areas are preserved. [17]

seeking contradictions A process for proving a conjecture that involves first making an assumption, then examining the assumption for its validity by carrying it to its (contradictory) conclusion. Also known as **indirect proof** or **proof by contradiction.** [3]

sequence A set of numbers ordered by a consistent rule, or set of rules, that determine all the terms. [5]

set A collection of particular things. Each thing is an **element,** or **member.** A set is composed of one or more elements, except an **empty** set or "null set," which has zero elements. A **subset** is a set within a set. [12]

slope The ratio of the change in the dependent variable to the change in the independent variable, also expressed as the change in y over the change in x. For a linear graph, slope measures steepness and the direction of slant. [17]

solution An answer to a problem and an explanation of how the answer was reached. [Intro]

Spatial Organization An overall theme to several problem-solving strategies in which the spatial layout of the problem and solution information is intrinsic to the strategy. [7, 17]

square number A number that is the square of a whole number. Examples: 1, 4, 9, 16, 25, . . . [5]

step function A function whose output is not continuous, but "steps" from one value to another. [17]

strategy A method for solving a problem. A **substrategy** is a smaller part of a strategy, used with other substrategies to solve the whole problem. [Intro]

subproblem A smaller problem that is part of the whole problem. [7]

subset See **set**.

substitution A substrategy where one piece of information can be used to replace other connected information. Used frequently in matrix logic problems. [4]

sum An indicated addition or the result of adding two or more numbers together. For example, $14 + 3$ and $x + 3y$ are sums, and 10 is the sum of 7 and 3.

supplementary angles Two angles whose sum is 180°.

system A procedure for performing a task or tasks in an organized, methodical way. [2]

systematic list A list generated by organizing elements. [2]

term A member of a sequence. The *n*th term is a generic description used to refer to any member of the sequence. [5]

think outside the box A popular expression that means to think of ideas that do not fit any preconceived or standard notions. [16]

tree diagram A type of systematic list in which options are shown as branching diagrams. [2, 15]

triangle See **polygon**.

triangular number A number that can be arranged as a triangle, of the form $n + (n - 1) + \ldots + 3 + 2 + 1$. For example: 10 is a triangular number, and this characteristic is used in the set-up of 10 bowling pins. Here, 1, 3, 6, and 10 are shown as triangular numbers. [5]

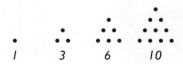

trigonometric function A function related to the ratio of the sides of a right triangle. [14]

unit analysis A method of organizing quantitative information that carefully keeps track of units during computation in order to produce the correct quantity (number and unit) in the answer. See **unit conversion.** [8]

unit conversion A method of changing units by **canceling,** or multiplying by a one-n-o with the original unit in the denominator and a different unit in the numerator. [8]

unit pricing The policy of many stores to give prices in terms of one ounce or one of some other quantity. [8]

universal set In a given problem, the set of all possible elements. [12]

variable An unknown quantity in a problem. For a function, the **independent variable** is the input and the **dependent variable** is the output. [17]

Venn diagram A diagram with bounded regions containing elements that share a characteristic. [12]

vertex A corner of a figure; a point where two or more sides meet. (plural: **vertices**) [9]

whole number Zero or one of the counting numbers; a member of the set {0, 1, 2, 3, 4, . . .}.

working backwards A strategy for solving a problem by reversing the steps of a process. [11]

y-intercept The value of the output when the input is zero. The point where a graph crosses the vertical axis. [17]

Bibliography

"Ask Marilyn." *Parade* (September 1990): 15.

Associated Press. "NASA Fixes Hubble's Antenna." *Sacramento Bee,* May 1, 1990.

Charles, Randall; Frank K. Lester; and Phares G. O'Daffer. *How to Evaluate Progress in Problem Solving.* Reston, VA: National Council of Teachers of Mathematics, 1987.

Cohen, Gilles, editor. *50 Mathematical Puzzles and Problems from the International Championship of Mathematics—Green, Orange, and Red Collections.* Emeryville, CA: Key Curriculum Press, 2001; and Paris: Éditions *POLE*, 1997.

Dell Math & Logic Problems and various other puzzle magazines. Dell Magazines, New York City, NY.

Dolan, D. T., and J. Williamson. *Teaching Problem-Solving Strategies.* Reading, MA: Addison-Wesley Publishing, 1983.

Dudeney, Henry Ernest. *536 Puzzles and Curious Problems,* edited by Martin Gardner. New York: Charles Scribner's Sons, 1967 (originally published in approximately 1931).

Dudeney, Henry Ernest. *Amusements in Mathematics.* New York: Dover Publications, 1958 (originally published in 1917).

Fixx, James. *More Games for the Superintelligent.* Garden City, NY: Doubleday, 1976.

Fixx, James. *Games for the Superintelligent.* Garden City, NY: Doubleday, 1972.

Games Magazine and Games World of Puzzles Magazine. Blue Bell, PA: Kappa Publishing Group.

Gardner, Martin. *The Scientific American Book of Mathematical Puzzles and Diversions.* New York: Simon & Schuster, 1959.

Kordemsky, Boris A. *The Moscow Puzzles: 359 Mathematical Recreations,* edited by Martin Gardner. New York: Dover Publications, 1972.

Loyd, Sam. *Mathematical Puzzles of Sam Loyd,* volume 2, selected and edited by Martin Gardner. New York: Dover Publications, 1960 (originally published in 1914).

Meyer, Carol, and Tom Sallee. *Make It Simpler: A Practical Guide to Problem Solving in Mathematics.* Reading, MA: Addison-Wesley Publishing, 1983.

Mira, Julio. *Mathematical Teasers.* New York: Barnes and Noble Books, 1970.

Newman, James R. *The World of Mathematics,* 4 volumes. New York: Dover Publications, 2000.

Penny Press Magazines – various puzzle magazines. Norwalk, CT.

Pólya, George. *How to Solve It: A New Aspect of Mathematical Method,* 2nd edition. Princeton, NJ: Princeton University Press, 1945.

Smullyan, Raymond. *The Lady or the Tiger and Other Logic Puzzles.* Oxford, New York: Oxford University Press, 1991.

Stewart, Ian. "A Puzzle for Pirates." *Scientific American* (May 1999): 98–99.

Stewart, Ian. "The Riddle of the Vanishing Camel." *Scientific American* (June 1992): 122–124.

Summers, George. *New Puzzles in Logical Deduction.* New York: Dover Publications, 1968.

Summers, George. *Test Your Logic: 50 Puzzles in Deductive Reasoning.* New York: Dover Publications, 1972.

Summers, George. *The Great Book of Mind Teasers and Mind Puzzlers.* New York: Sterling Publishing Company, 1986.

University of Mississippi (Ole Miss). Problem of the Week website: http://www.olemiss.edu/mathed/contest.

Wikipedia.org—the free online encyclopedia.

Index of Problem Titles

General Index

Page numbers in **bold** denote glossary terms.

Bingham, John, 281
Borchard, Monica, 317
bouncing. *See* cross-correlations
Brickner, Brianne, 316
Bridges, Josh, 346

C

calculators, using, 203, 213
Caler, Judi, 196
canceling, 205, **506**
changing focus, **176**, 443–468, **506**
  advantage in, 448
  applied problem solving, 457–458
  complementary problem, solving,
    450–454
  least common multiple, 449
  point of view, changing, 444–450, **506**
  representation, changing, 454–457, **506**
  thinking outside the box, 445
  ways to, 443–468
charts
  in easier related problem (ERP), 236
  use, 146–148
Chew, Jeremy, 313
circle(s)
  area of, 452
  circumference of, 264
circumference, 264, **506**
clues, 49–50
coefficient, **382, 506**
combining clues, in matrix logic, 90
common denominator, 243–244
common difference, 383–385, 389, 394
complement, defined, 450, **506**
complementary angles, **450, 506**
complementary problem, **450**
  solving, 450–454
compound units, **218**–223, **506**
  area, 218–219
computer programming, 222–223

cone, 193
consecutive pairs of numbers, relationship
    between, 119
contradiction, 95, 418
conversion, unit conversion, 204–218
Cook, Marcy, 275
correlating, **506**
Craig, Cory, 137
cross-correlations, 91, **506**
  with negative information, 91
  with positive information, 85, 91–92
cryptarithms, 52, **54, 506**
  important points, 59
  solving, 59
cube, 273–274
cubic equations/function, 406, **506**
cylinder, volume of, 264

D

Daniel, Bob, 311
Davis, James, 341
decimals, 245–247
degree of a polynomial, **381, 506**
Dell Math and Logic Problems, 74
denominator, 205, **243, 506**
dependent variable, **471, 506**
diagonal, **258, 506**
diagram(s), 11–26, **506**
  in algebra, 368
  drawing, 11–26
    advantages, 20
    area, **17**
    perimeter, **17**
    tree diagram, **31**, 41
  infinite differences, 395
  matrix logic and, 78–79
  occupational, 20
  physical representations compared
    to, 266
  pipelines, **214**–215

systematic lists compared to, 33

tree. *See* tree diagrams

Venn. *See* Venn diagrams

working backwards and, 306–309

See also graphs; scale drawings

diameter, **193**, **507**

difference of the differences, 119

digit, **507**

dimensional analysis. *See* unit analysis

disjoint sets, **325**, 507

divisibility rules, 5, 50

divisors, 243–246

dog icons, about, 5

dogs, 2, **507**

Donohoe-Mather, Carolyn, 210

drawing to aid comprehension, 190

See also diagram(s)

drug testing, 427–428

Dudeney, Henry, 69, 107, 261, 289, 373

E

easier related problem (ERP), **507**

creating, common ways for, 236–255

divisors, 243–246

reciprocals, 243–246

solving, 233–264

breaking big problems into
subproblems, 234

changing focus strategy, 236

chart, organizing the information
in, 241

eliminating unnecessary information,
238

look for a pattern, 240

reading the question carefully, 239

specific, easier example, working
on, 240

using easier numbers, 237

subproblems versus, 255–257

element, **322**, **507**

See also sets

eliminating possibilities, 47–73, **507**

compound statements, analyzing, 53

cryptarithms, 52, 54

guess-and-check strategy, 51

hidden clues, 64

indirect proof, **52**

last digit of product, finding, 63

for organizing information, 51

seeking contradictions, 52

systemic list use in, 64

working in groups, benefits of, 60–64

empty regions, in venn diagram,
325, 327, **507**

English units, 213–214

abbreviations, 211

conversions, 213–214

See also unit analysis

equations, **355**, **507**

algebra and. *See* algebra

coefficients, **382**

cubic, 388, 406

exponential, 381

finite differences and. See finite
differences

letters used in, 381

linear, 381–382, 387–388

quadratic, **381**–390

slope-intercept form, 381

valid, 355

writing of, 350, 369

See also functions; polynomial
functions

Erickson, Jessica, 109

ERP. *See* easier related problems (ERP)

even numbers, 61, **507**

Evenson, Katie, 316

exercise, **3**, **507**

expected value, **430, 507**
exponent rules, 247
exponential functions, **381, 507**
exponents, 246, **507**
expressions, **355, 507**

F

factorials, **133, 507**
factors, **446, 507**
Family family, **507**
Fibonacci number, **129**
Fibonacci sequence, 128–129, **507**
finite differences, 379–412, **507**
  aligning charts, 396
  calculating the differences, 404–405
    data, 405
    finite differences do not apply to all
      patterns, 404
  compute first differences, 389
  compute second differences, 390
  correct general chart, using, 390
  determine degree, 390
  diagram for clarity, 395
  evaluating, 379–412
  look for patterns, 395
  matching corresponding parts of two
    charts, 387, 391
Fitzpatrick, Katie, 316
Fixx, James, 194
fractions, 5, 8
  numerator and denominator, 201, 205
  See also ratios
freezing, **31, 507**
functions, **380, 507**
  cubic, **381**, 400–406
  exponential functions, **381**
  finite differences and. *See* finite
    differences
  linear, **381**, 383
  polynomial functions, **381**

quadratic functions, **381**, 406
step functions, **477**
trigonometric, 381, **510**
See also equations; polynomial
  functions; quadratic equations
536 Puzzles and Curious Problems, 69,
  107, 373

G

Games and World of Puzzles, 74
games of chance, 428, 430
Games magazine, 74
Gardner, Martin, 68–69, 107, 140, 141,
  170, 229, 260, 314, 373, 436, 463
Gauss, Carl Friedrich, 234–235
gender, 93
Gernes, Don, 463
Gonzales, Aletta, 315
Graduate Record Exam (GRE), 176
graphs, **470**–484, **507**
  advantages, 476, 489–497
  easier related problem, creating, 480
  effectiveness, 483
  linear equations, 382, 385–388
  maximum value of, **483**
  nomographs, **484**–485
  quadratic equations, **381**–382
  slope, 381, **471**–473, 475–476
  step functions, **477**
  systematic list in, 478
  y-intercept, **471**–472
  See also diagram(s); scale drawings
groups, working in. *See* team work
guess-and-check strategy,
  145–174, **507**
algebra and guess-and-check
  table, 355, **363**
algebraic equations, 160
organized table use in, 164
patterns in, 163

Loyd, Sam, 5, 24, 45–46, 140, 170, 229, 314, 373
Lucas, Edouard Anatole, 129, 140

M

magic squares, 282–284
Make It Simpler: A Practical Guide to Problem Solving in Mathematics, 197
making an assumption. See assumptions
manipulatives, **508**
  in physical representation, **266**–283
  for unit analysis, 206–209
  working backwards and, 304
marking traits, 82–83
marking traits, in matrix logic, 88–89
Mason, Alyse, 283
Mathematical Association of America sponsors, 432
Mathematical Puzzles of Sam Loyd, 24, 45, 46, 140, 170, 229, 314, 373
Mathematical Teasers, 494
Mathletes, 420
matrix/matrices, 75, 78, 92, 97, **508**
matrix logic, 73–113, **508**
  clues, reading, 76
  cross-correlations, 85
  subscripts, 87
  substrategies, 87–98
    adjunct lists, 81, 87
    category labels, 98
    combining clues, 90
    making assumptions, 93–98
    marking traits, 88–89
    repeated categories, 98
    seeking contradiction, 95
    substitution, 89
    using the introduction, 92–93
  things to remember, 77

two-dimensional chart matrix logic, 98
  writing, 86, 91
maximum value, **483, 508**
member, **322, 508**
metric units, 213–214
  conversions, 5, 215, 224
    See also unit analysis
Meyer, Carol, 197
Migliore, Ed, 416
miles per gallon (mpg), **204, 508**
miles per hour (mph), **204, 508**
Mira, Julio, 494
Mission: Impossible, 444
mixture problems, **184**–190, 364, **508**
models, 92–93, 266, 270–280, **508**
multiple, **508**

N

NASA, 279–280
nautical miles, **227, 508**
negative number, **508**
Nelson, Matthew, 316
Ngo, Hao, 306
no (as term), 322
nomograph, **484, 508**
*n*th term, **508**
numbers
  consecutive pairs of, 119
  even numbers, **61, 507**
  integer, **65, 508**
  negative, **508**
  odd, **61, 508**
  positive, **112, 509**
  prime. See prime numbers
  rounding, 199
  square numbers, **120, 510**
  triangular numbers, **120, 510**
  whole numbers, 44, **511**
numerators, 205, **244, 508**

Photo Credits

Chapter 1 © nullplus/iStockphoto; Chapter 2 © Ingram Publishing/ Superstock; Chapter 3 © Blend Images/ERproductions Ltd/Getty Images, Inc.; Chapter 4 © Ocean/© Corbis; Chapter 5 © NASA/ JPL-Caltech/NASA Jet Propulsion Laboratory Collection; Chapter 6 © Ryan McVay/Getty Images, Inc.; Chapter 7 © Chad Ehlers/Alamy; Chapter 8 © Masterfile; Chapter 9 © Ken Wramton/Getty Images, Inc.; Chapter 10 © Masterfile; Chapter 11 © moodboard/Getty Images, Inc.; Chapter 12 © Les and Dave Jacobs/Getty Images, Inc.; Chapter 13 © Andrew Peacock/Getty Images, Inc.; Chapter 14 © Andy Newman/AP/Wide World Photos; Chapter 15 © MCT via Getty Images; Chapter 16 © F1online digitale Bildagentur GmbH/Alamy; Chapter 17 © Eliane Sulle/Getty Images, Inc.

Answers to More Practice Problems

Chapter 1

1. Apartment Building

   5th floor

2. Movie Line

   40 people

3. Backboard

   131.22 feet from backboard to fence. Total distance of 418.828 feet

4. Working Out

   882 feet

Chapter 2

1. Pizza

   10 ways

2. Tomato Sauce

   10 ways

3. Odds to 22

   18 ways

4. Blackjack

   17 ways

Chapter 3

1. Mandarins

   87 mandarins

2. Blocks

   56 blocks

3. Missing Digits

   $1,694 + 8,728 + 9,493 = 19,915$

4. Interesting Number

   21,978

Chapter 4

1. Book Club

   Teddi – *The Kite Runner*
   Sherwood – *The Poisonwood Bible*
   Luann – *The Little Book*
   Kathi – *Water for Elephants*

2. Muscle Cars

   Tesha – Camaro
   Cigi – Challenger
   P-Dawg – Charger
   Karrin – Mustang
   Garrett – Chevelle

3. Lunch

   Nattalie – cheese quesadilla
   Max – steak
   Andrea – salad
   Savanah – chili
   Stephanie – spaghetti

5. Sunshine Café

   Shonda – blue, 1
   Mace – purple, 2
   Jaime – yellow, 3
   Trixie – green, 4

4. Second Grade Recess

   See-saw – Alyssa and Juanita
   Swings – Sean and Isabella
   Hopscotch – Jessica
   Slide – Ryan
   Monkey bars – Pedro
   Detention – Nicholas

Chapter 5

1. Find the Next Three

   a. 77, 98, 122
   b. 87, 100, 117
   c. 29, 31, 35
   d. 343, 512, 729
   e. 70, 14, 80
   f. 219, 334, 485
   g. 60, 97, 157
   h. 1,100, 1,101, 1,110

2. Units Digit of Large Power

   The units digit is 3.

3. Born on a Friday

   It will be Monday.

Chapter 6

1. Stock Profit

   $153.60 per share

2. Pay Cut or Pay Raise?

   $522

3. Pocketful of Money

16 hundreds, 24 fifties, 38 twenties

4. Poker

81 dollar bills

5. Dancing

124 total dancers

6. Room Key

Mellow walks 124.5 yds/min, Rich walks 130.5 yds/min

7. Cross-Country Flight

2,314 miles

8. Morning Commute

53 miles per hour

Chapter 7

1. Grazing Business Part 1

2,000 square yards

2. Grazing Business Part 2

214.5 square yards did not get eaten.

3. Respond, Please

No, 5 people did not get chairs.

4. Paycheck Part 1

$267.30

5. Paycheck Part 2

$73.06

6. The Palmetto Grapefruit

a. $0.76
b. $1.42
c. $4.65
d. 5.25 in.

Chapter 8

1. Racine to Chicago

a. 3.75 gal
b. 1.5 hours
c. $3.96/gal
d. 73.3 mi/hr
e. 16.5 cents/min
f. $4.95/passenger
g. 120.7 km
h. 225 passenger-mi
i. 60 passenger-mi/gal
j. 6.6 cents/passenger-mi

2. Starbucks Part 1

21.7 pounds

3. Starbucks Part 2

She will only gain 5.3 pounds, so 16.4 pounds fewer than in problem 2.

4. Pedometer Part 1

1,267 steps. My stride length is 35 inches.

5. Pedometer Part 2

1,478 steps

6. Pedometer Part 3

   0.6 miles

7. Pedometer Part 4

   About 4 miles/hour

8. Pedometer Part 5

   Approximately 10 minutes and 30 seconds

9. Pedometer Part 6

   2.33 strides/second

Chapter 9

1. Sisters and Brothers

   5 more sisters than brothers

2. Sum of Evens

   The sum is 9,003,000.

3. Big Difference

   The difference is 405,450.

4. Sequence Sum

   The sum is 12,497,500.

5. Holey Triangles

   485 triangles

6. Window Panes

   760 wooden insets

Chapter 10

1. Crossing a River with Different-Size People

   21 crossings

2. Square Table

   Kimbuck is the accountant. Kimi is the web designer.

3. College Reunion

   Summer lives in New Hampshire, January lives in Oklahoma, Autumn in New York.

4. Wedding Picture

   Left to right: Aura – maid of honor – bride's sister; Ayelet – flower girl – bride's niece; Peggy – bride's mother; Miriam – bride; Jane – bride's aunt; Aurelia – bridesmaid – bride's cousin

5. Which Square Stays the Same?

   F is 4.

Chapter 11

1. Painting Nails

   She bought 24 bottles.

2. Guadalupe's Homework

   53 problems

3. Block Party

   54 blocks

4. Bonus!

   $200

5. Diego and the Peanut Trail

   56 peanuts

6. Pokémon

   19 at start, 17 cards now

Chapter 12

1. Color My World

   10 liked only yellow. 4 didn't like any of these three colors.

2. Beauty Salon

   52; 25; 5; 12

3. Magazine Survey

   8; 30; 25; 80

4. School's Out

   30, 7, 5, 13, 3

5. Cup of Joe

   35, 12, 2

Chapter 13

1. Triangle

   Missing side is 29.75 cm.

2. Wagons and Trikes

   14 trikes

3. Home Field Advantage

   2,328 home fans

4. All the Queens' Children

   792 ants, 884 wasps, and 1,584 bees

5. State Insects

   17 honeybee states

6. Every Little Bit Counts

   78 bottles

7. Buying Fruit

   Mangoes cost 69 cents and melons cost $3.15.

8. A Lovely Bunch of Coconuts

   She can buy 7 pineapples and 12 coconuts, so she has to leave 2 coconuts behind.

9. Hourly Workers

   Janeen makes $9.23 per hour and she would have to work about 22.7 hours per week to make the same weekly amount as Shari makes.

10. Charlotte to Charleston

   229.475 miles and it took 3.425 hours (or 3 hours 25.5 min)

11. Jim's Juice

   Add 2.34 quarts of pure cranberry juice.

Chapter 14

1. A Few More Functions

   a. $y = 7x + 2$ $y = 3,649$ when $x = 521$
   b. $y = x^2 + 6x + 4$ $y = 416,020$ when $x = 642$
   c. $y = -6x + 50$ $y = -19,090$ when $x = 3190$
   d. $y = 4x^2 - x + 3$ $y = 88,658$ when $x = 149$
   e. $y = x^3 - 5x^2 + 3x + 8$ $y = 1,071,104$ when $x = 104$
   f. $y = 2x^2 + x + 5$ $y = 152,633$ when $x = 276$
   g. $y = 9x - 22$ $y = 779$ when $x = 89$
   h. $y = 2x^3 + x^2 - 6x - 4$ $y = 3,736,121$ when $x = 123$
   i. $y = 3x^2 + 7x - 2$ $y = 11,588$ when $x = 61$
   j. $y = 4x^3 - x^2 - 9x - 1$ $y = 1,681,199$ when $x = 75$

2. Dots in Rows

   $y = x^2 + 2x$ 8,099 dots in the 89th picture

3. Turning Squares

   $y = \frac{1}{2}x^2 + \frac{3}{2}x + 1$ 4,851 dots in 97th picture

Chapter 15

1. Which Digit Is Repeated?

   The two cubes are $27^3 = 19,683$ and either $35^3 = 42,875$ or $38^3 = 54,872$. The repeated digit is 8.

2. Room Rates

   Two people paid $109 for their room. The rates were $89, $99, and $109 on the south side, then $178, 132, and $109 on the north side.

3. Double Input Function

$f(x, y) = 2x^2 + 5y + 3$

4. The Dilly Deli Sandwich Spot

480 days

5. Abby's Spinner

a. 1/27   b. 19/27   c. 8/27

6. Shoveling Manure

Expected value is $14.375, so slightly worse.

7. Red, White, and Blue

1/2

8. Protein Shake

18%

9. Carnival Game

Don't play. Playing 15 games would cost $41.25 and pay back $40. To look at it another way, the expected value per game is $2.67, but it costs $2.75 to play.

10. Flip to Spin or Roll

Expected value is $2.78, so it isn't worth it.

11. Rochambeau

Answer 1: Armand wins 1/9 of the time, Barbara wins 2/3 of the time, Carlos wins 2/9 of the time, so Barbara is the most likely to win.

Answer 2: Armand gets 6 points when he wins, Barbara gets 1 point when she wins, and Carlos gets 3 points when he wins.

12. Cat and Dog Lady

4/9

13. Pigeons on the Roof

6/19

Chapter 16

1. More Dots

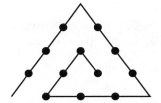

2. Sum of Digits

23/90

3. Interesting Sets of Three

85, 86, 87 is one set and 93, 94, 95 is the other.

4. Movie Reviews

Christy can choose to review 4 out of the 6 movies in 15 ways, which is equivalent to the complementary problem of choosing 2 out of the 6 movies to not review. The complementary problem is easier to solve than the original problem by either making a systematic list or drawing a diagram.

5. Rail Pass

21

6. Sprinklers on a Rectangular Lawn

139 square feet of lawn not watered by either sprinkler

7. Washing Dishes

About 60% (59.8)

8. Soccer Rotations

They will execute the switch 30 times until everyone is back in their original positions, so practice will last for 120 minutes.

Chapter 17

1. Sports Car Competition

About $3\frac{1}{2}$ years

2. Science Fiction Book Series

Book 7 should be about 530–535 pages.

3. TV Ratings

In year 8 or 9 the ratings will drop below 12 million people per week, so it will be canceled before the next year.

4. Broken Telephone Pole

31 feet

5. Rooftop Antenna

   He needs 3 spools (about 290 feet).

6. Lake Trip

   $8\frac{1}{2}$ hours due southwest and 4 hours due south for a total of $12\frac{1}{2}$ hours

7. The Neighbor's Tree

   180–185 feet for my tree and 160–170 feet for my neighbor's tree, so my tree is taller by 10–25 feet.